Bioarchaeology and Social Theory

Bioarchaeology and Social Theory aims to publish research grounded in empirical and scientific analysis of human skeletonized remains (referred to as bioarchaeology) from a wide variety of ancient, historic and contemporary contexts. The interpretations utilize social theory to frame the questions that blend cultural, environmental and social domains so that an integrated picture emerges. In this series, scholars have moved bioarchaeology into new methodological and theoretical areas.

Bioarchaeology is an exciting, innovative and relevant subdiscipline of anthropology and it is experiencing a fluorescence of application that has never been seen before. Bioarchaeologists are producing a body of scholarship that demonstrates the relevance of this kind of work for not only the unknown ancient past, but also for the present. Bioarchaeologists featured in this series are producing a body of scholarship that demonstrates the relevance of this kind of work for not only the unknown ancient past, but also for, in this case, known historical and contemporary practices. The biocultural approach encourages the use of multiple lines of evidence and this produces a more compelling and nuanced way of understanding human behavior in all of its complexities.

Social theory bridges biological data with cultural processes such as power, ideology, symbols, meaning, social structures, agency, and identity. The series promotes studies that link past understandings with present-day problem solving. In addition, ethical and critical considerations of bioarchaeological research are also emphasized.

Topics examined in this series include (but are not limited to) case studies that examine:

- identity, human variation, racism, gender and sex
- captives and slavery, warfare, conflict and violence, inequality and hierarchy,
- colonization and marginalization, industrialization and urbanization,
- social control, imperialism and subordination, migration and climate change.

This book series is indexed in SCOPUS.

For inquiries and submissions of proposals, authors may contact Series Editor, Debra Martin at Debra.Martin@unlv.edu or Publishing Editor, Christi Lue at christi.lue@springer. com

Anna Osterholtz

The Bodies Poetic

Identity and Interaction with the Dead

Springer

Anna Osterholtz
Department of Anthropology and Middle Eastern Cultures
Mississippi State University
Mississippi State, MS, USA

ISSN 2567-6776 ISSN 2567-6814 (electronic)
Bioarchaeology and Social Theory
ISBN 978-3-031-85997-7 ISBN 978-3-031-85998-4 (eBook)
https://doi.org/10.1007/978-3-031-85998-4

This Springer imprint is published by the registered company Springer Nature Switzerland AG
The registered company address is: Gewerbestrasse 11, 6330 Cham, Switzerland

For my mother, whose curiosity, compassion, and kindness are inspirational

Also, for Joe and Roger—I wish you two could've met

Series Foreword

The unique and insightful approach to death taken by Anna Osterholtz is one of those books that is so full of factoids and astounding bits of historical and contemporary information about death with all its vast meanings and expressions that it is an easy read. The author, using a biocultural bioarchaeological approach, has pulled from a range of disciplinary studies to highlight what can be learned about the dead for the living as well as vice versa. Using the theoretical framework of poetics of the body, the author brings us closer to understanding the meaning of death from a cross-cultural perspective. Like so many other volumes in this series, the bioarchaeological approach is both biocultural and anthropological. The body is interrogated through a prismatic lens of biological, cultural, and environmental factors focusing on symbolism, ideology, identity, and all the other things that humans use to make meaning of their lives as well as those who are dead.

Focusing on the multitude of cultural practices around death, the examples and case studies make it clear that death is a site of cultural negotiation that includes the importance of ritual, embodiment, and material culture. These principles hold for both ancestral people from long ago to contemporary times. The author also makes it clear how bioarchaeologists and others who study those who died long ago also become part of the past and the larger narratives that surround how people come to discuss and understand death. This aspect of self-reflection and engaged scholarship runs throughout the volume.

From Part I, readers will come to understand the usefulness in the poetics approach and the ways that this body of theory provides avenues for analysis in the reconstruction of the context of death. The case studies are culturally specific yet universally understood. The focus on the body as enshrouded in rituals and social interactions highlights the ways that identities are shaped and reshaped across the lifetime, and how that can be read through the body and its context at the time of death.

Understanding the power of the dead for the living is important for understanding who we are and how we identify ourselves. Subsequent sections of the volume deal with how the human body functions as material culture and is used to create and negotiate social relationships. Shifting to contemporary times, the author delves into topics such as cremation, memorialization, and how death and human remains are sometimes transformed into commemorative objects. There are chapters that take a deep dive into the political uses of bodies, and the ways that symbolism and politics play a role in how the dead are memorialized and the narratives that are perpetuated about those bodies. Finally, there is an exploration of the intersections between bodies and art, food and relics.

Department of Anthropology
University of Nevada
Las Vegas, NV, USA

Debra L. Martin

Preface

This volume explores how various processes contribute to the formation and negotiation of identity, with a particular focus on the material traces left by these processes, especially in relation to death and mortuary practices. I approach identity as a recursive, fluid process—one that is constantly evolving, negotiated, and renegotiated throughout an individual's life. The way we examine identity is shaped by the theoretical frameworks we adopt, the cultural contexts in which we are trained, and the personal experiences that spark our intellectual curiosity.

As a bioarchaeologist, my interest lies in understanding the processes that leave material evidence, particularly on bones and other elements of material culture. This focus has led me to examine the time surrounding death and the ways in which mortuary practices serve as tools for identity formation. As Robben (2004) notes, many of us who study death frame our understanding based on our own personal experiences. I acknowledge this influence and embrace it, recognizing that our views on death are shaped by our individual life histories.

I grew up in a household where death was not a taboo subject. My mother, a hospice nurse, worked closely with the dying and the bereaved, normalizing the idea of death as part of the life cycle. Her compassionate approach to the dying process helped me see death not as something to fear or avoid, but as a natural, inevitable part of human existence. This environment allowed me to explore questions about death from an intellectual standpoint without the cultural inhibitions that often surround the topic. It is in this spirit of openness and exploration that this volume was conceived.

In putting this volume together, I made a number of choices regarding structure and content. One of the central decisions was how to approach the discussion of mortuary practices among different ethnic groups in the USA. In recent scholarship, there has been a shift away from presenting the White, Colonial experience as the norm and framing African American (and other minority) experiences as deviations from this norm. This binary approach has its limitations, especially when discussing mortuary practices.

In this volume, I have chosen to present African American and White Colonial mortuary performances separately. I do so not to suggest that one experience is more meaningful or powerful than the other, but because these two traditions highlight different aspects of identity formation. African and African American mortuary performances often serve as mechanisms of resilience and identity formation in a "third space," shaped by historical oppression, survival, and resilience. In contrast, White Colonial mortuary performances tend to reflect processes of connection and the consolidation of power. These are distinct processes, and it is important to address them separately, while acknowledging that they each play an equally important role in identity formation.

Another key decision was the selection of case studies. The volume includes a mix of American examples, which reflects both my personal background and academic training. Growing up in the USA, I was exposed to historical sites and museums that fueled my early interest in death and memorialization within American culture. Many of the case studies in this book, such as those related to the Island of Cyprus, come from my own research, including my dissertation on Bronze Age tombs and the Committee on Missing Persons in Cyprus. These examples help us explore broader themes like mass graves, intentional forgetting, and cultural memory.

Additionally, some case studies are drawn from courses I teach, including the Anthropology of Death, Anthropology of Violence, and Death and American Culture. Over time, I've refined these examples based on student feedback and engagement, seeking those that spark interest while providing rigorous scholarly insight. For example, figures like Marie Curie, Frank Little, Marie Laveau, Emmett Till, and events like the atomic burials in Idaho serve as examples that reflect larger societal issues and transformations.

While the majority of these case studies feature men, this is partly due to the availability of popular sources, which tend to highlight men more than women. In Chap. 9, I discuss this issue further, reflecting on how both academic literature and popular literature often underrepresent women in death studies.

One of the most difficult decisions I faced was whether to include a discussion of the COVID-19 pandemic and its impact on mourning. The profound and far-reaching effects of COVID-19 on our relationship with death deserve more than a brief section or chapter. I have chosen to set this topic aside for future research, as the long-term consequences of the pandemic are still unfolding. The impact on mourning rituals, cultural memory, and the way we engage with death is a topic that will require more time to fully unpack. I believe that a poetics approach will be invaluable in understanding these shifts, but the time to fully explore this is not yet here.

This volume is not just an intellectual exercise; it is a reflection of my own experience as a bioarchaeologist and as someone who has spent much of their career studying the intersection of death, memory, landscape, and identity. I hope that the case studies and discussions presented here prompt readers to think critically about how we remember the dead, how we form identities through the material traces left by death, and how the dead continue to influence the living.

As you read through this volume, I invite you to consider the broader implications of death, mourning, and memorialization as tools for identity formation. This exploration is not just about the dead, but about how the living engage with them—through memory, material culture, and ritual—to construct meaning and negotiate their place in the world.

Starkville, MS Anna Osterholtz

Acknowledgments

This book is the product of many years of academic research. First and foremost, I am indebted to my mentor, Dr. Debra Martin. In my time at UNLV, she fostered an approach of exploration and curiosity that allowed me (and the rest of her students) to explore what anthropology looks like for them. Her courses and conversations have always meant the world to me, and I treasure the relationship we continue to have. I am also grateful for the relationship (both personal and professional) that I have with my fellow UNLV alumni, particularly my bestie Dr. Ryan Harrod. Through collaborative research and writing projects, barbeques, and being the Man of Honor at my wedding, I couldn't imagine a better friend and colleague.

My family has been a source of support and love, and I'm grateful for all of you. I blame my mother for my initial interest in anthropology, taking me to volunteer at archaeological sites when I was a teenager. From there, my family has supported the decisions I've made for an undergraduate major, graduate school, and my ultimate career. I was fortunate to have amazing teachers and professors who promoted cross-disciplinary interests, allowing me to forge a path in bioarchaeology that has been both fun and fulfilling. Joe, my partner, has learned more about anthropology than he ever wanted to, and has been my constant sounding board. Joe, I couldn't imagine this without you.

Finally, I want to thank the anonymous reviewers of the first draft of this book. Their comments were remarkably helpful and constructive. Through their careful review of the book, they helped me to reorganize and reframe some elements that made this a far stronger volume.

Competing Interests The author has no competing interests to declare that are relevant to the content of this manuscript.

Contents

Chapter 1
Introduction

Abstract Death, dying, and memorialization are central to how we see ourselves and tell the world where we position ourselves in relation to others in society. This chapter introduces Neil Whitehead's Poetics model. Initially formulated to understand group-level violence, this model can also be used to understand how mortuary activities can be used for identity formation and social negotiation. This extends past body manipulation to processes of memorialization that involve the body either in physical or symbolic form. This chapter also sets out the structure of the volume, with short descriptions of the five parts of the volume. Part I (Chaps. 2–5) details the development of a model of Poetics in bioarchaeology. Part II (Chaps. 6–9) applies the model to common mortuary processes. Part III (Chaps. 10–12) details modern interactions with the dead through a Poetics lens. Part IV (Chaps. 13 and 14) examines bodies and politics, with a focus on the use of the body as a political medium. Part V (Chaps. 15 and 16) examines the body as material culture, particularly through the lens of post-mortem agency. The Poetics lens, as introduced here, is expansive, and provides a lens to understand the relationship between the living and the dead.

There is a box in the spare room in my home filled with certificates that once belonged to my father. They are beautifully framed and range from certificates of achievement to his diploma from medical school. I have no idea what to do with them, where to put them. I don't really want to hang them on the wall, but I can't throw them away. I am incredibly stuck when it comes to these certificates. In the periodic bouts of clearing out things and decluttering, my partner suggested that we put the certificates into a binder and repurpose or dispose of the frames. While this is a good compromise, I can't bring myself to open the frames to remove the certificates. They are a physical link to him and therefore have significant power. It feels somehow wrong to disturb them. So I move the box from one side of the room to the other. In our daily lives, we have these reminders of those who have passed and understanding their power and the links to the dead become important for understanding who we are and how we identify ourselves.

A. Osterholtz, *The Bodies Poetic*, Bioarchaeology and Social Theory,
https://doi.org/10.1007/978-3-031-85998-4_1

Would anyone in the next generation of our family (or two generations down the line) have such ties to these pieces of paper in frames? I think about these certificates whenever I see memorial items in thrift stores, or see artwork from the nineteenth century made from human hair in antique stores. How long do objects connected to the dead continue to have agency in the lives of the living? At what point is it *ok* to dispose of memorial goods, and how do we do that? Given the rising popularity of tattoos as memorials, and use of cremains in tattoo ink (discussed in Chap. 10) and the ability to preserve tattooed skin (discussed in Chap. 16), it is now conceivable that an individual could incorporate the cremains of a deceased friend or relative into the memorial tattoo for that deceased individual that they wear for their life and after their death for that memorial tattoo with cremain-laden ink to be preserved. The family member with the preserved tattoo may keep it for years, passing it down through generations until it ceases to have the same connection between the living and the dead. What would we think if we came across this in an antique store? Would we recognize it for the multilayered expression of love and grief and connection?

Death, dying, and memorialization are central to how we see ourselves and tell the world where we position ourselves in relation to others in society. The Poetics framework first put forth by Neil Whitehead is a useful framework within which to begin to understand the power of objects and human bodies to convey social messages through their manipulation. It is also a useful framework within which to understand the role of place and space in how and when we commemorate the dead (and when we stop commemorating them).

1.1 The Poetics Framework

This volume explores concepts of identity and memorialization through a Poetics framework. First introduced by Neil Whitehead (2004a, b, 2005), this model contextualizes activities performed by groups in front of an audience to promote group identity. Whitehead focused the application of this model on the role of ritualized violence, emphasizing the role of violent performance in creating and expressing identity. According to Whitehead, violence is not only destructive but also a part of human expression and identity formation (Whitehead, 2004b). Building on this foundation, this volume expands the Poetics framework beyond violence to explore how the treatment of the dead—specifically through rituals, memorialization, and the manipulation of human remains—can also serve as a medium for group identity and negotiation/renegotiation of social status.

Applied to bioarchaeology and mortuary contexts, the Poetics framework allows us to view the body not just as a biological entity but as material culture—something that can be altered, shaped, and imbued with meaning. The body becomes a key site for the negotiation and renegotiation of social roles and collective identities. This approach challenges traditional perspectives, allowing scholars to explore embodiment, materiality, and landscape as integrated aspects of human experience.

It also highlights the postmortem agency of human remains—the idea that the dead continue to exert power and influence over the living long after death.

The model of Poetics as used in bioarchaeology has effectively examined how identity can be formed through a synergetic experience of ritual activity performed in front of an audience (e.g., case studies in Campbell & Osterholtz, 2024; Martin & Osterholtz, 2015; Osterholtz, 2016, 2018a, b; case studies in Osterholtz, 2020). In this volume, I will set out a larger view of Poetics than presented previously, with special focus on the importance of identity formation, memory, land-scape, and embodiment.

Poetics gives us a way to integrate bioarchaeological and archaeological exploration through examining how we change the body through human interaction and how we impact human interaction with the body. Biological constrains dictate that those changes must follow specific patters, but overall patterns of alterations to the biological with indicate social structure and change (just as inclusions in lithic raw material will dictate how a stone tool can be made). Using a Poetics approach, we must examine the physicality and materiality of the human body along the same lines as any other class of material culture. Reframing the body as material culture allows us to employ archaeological theories of embodiment, materiality, phenomenology, and landscape in ways previously unopen to us. We can begin to address the transformative properties of the body itself, and its ability to transform society. But as we are reminded, the dead do not bury themselves (Parker Pearson, 1999), so this manipulation of the body (particularly in this case, the dead body) is carried out by the living for the purposes of living social structures and social change.

Ultimately, I am a bioculturally oriented bioarchaeologist. I am concerned with the analysis of human remains and their burial contexts from archaeological contexts as a way to reconstruct past lived experience. To do this, an understanding of what the body is and how it is constructed from both biological and cultural matter must be made. This volume represents an attempt to unpack some of the elements of identity formation that may become inscribed on the bones and other elements of material culture related to the physical body. I am also interested in the materiality and embodiment of cultural processes such as mourning and memorialization. This means that I must extend my interests past just the physical or individual body into an exploration of the other bodies: the Social Body and the Body Politic (Scheper-Hughes & Lock, 1987). Ultimately, this need to understand issues of mourning and commemoration leads to larger discussions than just those revolving around the physical body. In this volume, a large amount of space is devoted to modern methods of commemoration, primarily with a focus on the material culture produced from those activities.

The Poetics framework encourages a shift in how we under-stand death and memory. By treating the human body as material culture, this framework allows us to apply theories of embodiment and materiality to the study of death rituals and memorial practices. Rather than viewing death as an end, the Poetics model frames it as an ongoing process of social transformation. The manipulation of the body—whether through ritual, commemoration, or mourning—becomes an integral part of shaping social structures and cultural norms.

In addition, the volume extends the concept of postmortem agency to suggest that the dead continue to influence the living through their remains. This idea challenges us to rethink how the bodies of the deceased are used to shape social memory and cultural identity, both in ancient and contemporary contexts.

1.2 Structure of the Volume

This volume is separated into five main parts in addition to the Introduction and Conclusion.

- Part I (Chaps. 2–5): Building a Model of Poetics in Bioarchaeology. This section introduces the theoretical foundations of Poetics, blending ideas from archaeology, bioarchaeology, and social anthropology. It outlines how the body, through ritual and social interaction, becomes a vehicle for identity formation.
- Part II (Chaps. 6–9): The Body as Material Culture. This section focuses on how the human body functions as material culture. It examines how bodies are used to create and negotiate social spaces, as well as how they act as mediums for the formation and renegotiation of social relationships.
- Part III (Chaps. 10–12): Modern Interactions Between the Living and the Dead. Focusing on contemporary practices, this section explores cremation, memorialization, and how the remains of the deceased are transformed into commemorative objects. It highlights evolving attitudes to-ward identity, memory, and individualization in modern death practices.
- Part IV (Chaps. 13 and 14): Politics and Bodies. This section examines the political use of human remains, analyzing how bodies are employed as symbols of sovereignty, national identity, and political power. It explores the role of necropolitics—the politics of death—in the memorialization of bodies.
- Part V (Chaps. 15 and 16): Postmortem Agency and Human Remains as Art. The final section explores the postmortem agency of human remains, looking at how they continue to influence culture and identity. Topics include the use of human remains in art, food, and relics, as well as the ethical ques-tions surrounding these practices.

1.3 A Note on Sources

The volume draws on a range of sources, from peer-reviewed academic articles to popular media (e.g., newspapers, blogs, company websites). This combination of sources is crucial for understanding how death rituals and memorialization practices are experienced in both academic and popular contexts. For example, examining the burial of Elvis Presley requires a review of both government re-ports (such as the National Park Service's application for landmark status) and media coverage (e.g.,

Rolling Stone articles), as both contribute to shaping our understanding of the cultural significance of his burial. Similarly, discussions of Arlington National Cemetery draw on both official documentation and academic interpretations to contextualize the role of such sites in collective memory.

1.4 Conclusion

The Poetics framework offers a comprehensive lens for exploring the materiality of the body and its role in shaping identity, memory, and social structures. By viewing death practices as a site of cultural negotiation, the frame-work emphasizes the importance of ritual, embodiment, and material culture in both ancient and modern contexts. As Oestigaard (1999, p. 358) states, "Death is not a non-existence; it is only another existence, which makes possible different forms of dialogues between the living and the dead through various performances and media." Anthropologists, when we study the past, become part of the story of the past. In effect, we act as witnesses who see an incomplete picture but are nonetheless transformed by our interactions with the material remains of the past. Haanstad (2020, p. 249) notes that "By reassembling the dead—human, plant, and animal—the inscribed representations of this conclusion continue to extend the meanings of death."

References

Campbell, R. A., & Osterholtz, A. J. (Eds.). (2024). *The poetics of violence in Afroeurasian bioarchaeology*. Springer Nature.

Haanstad, E. J. (2020). Conclusion: Poetic amplifications and extnesions, how to process the dead. In A. J. Osterholtz (Ed.), *The poetics of processing: Memory formation, cosmology, and the handling of the dead* (pp. 235–252). University Press of Colorado.

Martin, D., & Osterholtz, A. J. (2015). *The poetics of processing: Deviant performances and memory making in the ancient southwest*. Paper presented at the 114th American Anthropological Association Annual Meeting, Denver, CO.

Oestigaard, T. (1999). Cremations as transformations: When the dual cultural hypothesis was cremated and carried away in urns. *European Journal of Archaeology, 2*(3), 345–364.

Osterholtz, A. J. (2016). Patterned processing as performative violence. In A. J. Osterholtz (Ed.), *Theoretical approaches to analysis and interpretation of commingled human remains* (pp. 125–138). Springer.

Osterholtz, A. J. (2018a). Each one the same: Performance, demography, and violence at sacred ridge. In C. P. Anderson & D. L. Martin (Eds.), *Massacres: Bioarcheaology and forensic anthropology approaches* (pp. 82–93). University of Florida Press.

Osterholtz, A. J. (2018b). Interpreting and reinterpreting sacred ridge: Placing Extreme processing in a larger context. *Kiva, 84*(4), 461–479.

Osterholtz, A. J. (Ed.). (2020). *The poetics of processing: Memory formation, identity, and the handling of the dead*. University Press of Colorado.

Parker Pearson, M. (1999). *The archaeology of death and burial*. Texas A&M University Press.

Scheper-Hughes, N., & Lock, M. (1987). The mindful body: A prolegomenon to future work in medical anthropology. *Medical Anthropology Quarterly, 1*, 6–41.

Whitehead, N. L. (2004a). Introduction: Cultures, conflicts, and the Poetics of violent practice. In N. L. Whitehead (Ed.), *Violence* (pp. 3–24). School of American Research Press.

Whitehead, N. L. (2004b). On the poetics of violence. In N. L. Whitehead (Ed.), *Violence* (pp. 55–78). SAR Press.

Whitehead, N. L. (2005). War and violence as cultural expression. *Anthropology News, 46*(5), 23–26. https://doi.org/10.1525/an.2005.46.5.23.1

Part I
Building a Model of Poetics in Bioarchaeology

The following chapters (Chaps. 2–5) explore the elements that go into the Poetics model. Chapter 2 details Whitehead's Poetics model, tying together ritual, performance and identity. The model emphasizes that cultural actions, when performed in front of an audience, generate and allow for the negotiation of group identity. These acts are both degenerative and regenerative, both disruptive to and constructive of social order and social relationships. Chapter 3 examines Scheper-Hughes and Lock's Three Bodies Model in relation to bioarchaeological inference about lived experience in the past. Chapter 4 focuses on mortuary archaeology and ethnographic approaches, tracing the development of models used to interpret material culture has been used to infer social structure, inequality and cultural meaning. Processual, Marxist, and post-processual approaches are all discussed. Chapter 5 focuses on the interpretation of post-colonial construction of space and how mortuary performance plays a key role in the creation of new identities. This is told primarily through the history of African American funeral directors and the unique role of mortuary space in the creation of African American identity and resilience.

The chapters provide a rich and nuanced approach to understanding the intersection of bioarchaeology, mortuary archaeology, and cultural identity. By intertwining aspects of performance, ritual, and material culture, the Poetics model offers new ways of interpreting past societies, focusing on how cultural and social practices—especially related to death and the body—serve as mechanisms for constructing and negotiating identity, power, and social order. These chapters contribute to a broader understanding of how archaeology and bioarchaeology can illuminate the complex interplay between individual bodies, collective identities, and the material world.

Chapter 2
Poetics and Identity

Abstract This chapter introduces the application of Whitehead's Poetics model to bioarchaeology and mortuary archaeology, emphasizing its potential for understanding cultural practices and the formation of group identity. The model can be summed up as **performance + ritual = identity**, positing that cultural actions, when performed in front of an audience generate collective meaning and reinforce group identity. The chapter outlines how this framework can be used to interpret violent acts and mortuary rituals as social performances that simultaneously destroy and regenerate cultural and social order. Whitehead's Poetics of Violence model argues that violent performance, rather than a universal or physical phenomenon, should be understood as a contingent, culturally specific process, deeply embedded in and contributing to social meaning. This chapter extends the Poetics framework to mortuary practices, demonstrating how rituals surrounding death, such as the manipulation of human remains, grave goods, and tombs, play a crucial role in the creation and maintenance of collective identities. By applying Poetics to mortuary archaeology, this model contributes to a broader understanding of how identity is formed through the reciprocal interaction between individuals, communities, and the material world. Ultimately, this approach underscores the regenerative nature of social practices and provides a nuanced lens for interpreting the material records of past societies, as explored through case studies from diverse global contexts.

This volume is dedicated to the application of Whitehead's Poetics model to the subfields of bioarchaeology and mortuary archaeology. Therefore, it is important to lay out Poetics the way that Whitehead and others interpreted and operationalized it. One of the great strengths of the Poetics approach is that it allows for cultural activity to be seen as regenerative and meaningful, with meaning coming not only from the activity (ritual) but also from the involvement of the community or witnesses to the activity.

The basic model can be summed up as performance+ritual = identity. Whitehead's primary focus was on how violent actions conducted in front of an audience were key to collective identity, but this basic model can be used to look at other human activities as well. This volume is dedicated to looking at the performances of

A. Osterholtz, *The Bodies Poetic*, Bioarchaeology and Social Theory,
https://doi.org/10.1007/978-3-031-85998-4_2

violence as well as the performances of mortuary activity as important for collective or group identity.

Performance, as used here, should be viewed as an act conducted in front of an audience. The reciprocal feedback from the audience reinforces the underlying understanding of the act for all participants, even those who are not direct actors (i.e., the witnesses or audience). This meaning is not just a surface meaning of the act conducted but also the deep, culturally and historically contingent meaning embedded in the acts. The acts themselves can be thought of as the ritual part of the equation. Ritual acts involve repetition of acts, repetition can reinforce group identity. Bell (1992, pp. 19–21) notes that rituals are either (1) routinized, habitual actions that express conceptual orientation inspired and promoted by beliefs, symbols and myths; (2) mechanisms for integrating actions and thought, or (3) affirmations of communal unity, creating 'antistructures.' In other words, rituals cannot exist without belief, and so ritual is the embodiment and operationalization of beliefs through action. As Ryan Harrod and I found in cross-cultural ethnographic study of military initiation rites and drills, repetition of acts creates in-group and out-group identities that do not disappear when military service is concluded (Osterholtz & Harrod, 2013). But ritual does not require a performance for meaning to occur. For the Poetic "amplification of meaning" (Haanstad, 2020, p. 236) of ritual to occur, performance is needed. When heterogenous interpretations of social processes occur, groups may perform their own "rites of repair" (Diefendorf, 2012, p. 34) that solidify their own identity to the exclusion of others. In terms of religious rituals in France during the sixteenth century, for example, religious "rituals of repair were thus double-edged; whatever their avowed intention, they served more to heighten differences than to promote unity."

Whitehead (2002, 2004a, b, c, 2005, 2007) was primarily engaged in the study of violent interaction, essentially asked the question, "What is violence?". Whitehead argues that violence is an important cultural system, involving "recognition that violence is as much a part of meaning and constructive human living as it is an imagination of the absence and destruction of all cultural and social order (2007, p. 41). Violence is part of who we are and how we live our cultures. Through violent interaction, society and culture are simultaneously destroyed and created, being both *con*structive and *de*structive, *de*generative and *re*generative (Osterholtz, 2020a). Violence is contingent both culturally and historically, and this contingency itself makes violence difficult to understand within Anthropology (Whitehead, 2004b). Violence, instead of being a concept should be viewed as a structure or a process that must be interpreted within the system that gives it its phenomenological meaning. We cannot focus strictly on one form of violence, such as interpersonal violence, and hope to understand larger issues such as warfare since these different types of violent interaction serve different social functions and help to order society differently. According to Whitehead (2004b, p. 57), "these illusions—that somehow all acts of violence are related in virtue of their "being violent," that is it is possible to explain or interpret some essential core" present in all examples of violence, and that violence is only "physical hurt"—that prevent the integration of anthropological analyses and, more importantly, a credible explanation of violent expression.

Such reductive and misleading definitions of violence fail to address a wide variety of violence that has no immediate material correlates, such as sorcery or verbal aggression, because they use only physical injury as the linking element between examples of violence."

Violence is a cultural performance, but a performance that must be contextualized. Performance itself is structured through the interaction between the actors, in this case the victim, perpetrator and witness (Riches, 1986). Riches was focused on the individual experiences, but Whitehead argues that construction of meaning of the performance should be based on collectivities as well as individuals. Hidden meanings will not always be obvious in the acts themselves. Haanstad (2020) argues that Poetics is the amplification of these meanings to form collective identity, always mediated through individual experience. In this way, we perceive individual meaning that impacts how we receive the amplified collective message. The same performance can also be perceived by the audience in heterogenous ways based on intersecting elements of identity (Fig. 2.1).

Poetics has great interpretive power for understanding mortuary customs as well. The term *Poetics* has been used in the analysis of the interaction with dead before. While Langford (2016, p. 5 of 13) does not cite Whitehead in their conference paper, the interaction between the living and the dead amongst Khmer individuals in the United States is described as a "ghostly Poetics of ritual performance and bodily practice involving feasting, gifting, embodying, wrestling with and/or negotiating with the dead in the present." Danely (2018, p. 134) has described this ghostly Poetics approach as one that "animates uncertainty".

I would argue that the examination of violence as a Poetic process can be extended to other forms of human activity that create a material record, particularly as mortuary processing (Fig. 2.2). This was recently explored in an edited volume where contributors were asked to apply Whitehead's theoretical framework to

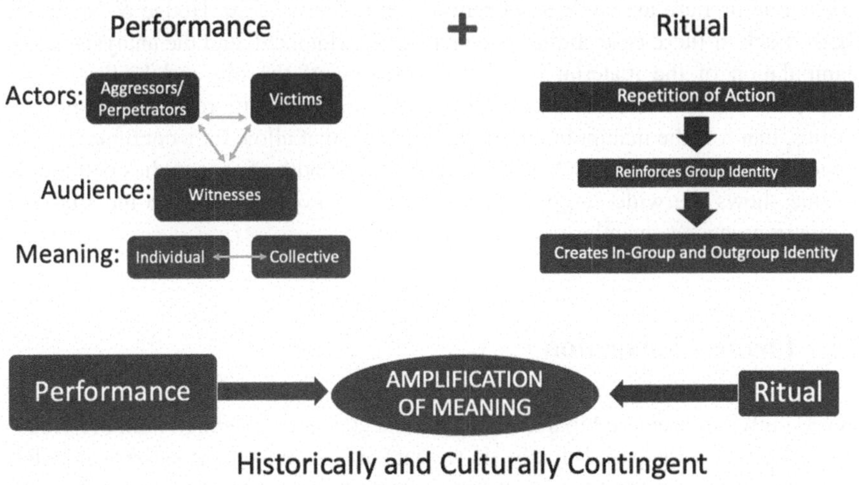

Fig. 2.1 Whitehead's poetics of violence model

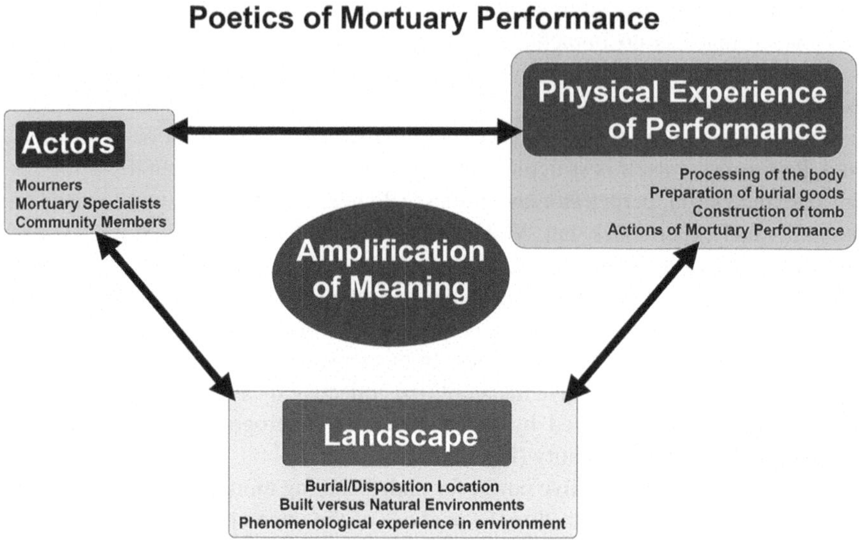

Fig. 2.2 The poetics of mortuary performance

mortuary processing around the world (Osterholtz, 2020b). Authors applied Poetics to examinations of feline trophy heads as grave goods from Peru (Scaffidi, 2020), processed individuals from Paquime, Mexico (Waller & Offenbecker, 2020), extreme processing in the southwestern US (Kuckelman, 2020; Martin & Osterholtz, 2020), tomb usage in Aksum, Ethiopia (Basanti, 2020), manipulation of power structures between the rulers and ruled in ancient Egypt (Campbell, 2020), social memory and embodiment of the house at Çatalhöyük, Turkey (Pilloud et al., 2020), tomb use and reuse at Nabatean Petra (Perry & Osterholtz, 2020), and structural violence in modern anatomical collections (de la Cova, 2020; Hodge & Nystrom, 2020). Each of these case studies is heavily contextualized, and the analysis of the manipulation of the material remains (consisting of the physical bodies, grave goods, and the graves themselves) allow for a view into the reciprocal nature of Poetics, that it is the manipulation of these objects that allow for a continual regeneration of society and social change to occur. The breadth of case studies within this volume shows the wide-ranging applicability of a Poetics model to the study of mortuary processes in anthropological research.

2.1 Identity Formation

Increasingly important to bioarchaeology is the concept of identity, both individual and collective, in the past. But why do we care about identity formation? And what do we mean by identity? Social identity is a useful concept here. Cipolla (2008, p. 196) defines social identity as "the ways in which people classify themselves and their social surroundings." Social identity is tied to memory in that memory is a social process (Halbwachs, 1992) and draws from both individual and group interpretations of the past. These interpretations are subject to change through the processes of iteration and citation (discussed below). Cipolla's definition of social identity is a relational one, relying on both the individual and the group. Effectively, this looks at the individual, social and political bodies (Scheper-Hughes & Lock, 1987); it is the relationship between these different identities that ultimately situates someone within the social, cultural, and physical landscape.

Cipolla's (2008) relational understanding of social identity utilizes the different forms of social capital utilized by Robert Putnam (1995, 2020). Putnam examines the difference between *bridging* and *bonding* social capital. Bridging social is inclusive, providing social ties between disparate groups (or different intersectional identities). Bonding social capitol is exclusive, performed through ritual that reinforces the ties between individuals of the same intersectional identity. As Putnam (2020, p. 23) states, "[b]onding social capital constitutes a kind of sociological superglue, whereas bridging social capital provides a sociological WD-40." Understanding the different types of connection (social capital) is useful in understanding the increasing rise of alternative mortuary performances discussed later in this book. Large-scale social structures at play, such as nationalistic identity provide bridging social capital, while bonding social capital is reinforced through mortuary ritual that highlights the smaller, closer bonds of "like-minded intimates" (Putnam 2020, p. 178), such as space burials or body composting.

Identity and identity formation will be discussed in upcoming discussions of mortuary archaeology and bioarchaeology and then applied through the rest of the chapters in this volume.

2.2 Summary

This chapter reframes Whitehead's Poetics model to bioarchaeology and mortuary archaeology, focusing on its potential to understand cultural practices and the formation/negotiation of group identity. The model (**performance + ritual = identity**), posits that cultural actions when performed in front of an audience creates collective meaning and allows for reinforces group identity. This chapter uses the Poetics framework to interpret mortuary rituals as social performances that are reflective of disruptions to social structure (acknowledging the death of an individual and the severing of those social bonds) and regenerative (the formation of new bonds through performative ritual). By extending the Poetics of Violence model to mortuary practices, this chapter highlights how rituals surrounding death—such as the modification of human remains, grave goods, and tomb construction/location—play key roles in shaping and maintaining group identities. The application of the Poetics model emphasizes the reciprocal relationship between individuals, communities, and the material culture created in performative ritual. Overall, the chapter underscores the regenerative nature of social practices and provides a framework for the interpretation of past material culture.

References

Basanti, D. S. (2020). Poetics of the house: Changing realities of body and person in aksumite mortuary practices. In A. J. Osterholtz (Ed.), *The poetics of processing: Memory formation, cosmology, and the handling of the dead* (pp. 103–122). University Press of Colorado.

Bell, C. (1992). *Ritual theory, ritual practice*. Oxford University Press.

Campbell, R. A. (2020). Smiting pharaohs: Violence and power in ancient Egypt. In A. J. Osterholtz (Ed.), *The poetics of processing: Memory formation, cosmology, and the handling of the dead* (pp. 123–144). University Press of Colorado.

Cipolla, C. N. (2008). Signs of identity, signs of memory. *Archaeological Dialogues, 15*(2), 196–215.

Danely, J. (2018). Mournign as mutuality. In A. C. G. M. Robben (Ed.), *A companion to the anthropology of death* (pp. 131–144). Wiley.

de la Cova, C. (2020). Processing the destitute and deviant dead: Inequality, dissection, politics, and the structurally violent legalization of social marginalization in American anatomical collection. In A. J. Osterholtz (Ed.), *The poetics of processing: Memory formation, cosmology, and the handling of the dead* (pp. 212–234). University Press of Colorado.

Diefendorf, B. B. (2012). Rites of repair: Restoring community in the French religious wars*. *Past & Present, 214*(suppl_7), 30–51. https://doi.org/10.1093/pastj/gtr017

Haanstad, E. J. (2020). Conclusion: Poetic amplifications and extnesions, how to process the dead. In A. J. Osterholtz (Ed.), *The poetics of processing: Memory formation, cosmology, and the handling of the dead* (pp. 235–252). University Press of Colorado.

Halbwachs, M. (1992). *On collective memory* (L. Coser, Trans. L. Coser Ed.). Chicago: University of Chicago Press.

Hodge, C. J., & Nystrom, K. C. (2020). Dissection as social process: Anatomical settings in the nineteenth-century United States. In A. J. Osterholtz (Ed.), *The poetics of processing: Memory formation, cosmology, and the handling of the dead* (pp. 189–211). University Press of Colorado.

Kuckelman, K. A. (2020). Ritual modification in the context of social unrest among ancestral pueblo Peoples of the Northern San Juan. In A. J. Osterholtz (Ed.), *The poetics of processing: Memory formation, cosmology, and the handling of the dead* (pp. 66–84). University Press of Colorado.

Langford, J. M. (2016). *Ghostly poetics*. Paper presented at the Penn Ghost Project, Philadelphia. Retrieved from https://d1wqtxts1xzle7.cloudfront.net/43605884/Ghostly_Poetics-libre.pdf?1457663987=&response-content-disposition=inline%3B+filename%3DGhostly_Poetics.pdf&Expires=1741280925&Signature=bBFM7EWMzP5ejlKJhVxM6oBhfM2aT44s~iapBZl-xzjqodlVgZtaavqPgB99OfENvKgTTZaaLYdi78QyYezB9A7ZUQLncKxHqqTs5~t1DoC8xWQ2fDm89r9t3uzKBbpq4OyFb50XcUJvHGIsCcTQnTK~8JQ9wZ4nYrSe~AjrzDOFW-l~5Gj3BTXu7DGqWx7AI15Kqq7cEPPWvUyzj8rIgrNFkf9XiEgWi9bg~qn4Xl-t4mLzqfCJcSuAuSVHPZkh53t3Nlep7WfJS0G6RYJZ7BoZ9foSzaIxomS8u7gTlfYp0zczOyixcg73BJyYVykbRIrkrSy9A1GlcKpFYe0KqA__&Key-Pair-Id=APKAJLOHF5GGSLRBV4ZA

Martin, D. L., & Osterholtz, A. J. (2020). The poetics of corpse fragmentation and processing in the ancient southwest. In A. J. Osterholtz (Ed.), *The poetics of processing: Memory formation, cosmology, and the handling of the dead* (pp. 85–102). University Press of Colorado.

Osterholtz, A. J. (2020a). Introduction: Processing and poetics, examining the model. In A. J. Osterholtz (Ed.), *The poetics of processing: Memory formation, cosmology, and the handling of the dead* (pp. 3–14). University Press of Colorado.

Osterholtz, A. J. (Ed.). (2020b). *The poetics of processing: Memory formation, identity, and the handling of the dead*. University Press of Colorado.

Osterholtz, A. J., & Harrod, R. P. (2013). *Warrior, soldier, big man: Warrior ethos, identity formation and the negotiation of social roles in multicultural settings*. Paper presented at the 78th Annual meeting of the Society for American Archaeology, Honolulu, HI.

Perry, M. A., & Osterholtz, A. J. (2020). Mingled bones, mingled bodies: Diversity in the poetics of death at Nabatean Petra. In A. J. Osterholtz (Ed.), *The poetics of processing: Memory formation, cosmology, and the handling of the dead* (pp. 166–188). University Press of Colorado.

Pilloud, M. A., Haddow, S. D., Knüsel, C. J., Larsen, C. S., & Somel, M. (2020). Social memory and mortuary practices in Neolithic Anatolia. In A. J. Osterholtz (Ed.), *The poetics of processing: Memory formation, cosmology, and the handling of the dead* (pp. 145–165). University Press of Colorado.

Putnam, R. D. (1995). Bowling alone: America's declining social capital. *Journal of Democracy, 6*, 65–78.

Putnam, R. D. (2020). *Bowling alone: The collapse and revival of American community (Revised and Updated)*. Simon & Schuster Paperbacks.

Riches, D. (Ed.). (1986). *The anthropology of violence*. Blackwell.

Scaffidi, B. K. (2020). Power, mediation, and transformation: Dismembered heads from Uraca (Majes Valley, Peru) and the Andean Feline-Hunter Myth. In A. J. Osterholtz (Ed.), *The poetics of processing: Memory formation, cosmology, and the handling of the dead* (pp. 15–40). University Press of Colorado.

Scheper-Hughes, N., & Lock, M. M. (1987). The mindful body: A prolegomenon to future work in medical anthropology. *Medical Anthropology Quarterly, 1*(1), 6–41. Retrieved from http://www.jstor.org/stable/648769

Waller, K. D., & Offenbecker, A. (2020). The politics and poetics of performance violence at Casas Grandes, Chihuahua, Mexico. In A. J. Osterholtz (Ed.), *The poetics of processing: Memory formation, cosmology, and the handling of the dead* (pp. 41–65). University Press of Colorado.

Whitehead, N. L. (2002). *Dark shamans: Kanaimà and the Poetics of violent death*. Duke University Press.

Whitehead, N. L. (2004a). Introduction: Cultures, conflicts, and the poetics of violent practice. In N. L. Whitehead (Ed.), *Violence* (pp. 3–24). School of American Research Press.

Whitehead, N. L. (2004b). On the poetics of violence. In N. L. Whitehead (Ed.), *Violence* (pp. 55–78). SAR Press.

Whitehead, N. L. (Ed.). (2004c). *Violence*. School of American Research Press.

Whitehead, N. L. (2005). War and violence as cultural expression. *Anthropology News, 46*(5), 23–26. https://doi.org/10.1525/an.2005.46.5.23.1

Whitehead, N. L. (2007). Violence & the cultural order. *Daedalus, 136*(1), 40–50. https://doi.org/10.2307/20028088

Chapter 3
Bioarchaeology and the Bodies

Abstract This chapter explores the intersection of bioarchaeology, cultural theory, and the study of the human body, focusing on how biological, cultural, and social factors shape lived experiences across time through the lens of the Three Bodies Model put forth by Scheper-Hughes and Lock. Bioarchaeological analysis provides a unique lens through which researchers can trace physical evidence of trauma, disease, and stress in skeletal remains, offering insights into historical health and social dynamics. The Three Bodies Model conceptualizes the body as an physical object as well as a social and political entity, offering a framework for understanding how cultural processes shape physical bodies. The Three Bodies Model is examined in relation to other concepts used in bioculturally engaged bioarchaeology. The concept of social buffering is introduced, illustrating how cultural norms and social structures can either protect or expose individuals to risks. Power dynamics are also examined with relation to the Three Bodies Model, Practice, and Structuration, exploring how individual agency and social structures shape lived experience and are inscribed on bone. The role of intersectionality is introduced as a critical framework for understanding how overlapping social identities (such as gender, age, social race, and socioeconomic status) influence both cultural practices and physical health. By integrating intersectionality into bioarchaeological analysis, scholars have argued that skeletal trauma and health disparities can be understood to be the product of complex, interwoven social and biological factors. Finally, the boundaries of the bodies are explored, with an emphasis on fluidity and the complex interplay between physical bodies, material culture, and social context. The interpretation of bioarchaeological data requires an understanding of the specific archaeological, historical, and cultural context in which human remains are found to provide a holistic understanding of lived experience in the past.

What is a body? In bioarchaeology, particularly a bioculturally focused model of bioarchaeology, the term body refers both to biological and cultural understandings and applications. One primary model that ties these two together is the three bodies model (Scheper-Hughes & Lock, 1987), discussed in detail below.

A. Osterholtz, *The Bodies Poetic*, Bioarchaeology and Social Theory,
https://doi.org/10.1007/978-3-031-85998-4_3

3.1 Why Study Bodies?

Elements of lived experience can become encoded or inscribed on our bones. In bioarchaeological analysis, the focus tends to be on paleopathological indicators of disease and stress during the life course, exposure to trauma, and changes in the skeletal system corresponding to muscle development. Bioarchaeological analysis allows for significant time depth in analyses, allowing researchers to chart diachronic changes over thousands of years. In my dissertation research, for example, I traced patterns of trauma and stress over a thousand year period on the island of Cyprus (Osterholtz, 2015).

There is a growing body of literature relating to the social construction of the body. Beginning with Foucault (2003 [1963]) and Wells (1964) and continuing through the work of bioarchaeologists in recent years (e.g., S. C. Agarwal & Glencross, 2011; Klaus et al., 2017; Zuckerman & Martin, 2016), there has been an increasing awareness of the role that culture plays in how the body looks and functions. This is explored by George Armelagos (2003) in his book chapter *Bioarchaeology as Anthropology*, which situates bioarchaeology as a discipline uniquely situated to incorporate multiple lines of evidence (e.g., biological, cultural, archaeological, and so on) and multiple theoretical models to understand lived experience through the study of human remains. In employing the biocultural model (as outlined by Armelagos), bioarchaeologists seek to contextualize the study of human remains historically and culturally.

One line of evidence employed is the study of the bones themselves through paleopathology and the establishing of demographic factors for individual bodies. Paleopathological analysis forms a key element to bioarchaeological reconstructions of lived experience. Much of paleopathological analysis is designed around the concept of stress. Goodman et al. (1984b) assign indications of stress to three main categories: indicators of general, cumulative stress; indicators of general, episodic stress; and indicators of stress associated with specific diseases or conditions. General stress indicators tend to be nonspecific responses since bone can only respond in a limited number of ways (Goodman et al., 1984a). Paleopathological data need to be examined by demographic factors such as age at death and estimated sex. The purpose of this volume is not to provide methodology, but the use of demographic factors is instrumental in developing a biological profile of an individual. Through the use of biological profiles, we can begin to understand if some individuals are more likely to experience specific disease (e.g., those who work with animals may be more likely to be exposed to disease via zoonotic vectors) or if males or females are dying younger or have fewer signs of systemic stress. In working with commingled assemblages, my coauthors and I have previously referred to this type of demographic data as "baseline data" (Osterholtz et al., 2014), meaning that we must understand the treatment of the body based on elements of identity that we can estimate such as age at death and sex. This same focus on baseline data and the intersections of those different elements of identity will be important when we examine mortuary processes as well.

3.2 The Three Bodies, Structure, Practice, and Power

In 1987, Nancy Scheper-Hughes and Margaret Lock (1987) introduced the Three Body Model. These can be thought of as the *Individual Body*, understood in the "phenomenological sense of the lived experience of the body-self"(1987, p. 7). These are the physical attributes including individual lived experience and "the ways in which the body is received and experienced in health and sickness"(1987, p. 7). The *Social Body*, then, refers the use of the body as a symbol. If the body can be seen as both a biological and cultural construction, the use of the body as a symbol of natural processes and vice versa can be used to understand both. Scheper-Hughes and Lock, in particular note the use of the human body to "domesticate the spaces in which humans reside"(1987, p. 20). The use of the body as symbol means that the body, and the characteristics of the body will impact the lived experience of the individual (e.g., different lived experiences for men and women in ancient Greece). Examination of the body can help elucidate the social constraints that the body was crafted under (Martin et al., 2013). This can be viewed as a bottom-up approach thinking about how the Individual Body comes to symbolize and embody the social structure in which it belongs. The *Body Politic* refers to the "regulation, surveillance, and control of bodies (individual and collective)"(Scheper-Hughes & Lock, 1987, p. 7). This can be viewed as a top-down approach. "When the sense of social order is threatened… the symbols of self-control become intensified along with those of social control"(Scheper-Hughes & Lock, 1987, p. 24).

In a Poetics sense, the three bodies are present in the amplified messaging. The amplified messaging can be seen as the Body Politic, the way the body is controlled, the individual experience of cultural performance is that of the Individual Body, mediated by the intersectionality (discussed in further detail below) of the Social Body.

3.2.1 *Bodies and Buffering*

The three bodies model has become useful within bioarchaeology as a way to conceptualize a view of the body that is both a physical and a cultural construct and as a way for us to begin to theorize how cultural processes may become inscribed in individual bodies and bones.

The interplay of the three bodies is important for interpretation of the physical remains we analyze in bioarchaeology. Identity, in itself, has an effect on health. We can see this as a relationship between identity and group inclusion leading to social buffering. Social or cultural buffering is a concept commonly employed in bioarchaeology (e.g., Bengtson, 2012; Dean et al., 1994; Geber, 2014; Goodman et al., 1984a, 1988; Reitsema & McIlvaine, 2014; Temple, 2019; Temple & Goodman, 2014). Social buffering should be seen as a set of rules that govern human-human, human-animal, and human-environment interactions. This definition reflects the complex nature of buffering. As noted by Temple and Goodman (2014, p. 188), "[c]ulture could be both a reducer or buffer of environmental constraint/stressors and a system to induce new stressors." An important element to how we view social buffering is that it not uniformly protective. Social rules and expectations may put some individuals at greater risk of infectious disease than others or may impact overall health. One way to envision this would be the individual as an egg, with the social buffering structures as bands of barbed wire wrapped around the egg (Fig. 3.1). Stone and Sanders (2020) explore the experience of being female during the Victorian era. For women of lower socioeconomic status, employment was difficult to come by. Women and girls were commonly employed in factories manufacturing artificial flowers and matchsticks, for example. They were sought for their perceived dexterity as well as the social convention of lower wages for female employees (Greenlees, 2007; Jowitt & McIvor, 1988; Stone & Sanders, 2020). This exposed these women and girls to numerous hazardous materials and toxins, but without employment they wouldn't have been able to survive in the urbanized environment of the Victorian era. Buffering, as this example shows, should not be seen as universally protective, but should be seen as a complex set of relationships based on an individual's lived experience.

Fig. 3.1 Representation of social buffering as both protective and exposing the individual to potential harms. The individual is represented by the egg, and the social buffering structures are represented by the bands of barbed wire

3.2.2 Structur(ation) and Practice

Practice theory is concerned with the identification of the "conduct of everyday life, tradition and routine" (Silliman, 2001, p. 191) and the relations of this routine to lived experience and power relationships. Ortner (1984, p. 148) asserts that "modern practice theory seeks to explain the relationship(s) that obtain between human action, on the one hand, and some global entity which we may call 'the system' on the other." In effect, the heart of practice theory as utilized by Sherry Ortner and others is the interplay between the individual and social bodies. Within practice theory, "social agents are frequently conceived of as individuals with goals, intentions and subjectivities and as historic moments in the negotiation of social structure, strategies and relations" (Silliman, 2001, p. 192).

These social agents (or individual bodies) have agency to make decisions within a culturally and historically contingent set of frameworks (structuration according to Giddens, 1984), meaning that there are some options that are more culturally appropriate than others and that these different pathways will vary in time and space. Structure and agency are mutually dependent upon each other for Giddens.

What constitutes normal practice? And what are the constraints placed upon social agents? Practice theory draws on the concept of and interplay between Bourdieu's *doxa* and agency. *Doxa* can be thought of as the shared givens inherent to social interaction (Bourdieu, 1977, p. 166) through which "the established cosmological and political order is perceived… as… a self-evident and natural order which goes without saying and therefore goes unquestioned." *Doxa* may also be seen as, "cultural knowledge that are so deeply taken for granted that alternatives are unimaginable"(Ortner, 2001, p. 271). *Doxa* encompasses the everyday experiences that we take as givens, without thought, something we may not even know to question. In some instances, these assumptions can be recognized and multiple alternatives can be followed (heterodoxy), while orthodoxy is viewed as an attempt to standardize and reinforce original assumptions. Agency informs the choices to either follow orthodoxic strictures or act to change them (through heterodoxic actions), making, unmaking, and remaking the world (Ortner, 2016, p. 63). Building on Giddens (1984), Silliman (2001, p. 192) approaches practice theory as a dialectic where "individuals often conducted their daily affairs in intentional and strategic *and* in routinized, nondiscursive or preconscious ways. To state the common expression, social agents are both constrained and enabled by culture [emphasis in original]."

Agency, in effect, is the explanation for deviations from orthodoxic social structures, and the root cause of all social change. Social change, though is mediated by social structures, and often centered around inequalities and power differentials (discussed more below). Silliman (2001, p. 194) adds in the concept of *practical politics* as well. This is defined as the "negotiation of politics of social position and identity in daily practices." This approach acknowledges social standing as a limiting or enabling factor present in the choices of any individual social agent (Individual Body). Practical politics can be viewed in relation to *capital* as utilized by Bourdieu.

Bourdieu looks at capital as economic, cultural, and social. Economic capital relates to the economic resources available to the individual social actor. Cultural capital can be embodied (embedded knowledge and practice including skills, accents, mannerisms that convey higher or lower capacities), objectified (material things with cultural significance) and institutionalized (symbols of knowledge and authority such as titles or degrees). Individuals sharing cultural capital can form collectives with similar backgrounds, and cultural capitol can form the basis for inequality. Social capitol is summarized as 'who an individual knows'(S. Schrader, 2019, p. 22), increasing with the higher the social capital of other people within their social network. For both Giddens and Bourdieu, society is relational, a dialectic between social structures and the individual, but they take slightly different viewpoints on how that relationship works and how conscious the individual is of their own practice. For Giddens, actions performed by individuals are both subconscious/unintentional and active/informed. For Bourdieu, doxa is mostly subliminal or unconscious, but for Giddens, individuals reflect on their actions and identities and use agency intentionally to alter *doxa* and *habitus* (see discussion in S. Schrader, 2019, pp. 24–25).

Doxa, *Agency*, *Capital*, and *Practical Politics* contribute to the every day activities that an individual experiences. All of these make up an individual's *habitus* (Bourdieu, 1977). Practice theory examines how individual agency impacts or reproduces *doxa* (Ortner, 2001). Put into a three bodies model, this interplay between the individual and social bodies may lead to different degrees or types of social buffering, differences in lived experience, and differential health outcomes that become inscribed in bone. I argue this is effectively what we view as *lived experience* in bioarchaeology, the day to day, embodied, inscribed-in bone-experience of an individual's identity (as identified by mortuary context, estimated age at death and sex, differential disease exposure/treatment, and trauma experience).

3.2.3 Practice and Power Relationships

Identity can also be viewed as a relational process placing an individual in society. This will typically involve power-relationships of some kind, even if that power is not permanent. We are used to thinking in terms of hierarchies that are more or less permanent such as higher or lower socioeconomic status. Hierarchies are structured relations in which some levels are subordinate to other levels (Crumley, 1979). Crumley (1995) breaks this into two different types of hierarchies: scalar and control. Scalar hierarchies allow for all levels to affect each other, while control hierarchies allow for effect in only one direction. Hierarchies provide social order, but human behavior and social standing is not static, but is constantly negotiated.

The concept of hierarchy, particularly in the relations between the Individual Body and the Body Politic. This top-down control of the Individual Body by the state or those in power has been the focus of anthropological theory for many years. Foucault (1977, 1980) brings this very much to the forefront with his analysis of the prison system. Lancaster (Lancaster, 2010, p. 63) builds on this in his analysis of the US prison system, noting that the "carceral state" (a phrase borrowed from Foucault), is not defined strictly by the number of individuals incarcerated, but one in which the "prison system supplies the norms for governance in general." This can be seen as an increased desire for security, something that trickles down to an increase in gated communities in what Caldeira terms an "aesthetic of security" (Caldeira, 2001, p. 291).

The control of the physical body by the Body Politic can be seen as a reinforcement of cultural views promoted by the Body Politic (whether these be heterodoxic or orthodoxic). Klein (2007, p. 7) discusses this in relation to the use of state-sponsored torture, quoting Uruguayan author Eduardo Galeano saying 'How can this inequality be maintained if not through jolts of electric shock?' As discussed above, practice theory gives us a way to discuss both the maintenance of orthodoxic cultural structures but also the mechanisms by which culture change can occur through resistance movements. Resistance to the doxa (exemplified in the Social Body) by the Individual Body's agency can be viewed as Ortner (2016, p. 63) does, "if we make the world through social practice, we can unmake and remake the world through social practice."

The concept of heterarchy may also be informative here. Crumley (1995, p. 3) defines heterarchy as "the relation of elements to one another when they are unranked or when they possess the potential for being ranked in a number of different ways." Essentially, Crumley (1995, p. 4) argues that power relations in human groups are complex, rankings may shift or be higher or lower in relation to what is being examined: heterarchical relationships among elements at one spatial scale or in one dimension (members of the same club) may be hierarchical at another (the privilege of seniority in decision making). Heterarchy is both a structure and a condition." Heterarchy may provide elements of social buffering while still allowing for cultural change (Crumley, 2015).

3.3 Embodiment and the Bodies

Embodiment theory is incredibly applicable to bioarchaeology. The archaeological foundations of embodiment theory are discussed in Chap. 4, here I want to provide a short discussion of the applications within bioarchaeology. At its heart, embodiment theory argues that humans are a product of both our biology and culture. Krieger (2005, p. 351) sees embodiment as a process, entailing "the temporal transformation of bodily characteristics as a consequence of animate beings' terms of engagement in their world." The body will physically change in response to and as a result of social processes, but those changes will depend on the identity (including intersectional elements of sex, gender, socioeconomic status, and so on) of the individual who experiences them.

Schrader and Torres-Rouff (2020, p. 15) argue that embodiment theory is "a conduit between the corporeal body and the social world with which the body engages." Critically, this viewpoint allows for the analysis of the physical body to understand how an individual engaged with the world around them, how their intersectional identities impacted their biology and how their biology impacted their lived experience. Human remains can be seen as "experiential, social, and agentive" (Schrader & Torres-Rouff, 2020, p. 15). Tung (2021, p. S125) argues that the body is both "brush and canvas", acting to convey information about social relationships, gender roles, and status. Daily activities and identities (that form the basis of lived experience and identity (see Identity formation section in Chap. 2) become inscribed on bone or "highlight the social nature of skeletal remains"(Schrader, 2019, p. 27).

As malleable pieces of material culture that are constantly changing and remodeling, human remains have the ability to shed light on both cultural and biological history, and how the two interact (Schrader, 2019). This has become a strong focus within bioarchaeology, with the examination of embodied social processes such as violent interaction (e.g., Duncan, 2012; Kenzler, 2019; Martin, 2016; Milner, 1995; Osterholtz, 2018a; Šlaus et al., 2012; Toyne & Narváez Vargas, 2014; Tung, 2007; Walker, 1997) structural violence (e.g., de la Cova, 2012, 2020a, b; Hodge et al., 2017; Hodge & Nystrom, 2020; Nystrom, 2011, 2016; Nystrom et al., 2017), class or ethnic identity (e.g., Klaus, 2008; Klaus et al., 2010; Tung, 2012a, b; Tung & Knudson, 2011), gender and gender-based violence (e.g., Geller, 2008; Harrod & Martin, 2010; Lorber & Martin, 2011; Martin & Akins, 2001; Martin et al., 2010; Osterholtz & Martin, 2017; Tung, 2021) and mortuary processes (Geller, 2004, 2012; Pérez, 2006), to name a few.

3.4 Intersectionality

Intersectionality is an important, emerging paradigm within bioarchaeology. This recognizes that social identity is multifaceted and will have overlapping categories to which an individual will belong (e.g., social race, gender, age, socioeconomic status). These intersectional identities are elements that create environments of structural violence and inequality (Collins, 2015; Crenshaw, 1989). Intersectionality theory originates from Black feminist and legal scholarship as a way to frame and understand experiences of Black women in the US (Crenshaw, 1989, 1991). Franklin (2001) called for the incorporation of Black feminist theory into social archaeology. Defining intersectionality is a difficult task, Collins (2015, p. 1) prefers to examine this more as a process than a term that provides an easy definition. She instead describes the "general contours" of intersectionality as race, class, gender, sexuality, ethnicity, nation, ability and age, noting that these "operate not as unitary, mutually exclusive entities, but as reciprocally constructing phenomena that in turn shape complex social inequalities." Intersectionality, at its heart, is an understanding that social identity is shaped by cultural factors (e.g., social race and gender) as well as biological factors (e.g., age). This is similar to the way Basanti (2020) views the body and identity.

In discussing identity, Basanti (2020, p. 103) introduces a biological metaphor, stating that "the body may not be a simple shell and a person may extend beyond their ostensible confines to create other kinds of entities, such as collectives, hybrids, ghosts, ancestors, nations, deities, and so on. Since these entities defy our standards of skin as a marker of the real, we may see them as only 'socially defined' (i.e., not really 'real') but in the processes of their formation they are not dissimilar from the relationships between swarms of cells, constituent proteins and amino acids, bacteria, and other living forms that compose the body. The body, while seemingly whole, is actually several creatures interacting and held together by the skin (which is itself a swarm of cells constantly replaced). The interactions of the body become so tightly wound that their constituents lose their independence, allowing a new greater entity to emerge." It is through the use of ritual acts that societies bind the body's "buzzing swarms into different creatures based on the relationships they most value."

Due to the history of intersectionality, developed in modern contexts by modern scholars to understand modern lived experience, this paradigm is well-situated for application to understand modern social phenomena. While it may be very difficult to operationalize this as effectively when examining social identity in the past (particularly during prehistory), a Black feminist approach pushes us to acknowledge the combined social and biological elements that make up social identity will have varying effects on the expression of lived experience embodied in the skeletons we study. Mant et al. (2021) argue that intersectional identities are "written in bone", that "intersecting factors and structural inequalities can impact not only physical health, but osteological health." Fractures, for example, can be viewed as a complex representation of intersectional identities. Studies of bone health and structure are impacted by both age and sex. Osteon size and frequency decrease with age (Stout et al., 2019), remodeling rates are also lower (Martin, 1993). Accumulation of microdamage in aging bone affects fracture risk, so older individuals are more prone to fracture (Stout et al., 2019). Other elements that impact bone health include lifestyle choices such as amount of exercise, smoking, and alcohol consumption (Agarwal, 2008). Psychosocial stress also plays a part in the likelihood of fracturing. Jiang et al. (2018) found a relationship between post-traumatic stress disorder and increased postcranial fracture risk.

de la Cova (2020a) argues that intersectionality can be used to investigate skeletal remains if four steps are followed:

1. research into and description of the cultural, social, political, and historical context through an analysis of material culture, historical sources, and laboratory analyses.
2. 'blinded' biological and paleopathological assessment of skeletons through skeletal analysis.
3. Uniting contextual information with the skeleton; and
4. Paleopathological and statistical analyses, and examination of lesion location/type/frequency with respect to cultural and biological intersections.

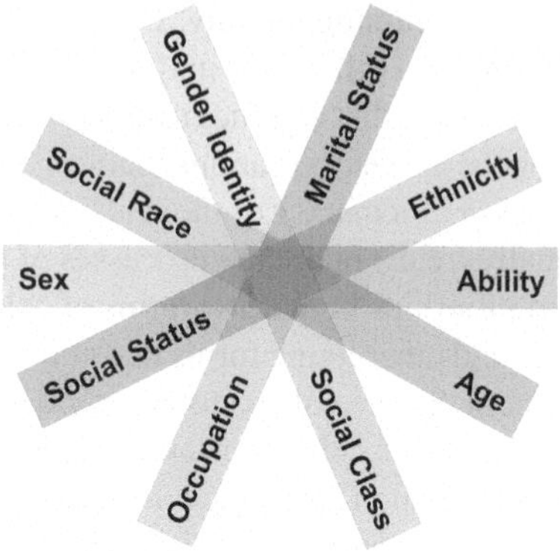

Fig. 3.2 Representation of how different elements of identity can intersect and impact lived experience of an individual

3.4.1 Intersectional Trauma, Intersectional Processing

Mant et al. (2021) provide an excellent discussion of the intersectionality of trauma. They argue that since traumatic lesions are effectively permanent markers, they are excellent ways to examine intersectionality. By examining the intersectional elements of identity with respect to traumatic lesion frequency and location, the role of violence within society may be postulated. Some intersectional groups will have greater or lesser exposure to different types of violence than others. This has been examined through use of the biocultural model by scholars such as Debra Martin in her analysis of La Plata. Scholars such as de la Cova, Mant, and Martin show how the intersectional model provides a critical lens and highlights the different lived experiences of these intersectional groups (Fig. 3.2).

Trauma may be acute (a single incident) or recidivist (meaning repeated insults). When contextualized with historical elements, intersectionality can be a lens through which structural violence can be examined as well. Martin and Harrod (2015) are contextual in their approach to gender and violence. Martin et al. (2010) combine a study of traumatic lesions, patterns of muscle marker development, and mortuary context to understand the subaltern group of females at La Plata. This is the expression of structural violence writ large, these individuals were not given a standard mortuary treatment and were subjected to heavy workloads as well as having an elevated frequency of antemortem traumatic lesions.

Martin and colleagues (2021) work is illustrative of how intersectional identities can impact mortuary processing as well. The different elements of the social identities impacted where they were buried and with what burial goods. As noted in Chap. 4, the social identity of individuals has long-been inferred based on mortuary context (e.g., Binford, 1971; Saxe, 1970). Intersectionality theory gives the opportunity for a nuanced approach to how those intersections may be influenced by biological and social intersections.

3.5 Bodies and Boundaries

In an archaeological context, the physical body is part of a greater complex of items in time and space. As Sofaer (2006, p. 50) states "The boundary of the body is extended and circumscribed by both corporeal (human) and non-corporeal (technological) elements. The body is, in this sense, 'technomorphic', and can grow and change shape by the addition of new technological aftefacts. The boundary of the body is unclear." It is in this sense, that the archaeological context must be part of the interpretation of bioarchaeological data: without the specific context (archaeological, historical, and cultural), bioarchaeological data cannot truly tell the story of the lived experience of the individual buried. "The grave constructs and restricts, forcing the person and objects into association. It envelopes the body and, as a human creation, is itself a form of material culture"(Sofaer, 2006, p. 50).

Scholars working with Maya remains have taken an approach of permeability and dividuality when examining personhood, identity, and boundaries. Duncan and Schwarz (2014) argue that the body can be seen as permeable. "Permeable bodies have porous boundaries with the outside world and may gain or lose animating essences or aspects of personhood throughout life" (p. 149). They also argue that bodies are partible, or internally divided. Different parts of the body have different "animating essences" (p. 149), indicating that different body parts may be removed from the body for different purposes. The decisions made by the living during processing should be seen as intentional choices that communicate social messaging. This takes the view that the sum of the body's parts may be greater than the whole.

The approach taken by these Maya researchers informed a view used in the analysis of processed human remains at Sacred Ridge (Martin & Osterholtz, 2020; Osterholtz, 2012, 2013, 2014, 2018a, b; Stodder et al., 2010). Essentially, we have argued that the act of *creating* the commingled assemblage (i.e., the processing of the bodies) was integral to the social meaning of the assemblage. It was the creation of parts that were greater (in terms of social meaning) than the individual bodies could ever have been. These parts change the relationship between the living and the dead and are transformed into relational bodies (as the term is used by Duncan and Schwarz (2014, p. 149)) to be bodies that are defined "in terms of their relationships with other people and objects and, as such, may well be quite fluid in definition and composition."

3.6 Summary

Bioarchaeology tends to focus on the physical body as a primary way to understand larger social structures based on the understanding that the physical body is mediated by the Social Body and the Body Politic. The way that we understand those relationships changes with the different approaches that we take within bioarchaeology, applying different theoretical models and heuristic devices such as the biocultural model to organize our observations into inference. Without individual bodies, we cannot understand the greater social processes that are present. By examining the individuals, we build a model for the larger social relationships evident in how the living interact with each other and the dead.

References

Agarwal, S. (2008). Light and broken bones: Examining and interpreting bone loss and osteoporosis in past populations. In S. R. Sanders & M. A. Katzenberg (Eds.), *Biological anthropology of the human skeleton* (pp. 387–410). Wiley-Liss.

Agarwal, S. C., & Glencross, B. A. (Eds.). (2011). *Social bioarchaeology*. Wiley-Blackwell.

Armelagos, G. J. (2003). Bioarchaeology as anthropology. In S. D. Gillespie & D. L. Nichols (Eds.), *Archaeology is anthropology* (Vol. No. 13, pp. 27–41). Archaeological Papers of the American Anthropological Association.

Basanti, D. S. (2020). Poetics of the house: Changing realities of body and person in aksumite mortuary practices. In A. J. Osterholtz (Ed.), *The poetics of processing: Memory formation, cosmology, and the handling of the dead* (pp. 103–122). University Press of Colorado.

Bengtson, J. D. (2012). *A biocultural perspective on sex and gender in late prehistoric West Central Illinois: Growth patterns, Mississippianization, and intracemetery social differentiation*. (3521718 Ph.D.). Michigan State University, Ann Arbor. ProQuest Dissertations & Theses Full Text database.

Binford, L. R. (1971). Mortuary practices: Their study and their potential. In J. A. Brown (Ed.), *Approaches to the social dimensions of mortuary practices* (p. 25). Memoirs of the Society for American Archaeology.

Bourdieu, P. (1977). *Outline of a theory of practice*. Cambridge University Press.

Caldeira, T. P. R. (2001). *City of walls: Crime, segregation, and citizenship in Sao Paulo*. University of California Press.

Collins, P. H. (2015). Intersectionality's definitional dilemmas. *Annual Review of Sociology, 41*(1), 1–20. https://doi.org/10.1146/annurev-soc-073014-112142

Crenshaw, K. (1989). Demarginalizing the intersection of race and sex: A Black feminist critique of antidiscrimination doctrine, feminist theory and antiracist politics. *University of Chicago Legal Forum, 1*, 139–167.

Crenshaw, K. (1991). Mapping the margins: Intersectionality, identity politics, and violence against women of color. *Stanford Law Review, 43*, 1241–1299.

Crumley, C. L. (1979). Three locational models: An epistemological assessment of anthropology and archaeology. In M. B. Schiffer (Ed.), *Advances in archaeological method and theory* (pp. 141–173). Academic.

Crumley, C. L. (1995). Heterarchy and the analysis of complex societies. *Archaeological Papers of the American Anthropological Association, 6*(1), 1–5. https://doi.org/10.1525/ap3a.1995.6.1.1

Crumley, C. L. (2015). Heterarchy. In R. A. Scott & S. M. Kosslyn (Eds.), *Emerging trends in the social and behavioral sciences* (pp. 1–14). https://doi.org/10.1002/9781118900772.etrds0158.

de la Cova, C. (2012). Patterns of trauma and violence in 19th-century-born African American and Euro-American females. *International Journal of Paleopathology, 2*(2), 61–68. https://doi.org/10.1016/j.ijpp.2012.09.009

de la Cova, C. (2020a). Making silenced voices speak: Restoring neglected and ignored identities in anatomical collections. In C. M. Cheverko, J. R. Prince-Buitenhuys, & M. Hubbe (Eds.), *Theoretical approaches in bioarchaeology* (pp. 150–169). Routledge.

de la Cova, C. (2020b). Processing the destitute and deviant dead: Inequality, dissection, politics, and the structurally violent legalization of social marginalization in American anatomical

collection. In A. J. Osterholtz (Ed.), *The poetics of processing: Memory formation, cosmology, and the handling of the dead* (pp. 212–234). University Press of Colorado.
Dean, J. S., Doelle, W. H., & Orcutt, J. D. (1994). Adaptive stress, environment, and demography. In G. J. Gumerman (Ed.), *Themes in southwest prehistory* (pp. 53–86). School of American Research Press.
Duncan, W. N. (2012). Biological distance analysis in contexts of ritual violence. In D. L. Martin, R. P. Harrod, & V. R. Pérez (Eds.), *The bioarchaeology of violence* (pp. 251–275). University of Florida Press.
Duncan, W. N., & Schwarz, K. R. (2014). Partiable, permeable, and relational bodies in a maya mass grave. In A. J. Osterholtz, K. M. Baustian, & D. Martin (Eds.), *Commingled and disarticulated human remains: Working toward improved, theory, method, and data* (pp. 149–172). Springer.
Foucault, M. (1977). *Discipline and Punish: The birth of the prison* (A. Sheridan, Trans.). New York: Vintage Books.
Foucault, M. (1980). Eye of power. In C. Gordon (Ed.), *Power and knowledge: Selected interviews and other writing* (pp. 145–165). Pantheon.
Foucault, M. (2003 [1963]). *The birth of the clinic: An archaeology of medical perception* (A. M. Sheridan, Trans.). New York: Taylor & Francis.
Franklin, M. (2001). A Black feminist-inspired archaeology? *Journal of Social Archaeology, 1*, 108–125.
Geber, J. (2014). Skeletal manifestations of stress in child victims of the Great Irish Famine (1845–1852): Prevalence of enamel hypoplasia, Harris lines, and growth retardation. *American Journal of Physical Anthropology*, n/a-n/a. https://doi.org/10.1002/ajpa.22567.
Geller, P. L. (2004). *Transforming bodies, transforming identities: A consideration of pre-Columbian maya corporeal beliefs and practices*. (Ph.D. Dissertation). University of Pennsylvania.
Geller, P. L. (2008). Conceiving sex: Fomenting a feminist bioarchaeology. *Journal of Social Archaeology, 8*(1), 113–138.
Geller, P. L. (2012). Parting (with) the dead: Body partibility as evidence of ancestor veneration. *Ancient Mesoamerica, 23*(1), 115–130.
Giddens, A. (1984). *The constitution of society: Outline of the theory of structuration*. University of California Press.
Goodman, A. H., Brooke Thomas, R., Swedlund, A. C., & Armelagos, G. J. (1988). Biocultural perspectives on stress in prehistoric, historical, and contemporary population research. *American Journal of Physical Anthropology, 31*(S9), 169–202. https://doi.org/10.1002/ajpa.1330310509
Goodman, A. H., Martin, D., Armelagos, G. J., & Clark, G. (1984a). Indications of stress from bones and teeth. In M. N. Cohen & G. J. Armelagos (Eds.), *Paleopathology at the origins of agriculture* (pp. 13–49). Academic.
Goodman, A. H., Martin, D. L., Armelagos, G. J., & Clark, G. (1984b). Indications of stress from bone and teeth. In M. N. Cohen & G. J. Armelagos (Eds.), *Paleopathology at the origins of agriculture* (pp. 13–39). Academic.
Greenlees, J. (2007). *Female labour power: Women workers' influence on business practices in the British and American Cotton Industries, 1780–1860*. Ashgate.
Harrod, R. P., & Martin, D. L. (2010). *Gender inequality in the ancient southwest (AD 900–1350): Non-lethal head wounds as evidence of institutionalized violence*. Paper presented at the 109th Annual Meeting of the American Anthropological Association, New Orleans.
Hodge, C. J., Morgan, M. E., & Rousseau, J. L. (2017). Teachings of the dead: The archaeology of anatomized remains from Holden Chapel, Harvard University. In K. C. Nystrom (Ed.), *The bioarchaeology of dissection and autopsy in the United States* (pp. 115–142). Springer.
Hodge, C. J., & Nystrom, K. C. (2020). Dissection as social process: Anatomical settings in the nineteenth-century United States. In A. J. Osterholtz (Ed.), *The poetics of processing: Memory formation, cosmology, and the handling of the dead* (pp. 189–211). University Press of Colorado.

Jiang, T., Veres, K., Farkas, D. K., Lash, T. L., Sørensen, H. T., & Gradus, J. L. (2018). Posttraumatic stress disorder and incident fractures in the Danish population. *Osteoporosis International, 29*, 2487–2493.

Jowitt, J. A., & McIvor, A. J. (Eds.). (1988). *Employers and labour in the english textile industries, 1850–1939*. Routledge.

Kenzler, H. (2019). Post-medieval burial customs in Germany – An archaeological perspective on materiality and spatiality. *Mortality, 24*(2), 123–144. https://doi.org/10.1080/13576275.2019.1585781

Klaus, H. D. (2008). *Out of Light Came Darkness: Bioarchaeology of mortuary ritual, health, and ethnogenesis in the Lambayeque Valley Complex, North Coast Peru, A.D. 900–1750*. (Unpublished Ph.D. dissertation). Ohio State University, Columbus.

Klaus, H. D., Guro, M., & Centurión, J. (2010). Bioarchaeology of human sacrifice: Violence, identity and the evolution of ritual killing at Cerro Cerrillos, Peru. *Antiquity, 84*(326), 1102–1122.

Klaus, H. D., Harvey, A. R., & Cohen, M. N. (Eds.). (2017). *Bones of complexity: Bioarchaeological case studies of social organization and skeletal biology*. University Press of Florida.

Klein, N. (2007). *The shock doctrine: The rise of disaster capitalism*. Henry Holt and Company.

Krieger, N. (2005). Embodiment: A conceptual glossary for epidemiology. *Journal of Epidemiology and Community Health, 59*, 350–355.

Lancaster, R. (2010). Republic of fear: The rise of punitive Governance in America. In H. Gusterson & C. Besteman (Eds.), *The Insecure American: How we got here and what we should do about it* (pp. 27–44). University of California Press.

Lorber, J., & Martin, P. Y. (2011). The socially constructed body: Insights from feminist theory. In P. Kvisto (Ed.), *Illuminating social life: Classical and contemporary theory revisited* (5th ed., pp. 183–206). Pine Forge Press.

Mant, M., de la Cova, C., & Brickley, M. B. (2021). Intersectionality and trauma analysis in bioarchaeology. *American Journal of Physical Anthropology*, n/a(n/a). https://doi.org/10.1002/ajpa.24226.

Martin, R. B. (1993). Aging and strength of bone as a structural material. *Calcified Tissue International, 53*, S34–S39.

Martin, D. L. (2016). Hard times in dry lands: Making meaning of violence in the ancient southwest. *Journal of Anthropological Research, 72*(1), 1–23.

Martin, D. L., & Akins, N. J. (2001). Unequal treatment in life as in death: Trauma and mortuary behavior at La Plata (AD 1000–1300). In D. R. Mitchell & J. L. Brunson-Hadley (Eds.), *Ancient burial practices in the American southwest* (pp. 223–248). University of New Mexico Press.

Martin, D. L., & Harrod, R. P. (2015). Bioarchaeological contributions to the study of violence. *American Journal of Physical Anthropology, 156*(S59), 116–145. https://doi.org/10.1002/ajpa.22662

Martin, D. L., Harrod, R. P., & Fields, M. (2010). Beaten down and worked to the bone: Bioarchaeological investigations of women and violence in the ancient Southwest. *Landscapes of Violence, 1*(1), 3.

Martin, D. L., Harrod, R. P., & Pérez, V. R. (2013). *Bioarchaeology: An integrated approach to working with human remains*. Springer.

Martin, D. L., & Osterholtz, A. J. (2020). The poetics of corpse fragmentation and processing in the ancient southwest. In A. J. Osterholtz (Ed.), *The poetics of processing: Memory formation, cosmology, and the handling of the dead* (pp. 85–102). University Press of Colorado.

Milner, G. R. (1995). An osteological perspective on prehistoric warfare. In L. A. Beck (Ed.), *Regional approaches to mortuary analysis* (pp. 221–244). Plenum Press.

Nystrom, K. C. (2011). Postmortem examinations and the embodiment of inequality in 19th century United States. *International Journal of Paleopathology, 1*(3–4), 164–172. https://doi.org/10.1016/j.ijpp.2012.02.003

Nystrom, K. C. (2016). *The bioarchaeology of dissection and autopsy in the United States*. Springer.

Nystrom, K. C., Sirianni, J. E., Higgins, R. L., Perrelli, D. J., & Liber Raines, J. L. (2017). Structural inequality and postmortem examination at the erie county poorhouse. In K. C. Nystrom (Ed.), *The bioarchaeology of dissection and autopsy in the United States* (pp. 279–300). Springer.
Ortner, S. B. (1984). Theory in anthropology since the sixties. *Comparative Studies in Society and History, 26*, 126–166.
Ortner, S. B. (2001). Practice, power and the past. *Journal of Social Archaeology, 1*, 271–278.
Ortner, S. B. (2016). Dark anthropology and its others: Theory since the eighties. *HAU: Journal of Ethnographic Theory, 6*(1), 47–73.
Osterholtz, A. J. (2012). The social role of hobbling and torture: Violence in the prehistoric Southwest. *International Journal of Paleopathology, 2*(2–3), 148–155. https://doi.org/10.1016/j.ijpp.2012.09.011
Osterholtz, A. J. (2013). Hobbling and torture as performative violence: An example from the prehistoric southwest. *Kiva, 78*(2), 123–144.
Osterholtz, A. J. (2014). *Patterned processing as performative violence at sacred ridge*. Paper presented at the 79th Annual Meeting of the Society for American Archaeology, Austin, TX.
Osterholtz, A. J. (2015). *Bodies in motion: A bioarchaeological analysis of migration and identity in bronze age cyprus (2400–1100 BC)*. (Ph.D. Dissertation). University of Nevada Las Vegas, Las Vegas, NV.
Osterholtz, A. J. (2018a). Bioarchaeological approaches to southwestern violence. *Kiva, 84*(4), 399–402.
Osterholtz, A. J. (2018b). Each one the same: Performance, demography, and violence at sacred ridge. In C. P. Anderson & D. L. Martin (Eds.), *Massacres: Bioarchaeology and forensic anthropology approaches* (pp. 82–93). University of Florida Press.
Osterholtz, A. J., Baustian, K. M., Martin, D. L., & Potts, D. T. (2014). Commingled human skeletal assemblages: Integrative techniques in determination of the MNI/MNE. In A. J. Osterholtz, K. M. Baustian, & D. L. Martin (Eds.), *Commingled and disarticulated human remains: Working toward improved theory, method, and data*. Springer.
Osterholtz, A. J., & Martin, D. L. (2017). The poetics of annihilation: On the presence of women and children at massacre sites in the ancient southwest. In C. Tegtmeyer & D. L. Martin (Eds.), *Bioarchaeology of women and children in times of war: Case studies from the Americas*. Springer.
Pérez, V. R. (2006). *The politicization of the dead: An analysis of cutmark morphology and culturally modified human remains from La Plata and Peñasco Blanco (A.D. 900–1300)*. (Doctor of Philosophy Dissertation). University of Massachusetts Amherst.
Reitsema, L. J., & McIlvaine, B. K. (2014). Reconciling "stress" and "health" in physical anthropology: What can bioarchaeologists learn from the other subdisciplines? *American Journal of Physical Anthropology, 155*, 181–185.
Saxe, A. A. (1970). *Social dimensions of mortuary practices*. (Doctor of Philosophy). The University of Michigan, Ann Arbor, MI.
Scheper-Hughes, N., & Lock, M. M. (1987). The mindful body: A prolegomenon to future work in medical anthropology. *Medical Anthropology Quarterly, 1*(1), 6–41. Retrieved from http://www.jstor.org/stable/648769.
Schrader, S. (2019). *Activity, diet, and social practice: Addressing everyday life in human skeletal remains*. Springer.
Schrader, S. A., & Torres-Rouff, C. (2020). Embodying bioarchaeology: Theory and practice. In C. M. Cheverko, J. R. Prince-Buitenhuys, & M. Hubbe (Eds.), *Theoretical approaches in bioarchaeology* (pp. 15–27). Routledge.
Silliman, S. W. (2001). Agency, practical politics and the archaeology of culture contact. *Journal of Social Archaeology, 1*, 190–209.
Šlaus, M., Novak, M., Bedić, Ž., & Strinović, D. (2012). Bone fractures as indicators of intentional violence in the eastern Adriatic from the antique to the late medieval period (2nd–16th century AD). *American Journal of Physical Anthropology, 149*(1), 26–38. https://doi.org/10.1002/ajpa.22083

Sofaer, J. R. (2006). *The body as material culture: A theoretical osteoarchaeology*. Cambridge University Press.

Stodder, A. L. W., Osterholtz, A. J., Mowrer, K., & Chuipka, J. P. (2010). Processed human remains from the sacred ridge site: Context, taphonomy, interpretation. In E. M. Perry, A. L. W. Stodder, & C. A. Bollong (Eds.), *Animas-La Plata Project: XV-Bioarchaeology* (pp. 279–415). SWCA Environmental Consultants.

Stone, P. K., & Sanders, L. S. (2020). *Bodies and lives in Victorian England: Science, sexuality and the affliction of being female*. Routledge.

Stout, S. D., Cole, M. E., & Agnew, A. M. (2019). Histomorphology: Deciphering the metabolic record. In J. E. Buikstra (Ed.), *Ortner's identification of pathological conditions in human skeletal remains* (3rd ed., pp. 91–167). Elsevier Academic Press.

Temple, D. H. (2019). Bioarchaeological evidence for adaptive plasticity and constraint: Exploring life-history trade-offs in the human past. *Evolutionary Anthropology: Issues, News, and Reviews, 28*(1), 34–46. https://doi.org/10.1002/evan.21754

Temple, D. H., & Goodman, A. H. (2014). Bioarcheology has a "health" problem: Conceptualizing "stress" and "health" in bioarcheological research. *American Journal of Physical Anthropology, 155*(2), 186–191. https://doi.org/10.1002/ajpa.22602

Toyne, J. M., & Narváez Vargas, A. (2014). The fall of Kuelap: Bioarchaeological analysis of death and destruction on the eastern slopes of the Andes. In A. K. Scherer, J. W. Verano, & J. Pillsbury (Eds.), *Conflict, conquest, and the performance of war in pre-Columbian America* (pp. 345–368). Dumbarton Oaks Research Institute.

Tung, T. A. (2007). Trauma and violence in the Wari Empire of the Peruvian Andes: Warfare, raids, and ritual fights. *American Journal of Physical Anthropology, 133*, 941–956.

Tung, T. A. (2012a). Violence against women: Differential treatment of local and foreign females in the heartland of the Wari Empire, Peru. In D. L. Martin, R. P. Harrod, & V. R. Pérez (Eds.), *The bioarchaeology of violence* (pp. 180–198). University of Florida Press.

Tung, T. A. (2012b). *Violence, ritual, and the Wari Empire: A social bioarchaeology of imperialism in the ancient andes*. University Press of Florida.

Tung, T. A. (2021). Making and marking maleness and valorizing violence: A bioarchaeological analysis of embodiment in the andean past. *Current Anthropology, 62*, S125–S144.

Tung, T. A., & Knudson, K. J. (2011). Identifying locals, migrants, and captives in the Wari Heartland: A bioarchaeological and biogeochemical study of human remains from Conchopata, Peru. *Journal of Anthropological Archaeology, 30*(3), 247–261.

Walker, P. L. (1997). Wife beating, boxing, and broken noses: Skeletal evidence for the cultural patterning of violence. In D. Martin & D. W. Frayer (Eds.), *Troubled times: Violence and warfare in the past* (pp. 145–175). Gordon and Breach.

Wells, C. (1964). *Bones, bodies, and disease: Evidence of disease and abnormality in early man*. Thames and Hudson.

Zuckerman, M. K., & Martin, D. L. (Eds.). (2016). *New directions in biocultural anthropology*. Wiley.

Chapter 4
Mortuary Archaeology: What Do We Learn from Mortuary Contexts?

Abstract This chapter explores the diverse theoretical approaches to mortuary archaeology, examining how burial practices and mortuary rituals provide insight into social structure, inequality, and cultural meaning. Beginning with processual approaches, the work of Binford, Saxe, and Goldstein highlights how burial locations and grave goods can be used to reconstruct social status and status hierarchies. Marxist interpretations, as outlined by Patterson and Lull, emphasize the role of labor, conflict, and social inequality, with funerary practices reflecting power dynamics between elites and laboring groups. These approaches argue that mortuary rituals and material culture, such as grave goods, are expressions of a community's social relationships and economic conditions. The chapter also examines post-processual perspectives, which shift the focus toward the symbolic and ideological dimensions of mortuary practices. Post-processual scholars stress the importance of understanding death rituals within cultural and historical frameworks, where these rituals function as sites for negotiating social identity and power. The role of materiality and agency is central to this discussion, with objects seen as active agents in shaping human behavior and societal outcomes. Material culture is not merely passive but actively constructs social relations and meaning-making. Ultimately, the chapter emphasizes the importance of material culture in the reproduction and transformation of social power. By examining the ways in which objects and ritual practices shape and reinforce societal norms, archaeologists can uncover the complex processes through which power, identity, and social structures are created, negotiated, and sustained over time. The chapter concludes that the study of death and mortuary rituals, through both material culture and ethnographic perspectives, offers critical insights into the social dynamics that underlie an understanding of death.

Mortuary contexts have long been the focus of archaeological investigation. Throughout archaeological investigation, though, the focus on usually been on the mortuary context (e.g., tomb type, location) and associated grave goods. The inhabitant(s) of the mortuary contexts, though, were usually set aside for specialized analysis. Results of the bioarchaeological analysis of these individuals and

A. Osterholtz, *The Bodies Poetic*, Bioarchaeology and Social Theory,
https://doi.org/10.1007/978-3-031-85998-4_4

populations were seen as completely separate analyses, usually relegated to appendices of reports or separate chapters. In reviewing the role of the mortuary archaeology in the study of materiality, I will present a general review of different approaches that have had an impact on mortuary archaeology.

I will begin in-depth discussions with changes that occurred in the 1970s with the introduction of the New Archaeology, with a special focus on the role of burials and burial goods to infer status and group affiliation. Marxist approaches will also be presented, with its focus on labor and material culture. With the introduction of post-processual approaches, the analytical shift moves to one of meaning and identity. These approaches focus on the interaction between individuals and the embodiment of those relationships. Here, special focus is given to materiality and landscape. How we interact with the landscape and with each other in specific landscapes tells us important things about our relationship with the dead and who we are. The chapter finishes with a discussion of ethnographic approaches, and the importance of the inclusion of ethnographically derived theories such as Van Gennep's and Hertz's models of mourning and ritual.

The biocultural model of bioarchaeology relies on context—archaeological, historical, and cultural—to fully understand lived experience in the past. The mortuary archaeology record has been interpreted differently through time, and understanding those shifts helps us to understand the ways in which our understanding of bioarchaeology has also shifted.

4.1 The New Archaeology

Proponents of the New Archaeology, particularly Binford (1971), emphasized the link between artifacts and human behavior, viewing material culture as direct evidence of social actions. Mortuary contexts, in this framework, were seen as reflections of the societies that produced them, with social organization inferred from burial remains. Binford proposed that mortuary assemblages could reveal key aspects of a society's structure, such as whether it was egalitarian or stratified, and the degree of social complexity present. Grave goods were used to infer individual identity markers, such as age, sex, social affiliation, and status, as well as to understand the conditions and locations of death. The primary focus of these studies was on the objects found in burial contexts rather than the bodies themselves. Central to this approach was the assumption of hierarchy within social groups, with efforts in mortuary practices believed to correlate with an individual's social rank (Tainter, 1978).

The spatial placement of burials was also central to understanding human behavior and social organization. Saxe (1970) argued that burials, especially in contexts of resource scarcity, would be located near valuable resources, thus symbolically claiming territory for the group associated with the burial site. Goldstein (1980, 2006) refined this idea, suggesting that permanent Burial Grounds demarcated the territorial rights of a corporate group, with rights to resources passed down through lineages. More formalized burial spaces indicated a greater connection to corporate identity. While Saxe and Goldstein both focused on the economic aspects of burial location, Goldstein, as a bioarchaeologist, introduced a shift in focus. Unlike Saxe, her work emphasized the human remains themselves, marking a significant departure in bioarchaeological analysis aimed at understanding the lived experiences of past individuals, not just the material culture they left behind.

4.2 Marxist Approaches: Conflict and Meaning

Marxist approaches highlight conflict and meaning. Patterson (2007) argued that Marxist archaeologists saw the world as structured by unobservable processes, that social development was historically constituted, and that the processes observable in modern societies could be used to understand past cultural development. The approach highlighted the role of the majority of society, those doing the work and who were governed as opposed to the elite within society. The focus tended towards the examination of the material culture related to modes of production, socioeconomic formation, lifeways, and culture. The Marxist approach to mortuary archaeology takes the view that "[b]urial sites are deposits of social labour. Both when the society uses death as a mechanism for achieving integration, and when the ritual is intended to express the collective mentality, what is certain is that *the* decisive factor is the social labour as opposed to the individual" (Lull, 2000, p. 578).

Funerary practices are reflections of social relationships, particularly the relationships of inequality between individuals with the form and grave goods associated with burials indicative of a community's economic capacity as well as the social relations of production (Lull, 2000). In this way, communities that lack Economic surplus may still show care to the dead through pooling of labor in tomb construction. The Marxist approach forces an evaluation of labor: how were activities organized and by whom (Chapman, 2003). For Lull, three elements must be considered in a Marxist approach to mortuary archaeology: (1) the form and function of containers for the dead (the source of raw materials, how they're made, and so on); (2) a systematic study of human remains to understand diet, health, and demography); and (3) "archaeological research should establish the relative social value of the grave goods" to then be used to develop hypotheses about social structures in conjunction with an analysis of settlements and political conditions (Lull, 2000, p. 579). Differences in burial treatments between individuals are to be viewed as differences between the socioeconomic and socio-ideological groups responsible for their mortuary performances.

4.3 The Dead Don't Bury Themselves: Post-processual Paradigms

Many of the post-processual approaches developed are a direct critique of the processual approach to mortuary archaeology. Hodder (2004, p. 43) argues that a strict focus on economics as put forward by Saxe tends to "relegate ideology to an epiphenomenon of the assumed primacy of functional contingencies and does not adequately consider the particular symbolic meanings of the monuments and rituals. The Saxe hypothesis [referring to Saxe's Hypothesis 8] not only presents a relatively passive view of society, but also, and more clearly, it disregards the cultural context so central to ideology and ideological functions. When individuals act socially, and represent their actions to others, they necessarily do so within a framework of meaning, and this framework is relative and historically constructed. Without consideration of the cultural context one cannot hope to understand the effects of past social actions." Hodder, as a major proponent of post-processual interpretations, focuses on issues of meaning and identity to a degree that was not possible in previous paradigms. Shanks and Tilley (1987) also argue that the Saxe/Goldstein model is deterministic due to its formulation using cross-cultural ethnographic survey and that the value of such a study is dubious. At its heart, the Saxe/Goldstein model is an economic one without significant attention paid to ritual or non-economic reasoning for mortuary treatment and placement. According to post-processual scholars, burial and mortuary treatment is simply a larger issue than property transmission. In his analysis of the Saxe/Goldstein model, Morris (1991) focused on ideological manipulation of the dead in classical Greece and Rome. He found no consistent relationship amongst cemetery location and land tenure. Instead, he found that the dead were placed in areas to resolve conflicts over power or property, with mortuary ritual providing a space for negotiation between conflicting groups. He states (1991, p. 163) "at Rome, the very limited victories of the plebians and the continuing erosion of the power of the poor were intertwined with the uses of the dead which divided the citizen group at least as much as they united it, drawing a line between elite and commoners every bit as strongly as between citizens and non-citizens."

"The dead do not bury themselves" (Parker Pearson, 1999, p. 3); this simple and concrete idea pushes us to think about the relationship between the living and the dead. Mike Parker Pearson, in his various theoretical writings (e.g., 1982, 1993, 1995, 1998, 1999) shows that mortuary processes can be used to understand how the living relate to the dead. Burial is, in fact, not for the benefit of the dead. It is for the benefit of the living. Funerary practices "are products of 'political' decisions (or sequences of decisions) in which the corpse is manipulated for the purposes of the survivors"(Parker Pearson, 1993, p. 203). This concept of the manipulation of the dead is not restricted to anthropological research. Post-processual paradigms tend to focus on agency and the "emphasis on actors and agency required bodies"(Sofaer, 2006, p. 18). This will be explored in detail later in this volume, but the manipulation of the body can be examined as a political tool around the globe in modern contexts. For example, Verdery (1999) examines politicized corpses in Eastern Europe and the former Soviet Union. Verdery focuses on the return of the bodies of prominent individuals who had been buried elsewhere as a way to reclaim identity with the fall of the Soviet Union. Sometimes these reburials involved treating the remains (in some cases only body parts) as "quasi-religious relics"(Verdery, 1999, p. 17), movement of these individuals is seen as politically transformative, part of larger processes of identity formation. The bodies of these individuals take on particular meanings and become an embodiment of a political, social, and economic transition.

4.3.1 Space and Place: The Role of Landscape Theory and the Phenomenological Approach

Landscape theory in archaeology has been used to understand the use of space in human experience. More broadly, "a landscape of archaeology is an archaeology of place, not just defined in a set of physical nodes in space (cf. Binford, 1982) but in its lived dimensions: experiential, social, ontological, epistemological, emotional, as place and emplacement concern social identity, as much as they concern the economic and environmental aspects of life"(David & Thomas, 2016, p. 38). It is a relatively young interpretive framework, coming to prominence in the 1970s and 1980s (David & Thomas, 2016). In the 1970s, a focus on relationships between humans and their physical environments meant increased refinement in field methodologies and statistical analysis of site distributions across the landscape as well as detailed focus on formation processes. This led to developments in survey technology and methodology (e.g., Foard, 1978) as well as predictive modelling (e.g., Sabloff, 1981), geoarchaeology (e.g., Hassan, 1979), and paleoecology (e.g., Shawcross, 1968). These early projects in Landscape archaeology were based on how landscapes were used and changed by human interaction. More recent uses of landscape focuses on how humans interact with the landscape in a more reciprocal manner, i.e., how the landscapes changes humans even as humans change the landscape. Particularly evident in analyses of built environments, landscape theory has been used to examine the use of ritual space and negotiation of social hierarchy. In his work on the island of Cyprus, Kevin Fisher (2007, 2009a, b) uses the concepts of restrictive space to argue for the creation of social *Place* formation and negotiation of social hierarchy. Building upon the work of Ashmore, Fisher's use of Landscape is integral in the processes of socialization and memory making. Specifically, Ashmore (2004, p. 264) states "monuments and movement and channel access to spectacles of sound or light, manifestations of the sacred." This new view of Landscape is tied to an understanding that space is "actively inhabited, and that social relations and spatial structure are linked recursively" (Ashmore, 2002, p. 1172). Ashmore, in effect is advocating for a *spatial* approach within a *landscape* approach. Space is viewed as an active participant in human activities.

Space and place are experienced phenomenologically (Tilley, 2016, pp. 271–272), and it is therefore an embodied concept. Tilley's approach to space and place claims that "landscapes have *agency* [emphasis in original] in relation to persons. They have a profound effect on our thoughts and interpretations because of the manner in which they are perceived and sensed through our carnal bodies… This is an approach that stresses the *materiality* [emphasis in original] of landscapes: landscapes as real and physical rather than simply organized or represented. The physicality of landscapes acts as a foundation for all thought and social interaction. It profoundly affects the way we think, feel, move, and act…. Landscape is fundamental for human existence because it provides a medium for, and an outcome of individual and social practices." The phenomenological approach centers on the experience of living, moving bodies in relation to landscape, monuments or architectural space (Sofaer, 2006, p. 21). Bringing in what Strathern considers the *dividual self*, people are constituted through social relations with each other, combined with their interaction with landscape (1996).

In my dissertation work, I used this concept of space and place making throughout the Bronze Age on Cyprus to argue that ritual spaces and activities may have provided social structures for the integration of new populations into a third-space or post-colonial concept of what it meant to be Cypriot (Osterholtz, 2015). Specifically, I argued that the movement of burials into the living spaces during the Late Bronze Age and the association of tomb spaces with buildings associated with political and economic importance brought legitimacy to those enterprises through an association with ancestral groups. This is in line with Ashmore's (2004, p. 265) understanding of the power of landscape to "create memories about the proper manner of articulating the living and the dead."

This approach to landscape combines elements of ancestor veneration (e.g., Antonaccio, 1995), politicalization of the dead (Pérez, 2006), landscape and restricted space (Fisher, 2007, 2009a, b), identity theory (Joyce, 2005), materiality (Sofaer, 2006), and aspects of violence theory (Derriennic, 1972; Parsons, 2007). For example, Petts (2003), notes that some grave markers may be indicators of land ownership in early Medieval Wales, where ownership was linked to inheritance and maintained through tribute relationships with "land rights… embedded via kinship through a direct physical presence of the dead in the land. The grave stands as a testimony to a close link between land and kinship" (Petts, 2003, p. 205). The use of stones as markers decreases with the rise of a charter system divorcing land ownership from kinship. Changes to Christianity beginning in the eighth century reconceptualized the relationship between the living and the dead, with the fate of the soul of the dead more influenceable by the actions of the living. This led to a change in how stones were used as memorials. Before, the stones were linked to the bodies of the dead, but after the eighth century, the stones were seen as a place of interaction between the "reader" (p. 208) and the soul of the deceased. This is not a passive relationship, but one where the two groups could intercede with each other. The stones no longer needed to be placed near a grave, but could be separate from the body, still memorializing the dead while not needing the physical body.

Stones, in effect, could have multiple meanings and uses with respect to the dead. Some stones from the late ninth-early tenth centuries from York were examined by Thompson (2003). During this time, the state of the soul of the deceased was unclear, as this was before the codification of Purgatory. For those who wished for the well-being of the dead, and with the means to procure them, the stones performed numerous functions. They served as memorials, status-markers and works of art. Stylistically, they are interpreted as political statements portraying the elite as the inheritors of both English and Scandinavian traditions and power. The stones and their carvings "not only reassured the onlooker, but they also battled for the salvation of the corpse, both in their form, as recumbent slabs, and in their ornament, with the protective beasts and in some cases also, perhaps the crosses" (Thompson, 2003, p. 225). The material nature of monuments, cemeteries, grave markings, and so on provide a physical connection between the living and the dead. This physical connection is rooted in history and reflects social understanding of status and social relationships. Importantly, this is not a static relationship, but changes with time. Buelow (2010, p. 37) argues that "neither identity nor memory will ever be fully developed through the constructed form, because they change with re-telling. New stories are invented each time they are told."

4.3.2 Materiality and Agency

Agency encompasses two fundamental and intertwined concepts, materiality and social reproduction. What do we mean with the concept of materiality? Materiality, as defined by Van Dyke (2016, p. 279) "refers to the reciprocal interactions among humans and the material, embedded within a set of cultural relationships." Brown and Walker (2008, p. 298) specifically define object agency as "the causal consequences objects (artifacts, architecture and landscape features) have on the course of human activity, and includes animate objects as well as the performance characteristics of material things." Materiality is the interactions between objects and social actors, and agency is the power of those relationships using these definitions, both leading to social construction, negotiation, and reproduction. Dobres and Robb (2005) argue that "the material world is not just "central" to social reproduction, but that material culture actually *constitutes* social relations and meaning making. It is within the tightly woven web of material, symbolic, and social engagement that agency reproduces and transforms society" (Dobres & Robb, 2005, p. 162). Dobres and Robb (2005, p. 160) note that while heavily utilized in archaeological interpretation, the concept of agency is under-theorized, arguing that it in some ways too large a concept and therefore difficult to link to specific understandings in prehistory, noting that "if we consider agency to be a fundamental quality of human existence, then "The Archaeology of Agency" begins to look somewhat like "The Archaeology of Breathing"—a dynamic so universal and inescapable that, without further linkage to something more specific, it is difficult to see how it can illuminate particular aspects of the past."

Given the overall scope of human experience, operationalizing the concept of agency becomes important. They argue that the only way that agency in the past can be understood is how it "worked" (Dobres & Robb, 2005, p. 161) materially. In effect, they argue that the embodiment of objects is the only way to understand agency and the ability of agents to construct and negotiate social structure. Nielsen (1995, p. 49) reiterates that social structure is not static but is "recursively reproduced, negotiated, and transformed by action itself." Objects can be seen to have agency in that they alter activities through "performance characteristics (Konti & Walker, 2015, p. 1 of 24). In some cases, objects can take on their own personae when humans using them treat them as sentient actors. Konti and Walker (2015, p. 1 of 24) refer to these as "nonhuman beings." By shifting the focus of the study of agency onto objects instead of those that produce them, interactions between people ("interactors" in Konti and Walker's (2015, p. 3 of 24) usage) can be examined in the archaeological record. Behaviors inferred are based on the performances of interacting people and objects.

Some aspects of *doxa* are rote and mundane, but some are intentional due to shared motivations and life histories which lead to *habitus*. To the idea of life histories, we can add *death histories*, an examination of the post-mortem usage of material remains (explored in Chap. 5; Geller, 2012).

4.3.3 Materiality and Embodiment

Walker (1995) argues for analysis of the life-history of objects, and how life-history can help to reconstruct the use of objects. Artifacts in this view, are defined in material terms and "include any humanly made or modified object, device, structure, place, or system and is used more or less interchangeably with technology" (Walker & Schiffer, 2006, p. 68). Walker and Schiffer, in formulating and stating this definition, take an operational approach to objects, and define them in relation to their makers and their use. In this way, artifacts can only be viewed as the embodiment of human interactions, and can therefore have power in themselves. Neilsen (1995, p. 49) views social power in practice terms as "the ability of actors to pursue goals" by gaining and using "resources (objects, information, and other people's actions)." Resources, broadly, refer to "artifacts, behavior, or information necessary for the biological, material, and cultural reproduction of a group." Materiality forms the basis of activity, which is the unit of analysis for understanding social structure and social power. Activity, in turn is defined as "specific-artifact interactions of limited duration usually confined to a place" (Walker & Schiffer, 2006, p. 70). Activity is composed of human-artifact interactions. This embodiment is key to understanding ritual in the Poetics model. How do we see the remnants of ritual activity in the archaeological record? Like many of the topics to be discussed in this volume, ritual should be seen as a process, not a singular act.

Embodiment of behavior and relationships can be seen by examining the collections of the Museum of Broken relationships (https://brokenships.com). Begun as a museum in the city of Zagreb, Croatia by artists Olinka Vištica and Dražen Grubišić, the museum exhibits every-day objects that are emblematic of lost or ended relationships. It is composed entirely of objects left by individuals who have visited the museum or one of many travelling exhibits. Many of the objects commemorate broken romantic relationships, but can also be representative of other relationships, such as the death of parents and children (e.g., Fig. 4.1).

Each object is exhibited to show the tremendous meaning that can be wrapped up in simple every-day objects along with a story written by the anonymous donor of the object. Vištica (n.d., p. 8) describes the museum as "an ever-evolving, community-built collection composed of keepsakes, trinkets of no apparent value, either monetary or artistic, all of which bear witness to the end of a relationship. Rock solid as the concrete garden gnome from Ljubljana, delicate as a feather from

Fig. 4.1 "My mum's personal belongings", exhibit at the Museum of Broken Relationships (Zagreb, Croatia). Left: English translation of accompanying signage, right: clothing donated by a woman to the museum as an act of mourning. Photos by the author, 2018

San Fransisco, perishable as the wisp of hair from Skopje, edible as the Shanghai moose moulded in chocolate, they are all displayed anonymously using the subjective personal stories of their donors as the only text." The exhibits memorialize human relationships through the physical objects that become part of the story or experience of that relationships. Going through the exhibits, it is not uncommon to see visitors laughing at one exhibit (such as the "Toaster of Vindication" with a description reading "When I moved out and across the country, I took the toaster. That'll show you. How are you going to toast anything now?" (Anonymous, n.d., p. 190) and then wiping away tears at the next exhibit (such as "My mum's personal belongings" whose descriptions reads in part "if I wish to finally end my mourning and simply remember what was beautiful; if I want to live my life and not hers, I need to let go. This is my letting go" (Anonymous, n.d., p. 180; Fig. 4.1).

4.3.4 Materiality and Practice: Actors and Agents

As discussed above, practice theory revolves around the relationship between doxa and agency. How do we identify *doxa* (either heterodoxy or orthodoxy) in the archaeology record. Frankel provides one mechanism, through the concept of *habitus* and its inherent materiality. The role of *habitus* in archaeological interpretation is long standing. Frankel (2000) sees *habitus* as the combined aspects of material culture and beliefs that make of an individual's social identity (or ethnicity in his words). Frankel views changes in material culture as reflective of changes in social identity, drawing back to Bourdieu's concept of *habitus* as a set of "durable dispositions" (Silliman, 2001, p. 192). *Habitus* is acquired through embodied practice through socialization, learned habits, skills, and so on (Schrader, 2019). Practice also builds on what Giddens (1984) described as *structuration*. For Giddens, societies should be viewed as "dialectics of structures and practices" (Bernbeck, 2008, p. 396) and not just as large-scale structures. Structure, to Giddens was the "intersection of presence and absence [of social relations or social phenomena]; underlying codes have to be inferred from surface manifestations" (1984, p. 16), drawing an association back to *doxa* as identified by Silliman. Social structures formed through practice and form practice in a feedback loop.

Joyce and Lopiparo (2005, pp. 365–366) argue that structuration is "simultaneous the exercise of agency and the constitution of society." This is in line with Walker's (2009) use of practice theory, culture should not assume that everyone within that social system shares exactly the same beliefs and ideology. Culture is not a construction of the mind but is experiential, relational, and based on practices. In effect, culture *does*, not *is*. Key to this is the idea of agency. Culture is viewed as a process, one in which meaning and embodied relationships are encoded in material objects by social agents. Joyce and Lopiparo (2005) argue that in order for culture to *do*, it must *do repeatedly*; shared practices must be repeated in order to be reflective of agency. Dobres and Robb (2005) argue that it is through agency that social reproduction occurs. It is important to note here that social reproduction is not monolithic, because social agents will have different interactions with social structures. Culture is complex. Joyce and Lopiparo (2005) argue that to employ a practice framework in archaeological interpretation, the archaeologist must focus on the commonalities of cultural experience over the differences introduced by the choices made by the various social agents (thereby gaining bridging social capitol along the lines of Putnam (2020)). While Joyce and Lopiparo acknowledge that the different choices made are important, they "often signal critical divergences in positions along the spectrum from more to less structural determination" (2005, p. 368). Overall, though, the emphasis on *habitus* and *doxa* suggests that most persons act without reflection on their actions.

The operationalization of agency as used by archaeologists is both phenomenological and repetitive. Archaeological approaches to agency rely on the repetitive nature of these to understand how social structures and therefore social identity is reproduced. Repeated activities create a material record that can be interpreted archaeologically. Joyce and Lopiparo (2005, p. 370) examine the way that the construction of space can restructure actions and interactions within those spaces, with a sense of place "enacted in the original construction…" and "reiterated in the movement of residents… in their daily life." For example, construction of architecture can be community-building, an expression of differing social standing manifest in stone, with some architectural construction "stratifying space and creating the potential for differential experience and knowledge. These building projects directly affected the bodily practices of community members, and through their embodied experiences community members incorporated… largely non-discursive dispositions"(Joyce & Lopiparo, 2005, p. 371). Construction of space and place, then, requires re-corporation of *doxa* either in an asserted tradition or consciously as innovation. Transformations of the *doxa* involve not only the planners of the project but also those who would perform the actual construction tasks (Joyce & Lopiparo, 2005). In this way, repetition of tasks already known to both parties can be a mechanism for both reinforcing cultural tradition and for the changing of those traditions through innovated architecture and new use of space.

4.3.5 Materializing Social Power

The preceding sections have dealt with how artifacts and people interact to create meaning and how that meaning is embodied within those artifacts. It remains, then, to understand how these relationships translate into the creation, negotiation, and renegotiation of structural social power. Social power can be seen as "the practices, by which one or more groups, that in fact influence a choice"(Walker & Schiffer, 2006, p. 73), while structural social power is the "socially defined authority to make choices affecting the artifact interactions of others"(Walker & Schiffer, 2006, p. 73). Social power can be linked to the Social Body (Scheper-Hughes & Lock, 1987) while structural social power can be linked to the Body Politic (Scheper-Hughes & Lock, 1987), but with a more material focus than that employed by Scheper-Hughes and Lock.

4.4 Ethnographic Approaches to Death and Mortuary Ritual

The preceding sections of this chapter dealt with changes in archaeological interpretation of mortuary contexts. Now, we switch gears to examine the role of ethnographic study into the role of death and mourning. Ethnographic approaches have and should inform archaeological interpretations, particularly with respect to ritual. The approaches described here include work into the role of ritual in terms of community identity as well as the understanding of death as a concept and the relationship of both individuals and society to the process of death.

This section is by no means meant to present a thorough understanding of all ethnographic approaches to death and mortuary ritual. Here, I am more concerned with touching on ethnographic and social anthropological research that contributes heavily to the Poetics model. Early work in anthropology looked for universals across cultures in death and mourning. These researchers studying death were primarily based in the western European tradition, and so it is worthwhile to trace Western European (and later American) conceptions of death in order to understand the underpinnings of the theoretical models they developed. Social theory, after all, is not developed in a social vacuum. Early western conceptualizations of death were based very much on the changing concepts of self. Ariès (1975) examines four basic themes: (1) people's sense of the individual; (2) the defense of society against unpredictable forces of nature; (3) belief in life after death; and (4) belief in the existence of evil. These in turn led to the development of several models of death. The 'tame death' model dominated between the eleventh and seventeenth centuries in western Europe. Death was not seen as a personal affair but was in fact a social event, attended by family and community and formulaic in approach, with rituals acting as reparative within society. In doing so, death was tamed and ritualized. Occurring concurrently with the 'tame death' model was the 'death of the self' model more prevalent amongst the elite and replacing the tame death entirely by the seventeenth century. This shift was due to changes in how identity was viewed within society. Individuality became more important as opposed to a collective identity. This shift emphasized a separation of body and soul, and examined the role of the soul in the afterlife. Fear of the corpse became more common.

By the eighteenth century, nature was regarded as partaking both in life and death (Robben, 2004a). This belief led to fear of being buried alive. "Death was neither tame nor hidden, but savage and mysterious" (Robben, 2004a, p. 4). During this period, the fear of death became a central theme in theoretical literature (typified in the work of Malinowski, Becker, and Lifton and Olson).

Fear of death shits between the eighteenth and twentieth centuries giving rise to the 'fear of the other' model (Ariès, 1975). In this model, people did not necessarily fear their own death but that of their relatives and friends and the ultimate separation that would occur. Within the twentieth century, the 'invisible death' model becomes dominant, with a separation of death from every-day experience, including the specialization of the death industry and increasing medicalization of the dying process. One very important conceptual framework that was developed during this time was that of Hertz (1960 [1907]). Hertz argued that the death of the body is not exclusively a biological reality or confined to the individual sorrow of the bereaved, but that death evokes moral and social obligations expressed through mortuary practices. Hertz is interested in the roles during the time between death and ultimate burial. First, there is an intermediary time during which the body is stored or temporarily buried. During this time, the soul of the deceased remains near the body, and the bereaved are separated from society by their bereavement and the mourning process. One important addition to our understanding of death brought about by Hertz is that death is not a single moment, but is a process. During the mortuary process, the deceased is still considered to be a part of society. It is only when the soul departs at the end of the formalized mourning process that the status of the deceased changes. If the mourning process is not followed to cultural specifications, the soul may not depart but may impact the living in negative ways. Harkening back to the 'fear of the other' model, mortuary ritual occurs in order to ensure proper disposition not only of the body but also of the soul.

The work of Hertz is often linked strongly to van Gennep's rites of passage (1960 [1909]). Important phases in life are marked by rites of passage from one status to another, e.g., the transition from adolescence to adulthood, or from being single to being a married person. Van Gennep analyzed funerals as rites of passage. Transitions such as mortuary ritual have three distinct phases. First, separation isolates the deceased and mourners from the rest of society. Mourners are separated from society through different forms of dress or actions they are either required to do or prohibited from doing. Drawing upon Turner's (1967) concept of liminality, the second phase is when ritual activity occurs that allows the actors involved to transition: the corpse to an ancestor and the mourners to have their status changed as well. Rituals such as manipulation of the body, feasting, and so on, may occur during this period. Finally, a post-liminal rite of incorporation introduces new statuses for both the deceased as an ancestor and the mourners back into society. Important to this framework is the analysis of *both* the deceased and the bereaved. Both of these groups consist of actors that have roles to play for the maintenance of society. Reincorporation rites may consist of scattering cremations, funeral rites, and so on. These should form the end of prescribed mourning. All actors are fundamentally changed by the rite of passage, for example transitioning from wife to widow. New social realities may exist for the widow that are not open to the wife, new social responsibilities may also be present. The widow, for example, may be expected to help shepherd others through the mortuary rituals of their own spouses. Through the liminal period, she has gained a new set of skills that must be used within society. In discussing death and mourning in Japan, Danely (2018, p. 134) describes mourning as less of a resolution and more of a change in perspective: "[d]eath in Japan does not bring closure but leaves a frayed and unfinished seam to the narrative of a relationship, calling to attention both the desire for continued connection and the recognition that the deceased is no longer quite the same."

Is the fear of death universal? Lifton and Olson (2004) argue for a universal fear of death and argue that the conflicting finality of biological death and the human search for meaning as a search for immortality. They break this into several types of immortality:

- biological immortality, or the continuation of life through offspring, a family name, or other group affiliation
- theological immortality (resurrection, reincarnation, rebirth, or spiritual life after death)
- natural immortality, the concept of an eternal universe and unendying cycles of nature
- experiential immortality, or altered states of consciousness such as ecstasy and enlightenment

Robben (2004b) examines the role of historical meaning and memory with relation to place, time, power, gender and the body. Meaning varies depending on context in this view while understanding that time will influence material culture and memory.

4.5 Summary

This chapter has focused on different approaches to understanding mortuary practices, both from archaeological and ethnographic points of view. Together, these approaches form the history of how we have interpreted mortuary activity and its relation to larger social issues. From a beginning understanding of the role of grave goods and cemetery location with respect to social identity and status (processual approaches) to a post-processual approach focused on symbolic and ideological importance of mortuary practices, we can chart our changing understanding of how these practices and their material culture correlates can be intertwined with ethnographic approaches. Mortuary archaeology in conjunction with ethnographic interpretation offer valuable insights into the negotiation of power, identity, and social dynamics. By examining burial practices and material culture, we can better understand how societies reproduce and transform their social structures over time.

References

Anonymous. (n.d.). The exhibits. In O. Vištica & D. Grubišić (Eds.), *Museum of broken relationships: A diary* (pp. 80–305). Museum of Broken Relationships.

Antonaccio, C. M. (1995). *An archaeology of ancestors*. Rowman and Littlefield.

Ariès, P. (1975). *Western attitudes toward death: From the middle ages to the present* (P. Ranum, Trans.). Baltimore: Johns Hopkins University Press.

Ashmore, W. (2002). "Decisions and dispositions": Socializing spatial archaeology. *American Anthropologist, 104*, 1172–1183.

Ashmore, W. (2004). Social archaeologies of landscape. In L. Meskell & R. W. Preucel (Eds.), *A companion to social archaeology* (pp. 255–271). Blackwell.

Bernbeck, R. (2008). Structural violence in archaeology. *Archaeologies: Journal of the World Archaeological Congress, 4*(3), 390–413.

Binford, L. R. (1971). Mortuary practices: Their study and their potential. In J. A. Brown (Ed.), *Approaches to the social dimensions of mortuary practices* (p. 25). Memoirs of the Society for American Archaeology.

Binford, L. R. (1982). The archaeology of place. *Journal of anthropological archaeology, 1*(1), 5-31.

Brown, L. A., & Walker, W. H. (2008). Prologue: Archaeology, animism and non-human agents. *Journal of Archaeological Method and Theory, 15*(4), 297–299. https://doi.org/10.1007/s10816-008-9056-6

Buelow, D. A. (2010). *Peripheral memory: New York's forgotten landscape*. Massachusetts Institute of Technology.

Chapman, R. (2003). Death, society and archaeology: The social dimensions of mortuary practices. *Mortality, 8*, 305–312.

Danely, J. (2018). Mourning as mutuality. In A. C. G. M. Robben (Ed.), *A companion to the anthropology of death* (pp. 131–144). Wiley.

David, B., & Thomas, J. (2016). Landscape archaeology: Introduction. In B. David & J. Thomas (Eds.), *Handbook of landscape archaeology* (pp. 27–43). Routledge.

Derriennic, J.-P. (1972). Theory and ideologies of violence. *Journal of Peace Research, 9*(4), 361–374. https://doi.org/10.2307/422517

Dobres, M. A., & Robb, J. E. (2005). "Doing" agency: Introductory remarks on methodology. *Journal of Archaeological Method and Theory, 12*(3), 159–166. https://doi.org/10.1007/s10816-005-6926-z

Fisher, K. D. (2007). *Building power: Monumental architecture, place and social interaction in late bronze age Cyprus*. (Ph.D. Dissertation). University of Toronto, Toronto.

Fisher, K. D. (2009a). Elite place-making and social interaction in the late cypriot bronze age. *Journal of Mediterranean Archaeology, 22*(2), 183–209.

Fisher, K. D. (2009b). Placing social interaction: An integrative approach to analyzing past built environments. *Journal of Anthropological Archaeology, 28*, 439–457.

Foard, G. (1978). Systematic fieldwalking and the investigation of Saxon settlement in Northamptonshire. *World Archaeology, 9*, 357–374.

Frankel, D. (2000). Migration and ethnicity in prehistoric Cyprus: Technology as habitus. *European Journal of Archaeology, 3*(2), 167–187. https://doi.org/10.1177/146195710000300202

Geller, P. L. (2012). Parting (with) the dead: Body partibility as evidence of ancestor veneration. *Ancient Mesoamerica, 23*(1), 115–130.

Giddens, A. (1984). *The constitution of society: Outline of the theory of structuration*. University of California Press.

Goldstein, L. (1980). *Mississippian mortuary practices: A case study of two cemeteries in the lower Illinois valley*. Evanston: Scientific Papers, No. 4, Northwestern University Archaeological Program.

Goldstein, L. (2006). Mortuary analysis and bioarchaeology. In J. E. Buikstra & L. A. Beck (Eds.), *Bioarchaeology: The contextual analysis of human remains* (pp. 375–388). Academic.

Hassan, F. A. (1979). Geoarchaeology: The geologist and archaeology. *American Antiquity, 44*, 267–270.

Hertz, R. A. (1960 [1907]). *Death and the right hand: A contribution to the study of the collective representation of death* (R. Needham & C. Needham, Trans.). Glencoe, IL: The Free Press.

Hodder, I. (2004). *Theory and practice in archaeology*. Routledge.

Joyce, R. A. (2005). Archaeology of the body. *Annual Review of Anthropology, 34*, 139–158. Retrieved from http://www.jstor.org/stable/25064880

Joyce, R. A., & Lopiparo, J. (2005). PostScript: Doing agency in archaeology. *Journal of Archaeological Method and Theory, 12*(4), 365–374. https://doi.org/10.1007/s10816-005-8461-3

Konti, K., & Walker, W. H. (2015). Animate shadows of bears and giants. In C. Papadopoulos & H. Moyes (Eds.), The Oxford handbook of light in archaeology. https://doi.org/10.1093/oxfordhb/9780198788218.013.10

Lifton, R. J., & Olson, E. (2004). Symbolic immortality. In A. C. G. M. Robben (Ed.), *Death, mourning, and burial: A cross-cultural reader* (pp. 32–39). Blackwell Publishing.

Lull, V. (2000). Death and society: A Marxist approach. *Antiquity, 74*(285), 576–580.

Morris, I. (1991). The archaeology of ancestors: The Saxe/Goldstein hypothesis revisited. *Cambridge Archaeological Journal, 1*, 147–169.

Nielsen, A. E. (1995). Architectural performance and the reproduction of social power. In J. M. Skibo, W. H. Walker, & A. E. Nielsen (Eds.), *Expanding archaeology* (pp. 47–66). University of Utah Press.

Osterholtz, A. J. (2015). *Bodies in motion: A bioarchaeological analysis of migration and identity in Bronze Age Cyprus (2400–1100 BC)*. (Ph.D. Dissertation). University of Nevada Las Vegas, Las Vegas, NV.

Parker Pearson, M. (1982). Mortuary practices, society and ideology: An ethnoarchaeological study. In I. Hodder (Ed.), *Symbolic and structural archaeology* (pp. 99–113). Cambridge University Press.

Parker Pearson, M. (1993). The powerful dead: Relationships between the living and the dead. *Cambridge Archaeological Journal, 3*, 203–229.

Parker Pearson, M. (1995). Return of the living dead: Mortuary analysis and the New Archaeology revisited. *Antiquity, 69*(266), 1046–1048. https://doi.org/10.1017/S0003598X00082648

Parker Pearson, M. (1998). Stonehenge for the Ancestors: The Stones Pass on the message. *Antiquity, 72*(276), 308–326.

Parker Pearson, M. (1999). *The archaeology of death and burial*. Texas A&M University Press.

Parsons, K. A. (2007). Structural violence and power. *Peace Review, 19*(2), 173–181. https://doi.org/10.1080/10402650701353612

Patterson, T. C. (2007). Social archaeology and Marxist social thought. In K. Kristiansen & M. Rowlands (Eds.), *A companion to social archaeology* (pp. 66–81). Routledge.

Pérez, V. R. (2006). *The politicization of the dead: An analysis of cutmark morphology and culturally modified human remains from La Plata and Peñasco Blanco (A.D. 900–1300)*. (PhD Unpublished Ph.D. dissertation). University of Massachusetts Amherst, Amherst.

Petts, D. (2003). Memories in stone: Changing strategies and contexts of remembrance in early medieval wales. In H. Williams (Ed.), *Archaeologies of remembrance: Death and memory in past societies* (pp. 193–213). Kluwer Academic/Plenum Publishers.

Putnam, R. D. (2020). Bowling alone: The collapse and revival of American community (Revised and Updated). New York, Simon & Schuster Paperbacks.

Robben, A. C. G. M. (2004a). Death and anthropology: An introduction. In A. C. G. M. Robben (Ed.), *Death, mourning, and burial: A cross-cultural reader* (pp. 1–16). Blackwell Publishing.

Robben, A. C. G. M. (2004b). State Terror in the Netherworld: Disappearance in reburial in Argentina. In A. C. G. M. Robben (Ed.), *Death, mourning, and burial: A cross-cultural reader* (pp. 134–148). Blackwell Publishing.

Sabloff, J. (Ed.). (1981). *Simulations in archaeology*. University of New Mexico Press.

Saxe, A. A. (1970). *Social dimensions of mortuary practices*. (Doctor of Philosophy). The University of Michigan, Ann Arbor, MI.

Scheper-Hughes, N., & Lock, M. M. (1987). The mindful body: A prolegomenon to future work in medical anthropology. *Medical Anthropology Quarterly, 1*(1), 6–41. Retrieved from http://www.jstor.org/stable/648769

Schrader, S. (2019). *Activity, diet, and social practice: Addressing everyday life in human skeletal remains*. Springer Cham.

Shanks, M., & Tilley, C. (1987). *Social theory and archaeology*. University of New Mexico Press.

Shawcross, W. (1968). An investigation of prehistoric diet and economy on a coastal site at Galatea Bay, New Zealand. *Proceedings of the Prehistoric Society, 33*, 107–131. https://doi.org/10.1017/S0079497X00014079

Silliman, S. W. (2001). Agency, practical politics and the archaeology of culture contact. *Journal of Social Archaeology, 1*, 190–209.

Sofaer, J. R. (2006). *The body as material culture: A theoretical osteoarchaeology*. Cambridge University Press.

Strathern, A. J. (1996). *Body thoughts*. University of Michigan Press.

Tainter, J. A. (1978). Mortuary practices and the study of prehistoric social systems. *Advances in Archaeological Method and Theory, 1*, 105–141.

Thompson, V. (2003). Memory, salvation and ambiguity: A consideration of some Anglo-Scandinavian Grave-Stones from York. In H. Williams (Ed.), *Archaeologies of remembrance: Death and memory in past societies* (pp. 215–226). Kluwer Academic/Plenum Publishers.

Tilley, C. (2016). Phenomenological approaches to landscape archaeology. In B. David & J. Thomas (Eds.), *Handbook of landscape archaeology* (pp. 271–276). Routledge.

Turner, V. W. (1967). Betwixt and between: The liminal period in Rites de Passage. In V. W. Turner (Ed.), *The forest of symbols: Aspects of Ndembu ritual* (pp. 46–55). Cornell University.

Van Dyke, R. M. (2016). Memory, place and the memorialization of landscape. In B. David & J. Thomas (Eds.), *Handbook of landscape archaeology* (pp. 277–284). Routledge.

van Gennep, A. (1960 [1909]). Les Rites de Passage. : University of Chicago Press.

Verdery, K. (1999). *The political lives of dead bodies: Reburial and postsocialist change*. Columbia University Press.

Vištica, O. (n.d.). A shelter for exiled loves. In O. Vištica & D. Grubišić (Eds.), *Museum of broken relationships: A diary* (pp. 6–25). Museum of Broken Relationships.

Walker, W. H. (1995). *Ritual prehistory: A Pueblo case study*. (PhD). University of Arizona, Tucson.

Walker, W. H. (2009). Warfare and the practice of supernatural agents. In A. E. Nielsen & W. H. Walker (Eds.), *Warfare in cultural context: Practice, agency, and the archaeology of violence* (pp. 109–135). The University of Arizona Press.

Walker, W. H., & Schiffer, M. B. (2006). The materiality of social power: The artifact-acquisition perspective. *Journal of Archaeological Method and Theory, 13*(67–88).

Chapter 5
The Poetics of Processing: Bodies as Cultural Media

Abstract This chapter examines the role of mortuary practices in shaping social identity, resistance, and cultural preservation, with a focus on African American communities and their historical burial rituals. It reimagines the human body as a significant form of material culture, arguing that both the living body and the body after death serve as dynamic sites for cultural expression, social negotiation, and identity formation. The chapter draws on bioarchaeology, postcolonial theory, and social history to explore how the body's manipulation through mortuary rituals reflects broader social structures, power dynamics, and cultural hybridity. Particular attention is given to the "third space" concept from postcolonial theory, which illustrates how marginalized groups, such as enslaved African Americans, developed new, hybrid cultural practices in response to forced migration and oppression. The chapter discusses how African American mortuary rituals, born out of a fusion of African traditions and new circumstances in the U.S., became acts of resistance and empowerment. Through the examination of burial grounds and funerals, the chapter argues that these spaces provided opportunities for these new communities to assert their identity and resist social marginalization, ultimately acting as culture keepers. The role of African American funeral directors is explored, from the Civil Rights Movement through the Black Lives Matter movement. Ultimately, the chapter underscores how mortuary practices are not only essential for cultural survival but also serve as powerful tools for social and political change.

In this chapter, I will first go explore the malleability of the body and the use of the body as cultural medium for messaging. The human body, while alive, experiences and contributes to social life. This quality makes the physical remains of those who once were alive uniquely able to act as a social messenger after death. Later in this chapter I will examine some examples of how bodies and the processing of those bodies (visible changes in mortuary performance) can be seen as indicative of social change and how mortuary performances (including the processing of the body) can act as a way in which identity can form and resilience can be performed, primarily through a discussion of the role of funerary processes and the development and resilience of identity in African American communities.

A. Osterholtz, *The Bodies Poetic*, Bioarchaeology and Social Theory,
https://doi.org/10.1007/978-3-031-85998-4_5

5.1 The Body as Material Culture

A divide has long been present in archaeological practice between material culture and the humans that created the objects, modified the bones, and built the architecture. Critically, the subfield of bioarchaeology bridges this divide by examining human remains as a specific form of material culture, the analysis of which can be used to understand lived experience in the past. In 2018, I argued that "skeletal analysis takes on additional meaning when properly contextualized. It is a feedback loop between the analysis of material culture and the analysis of the makers of that material culture: one cannot fully understand lived experience in the past without understanding both the skeletal remains *and* material culture" (Osterholtz, 2018, p. 401). Using Van Dyke's view of materiality as a process of interaction, and building on an understanding of intersectionality, the body should be seen as a form of material culture that embodies meanings and relationships, not a separate entity. Bodies as artifacts, according to Meskell (2000), form normative typologies which function within networks of social forces and at the same time have the power to produce and transform those social forces. Schrader (2019, p. 28) cautions that human bones are unlike other forms of material culture in that they are constantly in a state of remodeling and change, but this does not limit their role as material culture. Due to the constant change and remodeling of the material culture type examined (i.e., the skeletal remains), bioarchaeology has "the ability to address both the biological and social past."

A growing body of literature exists building upon Sofaer's (2006) book *The Body as Material Culture*. Across time and space, bioarchaeologists have taken to heavily contextualizing skeletal data to add to an overall understanding of lived experience in the past. Geller (2012) builds on Saul's (1972, 1976) concept of osteobiographies as *life histories* of individuals by examining their *death histories*, or how those bodies and body parts are used after death by the living. This approach effectively looks at the type of processing of bodies and how that processing is used by the living to negotiate social structure. As noted by Geller and Suri (2014, p. 501), "life and death histories facilitate recognition of bodies as a nexus for experiencing the world, as a focus of external cultural expectation and attention and sites for political engagement—both during life and after death. Bodies *qua* bodies are thus not only physical objects but also sites for less tangible meaning-making and expression. In this way, they are an ideal subject for scholars interested in examining the intersection of temporality, modernity and memory." Bongers (2019, p. 49), looking at the Middle Chicha Valley, Peru, examined the body as an "inscribable surface, a medium of expression, and a culturally produced artifact with a capacity to serve as a symbol and influence and enable social action."

Just as lithicists examine stone tools using a specialized set of methodologies and techniques, so do bioarchaeologists analyze human remains. Both are seeking to understand human lived experience in the past, just utilizing different data sets. The physical body should remain part of the archaeological picture. The story of ceramics and architecture cannot be told fully without including the makers of those ceramics and the builders of those temples and houses.

The Poetics framework integrates bioarchaeological and archaeological lines of evidence: it forces us to examine the recursive relationship between the physical body of the dead and the social structures of the living. As Parker Pearson (1999) reminds us, the dead do not bury themselves. The body is a transformative object, and manipulation of the bodies of the dead is carried out by the living for the ultimate end goals of the living. When this manipulation occurs as a performative act, it can be considered Poetic.

Van Gennep's (1960 [1909]) transitions exist for both mourners and for the deceased. Transitions for the corpse may include preparation for display, defleshing, mummification, cremation, provisioning the grave, and any other processing. These are all actions that require interaction with the physical body and the modification of that body *by the living*. The acts performed for the dead, by the living, are part of the acts that cause the transitions for the mourners. The physicality of touching, provisioning, and modifying the body are integral to the successful transition of both corpse and the community.

5.1.1 The Body as a Phenomenological Object

At its heart, the Poetics model as laid out by Whitehead is phenomenological. Phenomenology can be viewed as the various ways in which we are connected to the world around us and to one another. These connections are meaningful to the actors involved, and embody our way of being in the world (Merleau-Ponty, 2005 [1962]).

The model of **ritual + performance** (i.e., the Poetics model) necessitates an experience. This is about the experience, the ritual at the heart of community identity formation. This can be seen along the same lines as phenomenological landscape interpretation. This experiential element helps to reinforce and allow for the negotiation of social identity. As Tilley (2016, pp. 271–272) notes, this approach to landscape "stresses the *materiality* of landscapes: landscapes as real and physical rather than simply organized or represented. The physicality of landscapes acts as a foundation for all thought and social interaction. It profoundly affects the way we think, feel, move, and act. The phenomenologist is a figure immersed within the ground of landscape. Landscape is fundamental for human existence because it provides both a medium for, and an outcome of, individual and social practices. The physicality of landscapes grounds and orientates people and places within them, a physical and sensory resource for living and the symbolic construction of lifeworlds." Essentially, landscape forms one elements of lived experience that works in a reciprocal loop with other social and biological elements to form social identity. How do we see ritual in the archaeological record? Ritual processes leave material traces through the objects used and manipulated in during the acts. In this volume,

I use Walker's (1996) application of artifact life-histories to indicate ritual acts and their embodied artifacts.

The phenomenological aspects of body manipulation can be an intense part of any ritual experience involving the body. It is worthwhile to briefly discuss the processes of decomposition that most bodies will go through. Exposure to a decomposing body exposes those around it to a plethora of sights, sounds, and smells. This is a process that every body goes through, regardless of age, status, or gender, unless altered by artificial or natural interventions such as mummification or anatomical preparation.

Understood as taphonomic changes, decomposition produces a variety of sensory experiences that have significant impacts on the material qualities of the body. After death, the body goes through identifiable stages of decomposition unless this process is altered by cultural processes. Initially after death, the process of algor mortis (cooling of the body after death) begins. Livor mortis, or the pooling of the blood in the body due to gravity occurs. This occurs on the lowest points of the body, leading to discoloration in some parts of the body, particularly if the body is moved during this period. Typically, livor mortis is evident within hours of death and most pronounced approximately twelve hours after death. Rigor mortis is the constriction of the muscles of the body caused by depletion of adenosine triphosphate within the muscle tissue. This starts in the small muscles of the face and jaw and ultimately extends to the rest of the body. Rigor mortis sets in within the first several hours after death, peaks around twelve hours after death, and relaxes over the next twenty-four hours as the muscle tissue begins to degrade. The process of decomposition involves autolysis and putrefaction. Autolysis is the destruction of cells through the actions of their own enzymes, this is most rapid in the pancreas and stomach. Putrefaction is microbial deterioration of tissues due to the proliferation of bacteria within the body itself. Putrefaction leads to color changes in the body including various shades of green, purple, and brown during the breakdown of internal structures. Numerous gasses are produced during putrefaction, most notably within the abdomen, leading to the bloating of the body. Over time, these bloated areas will either rupture or the gasses will subside on their own. Skin slippage is caused by the deterioration of the junction of the dermis and epidermis and occurs during putrefaction as well (Christensen et al., 2014).

Environmental factors heavily impact the rate of decay, including temperature, soil composition, precipitation, predation, and humidity. For example, in areas of low humidity, natural mummification can occur in either hot or cold environments. This may turn the skin into a dark brown leathery texture. In a time before the medicalization of death (roughly beginning in the US in the nineteenth century), family members would have been present and experienced the changes associated with putrefaction and death.

5.2 Community and Identity in the Third Space

The Third Space is that space created after cultural contact, and is a fundamental element involved with postcolonial theory. Postcolonial theory focuses on understanding and deconstructing the ways that colonial histories shape contemporary social, political, and cultural realities. Said (1978) highlights how scholarship has tended to focus on non-western societies as the *other*, primarily as a mechanism to justify colonialism.

The form of postcolonial theory employed here relates to the work of population movements and cultural interactions as reflected in their mortuary performance, tomb locations, and changes in body processing. Central to this are the concepts of hybridization and migration, a way to move away from the dichotomy of colonizer and colonized that can sometimes be presented in postcolonial theory (Osterholtz, 2015; Voskos & Knapp, 2008). With hybridization, we can examine how cultures change each other when they come into contact with each other, no matter how they come into contact with each other. This concept is similar to Locard's principle in forensic sciences, the idea that transfer will occur when two objects come into contact whether the transfer is intentional or not (Locard, 1920).

Whether through forced migration (enslavement or dislocation) or voluntary migration, newcomers bring their own skillsets, beliefs, and performances from their home place. What emerges in the postcolonial space (the third space), is not what was occurring in their home place, but it changes the performances and society of their new place as well. Postcolonial theory challenges the dichotomy of *here* and *there*, instead allows for the identification of a "shared postcolonial terrain" (Mains et al., 2013, p. 133). Voskos and Knapp (2008) extend this concept of hybridity and cultural adaptation in the third space into the past, examining the role of migration in the Mediterranean region. This approach can be useful in other contexts as well, examining the role of population movements and interactions through an analysis of material culture. Bhabha's work on hybridity (2015) has an emphasis on its transgressive and transformative nature, allowing for the formation of multiple viewpoints to exist in the same Third Space. Key to this is the adaptation of new traditions that reflect the transgressive nature of hybridity.

Adoption of new traditions can be examined using a variety of methods, including the diffusion of innovations theory (E. M. Rogers, 1995), which examines who within society is more likely to be receptive of cultural innovations (both material changes such as the adoption of new farming methods) or ideas and practices (such as mortuary rituals). Early adoption of new traits may be linked to a desire to gain status, and so individuals seeking status may be more likely to be early adopters. Essentially, those willing to take the risks are those lacking significant prestige, such as enslaved peoples coming from multiple cultural groups encountering each other. For those groups, the hybridization of ritual is a way to form new communities. As Anwar (2014, pp. 25–26) notes, "The politics of place and the politics of identity and difference are co-instructive, a terrain on which claims of 'origin' and strategies of fixing identities becomes important for marginalized groups." Cohesive group identity can be formed through the hybridizing of ritual, and this can be used as a way to form resilient communities.

5.2.1 As a Way to Legitimize Cultural Change

The bodies of prior political or corporate entities can be manipulated by newcomers as a way to indicate access or ownership over land or resources. For example, Duncan and Schwarz (2014, p. 164) examine the idea of cemetery desecration in the Petén Lakes Region of central America. Desecration of an enemy cemetery created a secondary commingled assemblage. Some of the body parts were thrown into nearby lakes, some reburied, and some placed in the corner of the main ceremonial square "to make a public symbol from the enemy remains". This public display of enemies remains brings to mind the top-down control of bodies by the Body Politic (Scheper-Hughes & Lock, 1987). This idea of display of the body as punishment is discussed below, what is important for this section is the idea of manipulation of the remains themselves as a show of power of one group over the dead and the living attached to them.

The use of bodies as legitimizations for political change link new regimes or ruling lines to old ruling lines, creating a kind of fictive kin network. Geller (2012, p. 124) argues that community leaders "required the physical presence of original ancestors to legitimize their positions of power." Geller is specifically interested in the partiability of bodies in Belize during the Late and Terminal Classic Maya Periods (A.D. 650–900). She argues that partiability, the intentional selection of body parts for deposition started "ritual acts that serve to reconstitute decedents' identities. While there are certain guidelines that direct these rituals, the particulars of a decedents' individuality may also contour the living's response"(Geller, 2012, p. 122). She specifically identifies the importance of a *death history*, a concept that extends the idea of a life history to include the ongoing engagement with the bodies of the dead. "Such relations are materially discernible, symbolically meaningful and contextually contingent" (Geller & Suri, 2014, p. 501).

In my dissertation work, I examined the reuse of tomb space in the Late Bronze Age on Cyprus as a way for newly growing populations and potentially new corporate groups to legitimize control over resources (Osterholtz, 2015). While on the face of it, the use of mortuary space in this matter would seem to support Saxe's Hypothesis 8 (Saxe, 1970), I argue that it is the mortuary ritual, not simply the use of space that is important here. That ritual becomes embodied in the manipulation of human remains and other artifacts within the mortuary ritual. The different life (and death) histories of these remains due to this manipulation form the basis for corporate control over space or resources, not the mere presence of the remains in the space they're found in. If we take the idea that mortuary ritual is conservative, changing only with large-scale social changes, then the legitimization of these social changes through a holding-on of some elements of the past in the new ritual should be seen as an important element to legitimization of the new.

5.2.2 In African-American Contexts

In Chap. 6, I will give a brief overview of the White, Colonial way of death in the US. As discussed in greater detail there, I have made a choice to present the African American and White, Colonial mortuary processes separately due to the different roles that mortuary processes played within these communities. For Africans forcibly brought to the US and their descendants, the development of mortuary ritual occurred in a third space, one that combined numerous African traditions and created new ones as a show of resilience. The same cannot be said for the controlling White Colonial population. In effect, the White, Colonial power structures allowed for a plethora of mechanisms to perform and create collective identity that were denied to African Americans. For African Americans, mortuary performance in spaces set aside for this use became a space and place in which new identities and communities could form.

The role of mortuary ritual in African American contexts is an important one, though it can be more challenging to study from an academic standpoint. First, there is a historical bias towards studying European American mortuary processes and spaces. Efforts to address the lack of focus on African American death studies has been pushed brilliantly by Dr. Kami Fletcher. Fletcher spearheads the Collective for Radical Death Studies (CRDS, https://www.radicaldeathstudies.com). They list their mission as one to "interrogate the field of Death Studies to decolonize and de-center whiteness while calling to radicalize death practices, all in theory and in practice from a variety of angles, i.e. our research, writing, community work, and professional careers" (https://www.radicaldeathstudies.com/our-mission). The website acts as a gathering place, providing zoom gatherings every other month, links to bibliographies and educational materials and engaging videos.

Second, African-American mortuary processes during enslavement may have been clandestine or were restricted in time and space. This means that they left different material traces: they are not as easily found and more likely to be disturbed. John Sella Martin, a formerly enslaved individual described his mother's burial: "My mother died in 1852… she was buried in unconsecrated ground, outside the graveyard for the whites, and her grave was walked over every day by the beasts of the field" (quoted in Blassingame, 1977, p. 734). Cemetery spaces set aside for African American use was land that was seen as less desirable (see discussion on Potter's Fields, Chap. 7), usually near train tracks or waterways. It is more likely to be have been destroyed or flooded. There tends to be lower levels of monumentality leading to lessened ability to identify these spaces in the long run. Cemetery spaces, when well-marked were more likely to be the target of "resurrection men", targeted for the clandestine procurement of cadavers for medical education (see discussion in Chap. 11). Finally, African-American cemeteries, being located in more marginal areas, are more likely to be disturbed for urban expansion than more centrally located historic White cemeteries (Blanks et al., 2021).

From the earliest arrival of enslaved people to North America, mortuary performances can be seen as an important element to forging new relationships and new identities. The burial of enslaved individuals in Colonial America can be examined through work on the African Burial Ground in New York City. During construction of a new federal building in the early 1990s, the remains of approximately 15,000 people of African origin and their descendants were found. This cemetery was in use from at least 1712 but likely has origins in the late seventeenth century after Trinity church banned the burial of Africans within its churchyard in 1697. The cemetery was in use until around 1795 when the land was filled in and leveled. Aspects of mortuary performance can be gleaned through an analysis of the material culture. Individuals were buried within coffins (evidenced by coffin nails) and enshrouded (shroud pins). Coffin orientation was general East-West with the head to the west. This shouldn't be seen as indication of Christianity or non-Christian faith—burial in this orientation was common amongst west African groups as well. Kohrs (2021), in a talk on the Corhaven Graveyard in Virginia highlights commonalities between West African burial traditions and those identified at Corhaven, historically known as the Sam Moore Slave Cemetery in Quicksburg, VA. She notes that the Ibo (in modern day Nigeria) bury the dead with feet facing west, for example. Grave goods can also be used to infer identity and relationships. At the African Burial Ground in New York City, 145 beads were recovered from the grave contexts. These beads were used to establish the time frame of the burials as well as gain understandings of bead usage by the enslaved new arrivals to the US and their descendants. LaRoche (1994) ties the presence of necklaces and waistbands to indicators of spirituality, status, fertility, apotropaic (ritually protective) uses, and as heirlooms that were carefully curated and passed down until placed within the burials. In 1993, 419 sets of remains were removed to Howard University for analysis lead by Michael Blakey.

These remains were invaluable for understanding lived experience of these individuals, including the markers of identity such as filed teeth (Blakey, 1998; Blakey & Rankin-Hill, 2004; Mack & Blakey, 2004). The work on the African Burial Ground also highlights the importance of community engagement and the role of African American bioarchaeologists in the work (Blakey, 2001). These 419 individuals were returned to the site in 2003 and reinterred. In 2007, the African Burial Ground was declared a National Monument (Fig. 5.1) with the visitor center opening in 2010 (https://www.nps.gov/afbg/index.htm). Its museum holds approximately 8000 archival records detailing the history of New York's enslaved and free African population.

By the nineteenth century (and continuing until today), mortuary ritual was used as a method for creating community amongst diverse African American communities (Jamieson, 1995). For Vlach (1991, p. 109), "the graveyard was, in the past, one of the few places in America where an overt black identity could be asserted and maintained." These mortuary processes were shaped by multiple types of African beliefs and how those were interpreted by African Americans, Christian canons, and "experiences unique to black Americans" (Blakely & Harrington, 1997, p. 169) such as slavery and racism. "Institutions of slavery restricted the ability of African Americans to maintain cultural links or cohesive cultural identities" (Jamieson, 1995, p. 29) from Africa. Patterson (1982, p. 5) describes slavery as "a substitute for

Fig. 5.1 Aerial view of the African Burial Ground National Monument in New York City. Photo attribution: United States-National Park Service, Public domain, via Wikimedia Commons. https://upload.wikimedia.org/wikipedia/commons/3/35/African_Burial_Ground-aerial_view-NYC.jpg

death, usually violent death… Slaves differed from other human beings in that they were not allowed freely to integrate the experiences of their ancestors into their lives, to inform their understanding of social reality with the inherited meanings of their natural forebearers, or to anchor the living present in any conscious community of memory." As a result, a third-space, postcolonial identity is constructed for enslaved individuals and their descendants. Patterson (1982) refers to this as a process of *desocialization*, the lack of ability to import culture from their homelands and the prohibitions against traditional African ritual and cultural practices, created a space where resistance would need to be hidden and reinterpreted. African tradition is not lost in the new world, but it is reinterpreted through oral history and brings in elements of Christianity and other non-African traditions. This led to a synchronism in mortuary ritual built from numerous cultural traditions filtered through the lens of enslavement, racism, and forced migration in the Americas (Jamieson, 1995). Mortuary ritual amongst enslaved individuals was less tightly controlled by slave-owners, this lack of overt control over mortuary ritual allowed enslaved persons to perform their own post-colonial identity through the creation of new mortuary processes. Jamieson (1995), in fact, argues that cultural transmission between Africa and the Americas was strongest with funerary rituals since enslaved individuals were responsible for burying their own dead in their own cemeteries.

For first generation enslaved individuals, members of the same ethnic group may have been involved with burial practices for one of their members, but practices would shift as time passed and direct cultural connections became changed through syncretism. Different African groups had different burial customs, and these differences were transmitted and altered as well, depending on which specialists were present. For example, David's (1992) work in Cameroon indicate that male *transformers* were responsible for the funeral. If a *transformer* were enslaved and brought to the Americas, he would be able to bring that specialized knowledge to an American context, but if *transformers* were not enslaved and brought to the Americas, this knowledge would be lost in the Americas, potentially leading to a different mortuary process.

The Newton plantation on Barbados contains a cemetery of enslaved individuals that have been thoroughly analyzed since their excavation in the 1970s. Barbados was England's first American territory to depend on slave-based labor. Thousands of individuals were brought from West Africa to Barbados, and so in his reconstruction of ethnographic behaviors, Handler and colleagues (1995, 1996, 1997, 1983) draw associations between Western African cultural groups such as the LoDagaa, Ibibio, and Ga and their mortuary practices. The cemetery at Newton consists of thousands of graves of enslaved individuals who died between around 1650 and 1838 (Handler, 1995). In addition to studying the skeletal remains, they also analyzed the spatial patterning within the graveyard. They found two primary clusters of burials in the cemetery, one with 17 burials and the other with over 40 sets of remains, among the randomly scattered burials. Through analysis of dental nonmetrics, they determined that the clusters may represent family groupings.

Enslaved individuals within the Newton cemetery found ways of honoring individual as well as group identities with their dead. Burial 72 provides an excellent example (Handler, 1997). This is the burial of an adult male around 50 years of age at death. There was no coffin present, and he was buried in a supine extended position with arms extended along the sides of the body. Orientation, absence of a coffin and grave goods suggest burial between the late 1600s and early 1700s. Analysis of bone lead content may suggest birth on Barbados, as does the lack of culturally modified (filed) teeth. But the origin of this individual is ultimately unclear. This individual was buried with the greatest number of grave goods of any of the excavated burials within the cemetery and provides a clear African influence on grave goods (Jamieson, 1995). Artifacts recovered with this individual included an iron knife between the legs near the left hand, six metal finger rings (consisting of copper and silver as well as mixed metals), three copper and brass bracelets on both forearms (two on the left and one on the right), a short-stemmed earthenware pipe on the pelvic area, and a necklace of cowrie shells, drilled fish vertebrae, drilled canid canines, European glass beads, and a carnelian bead. According to Handler (1997), it is likely that at least some of the materials such as the pipe, necklace and metal jewelry were transported across the Atlantic during the middle passage. The metal jewelry, in particular is of a "general African style" but not sufficiently distinctive to permit identification with specific cultural groups (Handler, 1997, p. 111). All employ techniques common to African metal working, it is unclear if they were made in Africa or on Barbados by African metalworkers, though. Handler (1997) believes this individual may have been a healer/diviner or medicine man based on the material culture found with the skeletal remains based on ethnographic accounts of similar metal jewelry used as protective amulets or charms. In particular, one clay pipe found was very similar to those found in Gold Coast archaeological sites, with formal characteristics such as form, shape, stem formation and decorative features resembling numerous pipes from coastal Ghana from the latter half of the seventeenth century. The elaborate necklace containing animal teeth and fish bones likely functioned as a protective amulet or charm (Handler, 1997). This individual, likely born on Barbados and subject to restrictions in cultural expression during life (e.g., the lack of tooth modifications), but numerous expressions of African cultural connection were included in his grave. Together, this individual inhabits the third space created by enslavement and desocialization. They also embody the lived experience of resilience within this third space, with markers of cultural connection to a place they likely had never physically been included in their grave.

A different burial from Newton shows that social standing attained in life can be reflected in funerary processes. Burial 9 contains the remains of a young adult female, approximately 20 years of age at death (Handler, 1996). The remains had high levels of lead that may indicate that the individual was born on Barbados. Lead levels, in fact, were more than twice the mean for all age groups as well as her own age group. While like Burial 72, this individual was found in an extended position, but Burial 9 was interred in a prone position. Burial 9 is the only interment within Mound 1, this single interment differs from the more common multiple interments that represent long term usage of the same burial mounds. Mound 1 was constructed before the other mounds, but no other interments were included in Mound 1 despite the crowding that occurred in other mounds. This lack of reuse of the funerary mound sets Burial 9 apart. It is the only prone burial within the cemetery, and the individual was buried alone, all of which raises questions about "her status in the slave community" (Handler, 1996, p. 79). What about this individual's lived experience or identity changed the way she was seen and handled in death? Extremely high levels of lead in the skeletal remains suggest she may have had lead poisoning. According to Hernberg (2000), lead poisoning can lead to lead colic, characterized by muscle spasms, joint and muscle pain, and abdominal cramping. Also common with lead poisoning are kidney issues and neurological impairments. The effects of lead poisoning may have been very visible, involving involuntary movement, leading to different social interactions.

The LoDagaa of Northwestern Ghana reserve some methods of burial for community members considered to be dangerous (Goody, 1962). One such method involves the creation of a mound over the dangerous individual, including individuals who die via suicide or are considered to be witches. Prone burial also may also indicate unusual or dangerous ways of dying (such as suicide or during pregnancy or childbirth). While this is an ethnographically based interpretation, the association with a specific group such as the LoDagaa is uncertain and so the interpretation that Handler makes for this individual may be an interpretive over-reach. What can be said is that the relationship between Burial 9 and the living enslaved population was different than others in the community. Due to possible lead poisoning, this individual would have moved and acted differently than others, and ultimately this seems to have led to a differential or atypical burial.

While the Newton Plantation case studies discussed above give a good idea of how burial practices can be interpreted within an identified space, finding the burials of enslaved individuals can often prove to be difficult. Rainville (2009) notes that since most enslaved individuals were buried in unmarked plots, burial locations are rarely marked on United States Geodetic Survey (USGS) maps and/or property deeds, making their locations difficult to identify prior to accidental discovery. These "unrecognized graves"(Rainville, 2009, p. 196) have a greater possibility of destruction.

After the end of slavery, the segregation of cemeteries by social race continued. In some places, this was written into cemetery rules and regulations. For the Elmwood Cemetery in Birmingham, Alabama, regulations stated that "Cemetery lots shall be owned only by human beings of the *white and/or Caucasian race* and the said lots shall be used only for the burial of human bodies of the *white and or Caucasian race*"(Rules and Regulations of Elmwood Cemetery adopted in 1954 as quoted in Rogers, 2005, p. 1153). The family of an African American serviceman killed in 1969 challenged this as a violation of 1866 Civil Rights Act. They were ultimately successful in their suit, leading to the integration of the cemetery. Rogers (2005) relates that approximately 90% of public cemeteries nationwide had at least some restrictions based on race in 1969. This general degree of segregation, supported by cemetery regulations and generally supported by the courts until the late 1960s and 1970s, led to expressions of African American identity within the cemeteries themselves. African American cemeteries had long been the target for body procurement for medical school dissecting rooms and anatomical collections (discussed below in greater detail).

Richardson (1989, p. 305) examines historically Black Burial Grounds of the Washington, DC area. His historical work shows a history of differential access to burial areas as a "microcosm of larger inequities of Washington life between 1880 and 1919." Many of the cemetery spaces available for use by African Americans were marginal, swampy plots land of limited commercial value. Cemeteries were commonly segregated by thorn fences. The Columbian Harmony Society was established by Congress in 1796 to provide aid to free-born and formerly enslaved African Americans. These individuals were unable to legally incorporate businesses, so the Society solicited contributions to purchase lad plots for cemetery spaces. The first of these, "Harmoneon" was used for burial from 1829, with burial within the cemetery limited to members of the Society. This cemetery was relocated in 1856 due to a lack of space and city growth.

Other African American cemetery space was purchased by church groups and other aid societies such as the Female Ladies Union Band Society, a group formed by free African American women who desired a place for burial for free African Americans (Richardson, 1989). This group purchased a lot next to that purchased by the Methodists. The two Burial Grounds are collectively identified as Mount Zion Cemetery (Fig. 5.2). As a symbol of increased independence on the part of

Fig. 5.2 Mount Zion Cemetery in Washington, D.C. Photo credit: APK, CC BY-SA 4.0 <https://creativecommons.org/licenses/by-sa/4.0>, via Wikimedia Commons. https://upload.wikimedia.org/wikipedia/commons/0/08/Mount_Zion_Cemetery_%28Washington%2C_D.C.%29_-_1.jpg

Washington's African American community, it was an important place both physically and socially, serving as a station in the Underground Railroad, and continuing to survive today. In 1816, Mount Zion Methodist church was founded by 125 African American members of the Montgomery Street Church who were dissatisfied with the discriminatory treatment they'd experienced. They purchased a lot and built a church known as "The Meeting House." Records of the church form a narrative of the African American experience in Washington, including notations of sales of individuals during slavery as well as indications of movement of members to Liberia (Mitchell, 1984). The name of the church was changed in 1844 to Mt. Zion Methodist Episcopal Church, and by 1850 there was a call amongst the congregants for African American clergy. When the church was moved in the 1880s, African American artisans were responsible for the construction. This church was a seat of African American identity and power and the associated cemetery, Mount Zion Cemetery, was no different. Mitchell (1984) notes that the cemetery was integrated, containing members of both Mt. Zion and its parent church, the Montgomery Street Church.

Though numerous notables were buried within the Mt. Zion cemetery in the nineteenth century, the last recorded burials occurred in 1950 and the District of Columbia prohibited additional interments in 1953 because the cemetery was out of compliance with Health Department regulations (including sunken graves, trash, and overgrowth in the cemetery space). Heirs to the members of the Female Union Band petitioned the US District Court to allow disinterment of remains in 1964 and ultimately sold the land containing the cemetery for $144,000 plus the cost of disinterment and reburial in cemeteries with perpetual care in 1974. Disinterment was challenged by the Afro-American Bicentennial Corporation (ABC) after the sale. ABC was established to "stimulate and organize participation of blacks and other minorities in the two-hundredth anniversary of the nation's independence" and to develop programs to "assure the perpetual endurance of the contributions of blacks and other minorities to the American heritage"(Mitchell, 1984, p. 114).

It was in this role that they intervened in the quiet movement of individuals from an important African American cemetery. They argued that instead of moving the cemetery to make way for housing, the space should be celebrated and declared a national monument under the care of the US Department of Interior ("FEMALE UNION BAND ASSOCIATION et al., Plaintiffs, v. UNKNOWN HEIRS AT LAW, DEVISEES AND ALIENEES OF Joseph MASON, Deceased, et al., Defendants, and Neville Waters and Afro-American Bicentennial Corporation, Intervenors. Charles L. NORRIS et al., Plaintiffs, v. UNKNOWN HEIRS AT LAW et al., Defendants, and Neville Waters and Afro-American Bicentennial Corporation, Intervenors," 1975). On August 6, 1975, Mount Zion Cemetery was granted status as a Historic Landmark with the following justifications:

- Its history and that of the generations of Blacks both free and slave interred therein uniquely convey the quality and thrust of Black life and evolving free Black culture in the District of Columbia from the earliest days of the city to the present.
- It is one of the few physical reminders of the significant contributions of Black people to the development of Georgetown, which during the late nineteenth century was between 35 and 45 percent Black in population (*547 III. Efforts at Preservation).

The judge's decision in this case touches not only on the concept of disinterment, stating that it is not favored in any instance, but on the mass disinterment that the repurposing of the land would entail. He states, "Disinterment would not be one on one, but *en masse*. It would be executed not by a shovel, but by a bulldozer-like machine called a 'back-hoe.' No matter how reverently the back-hoe might be maneuvered, no matter how respectfully the bones might be placed into a pile, this kind of operation cannot but offend the sensitivities of a civilized people." He goes on to note the commitment in 1975 to monumentation and restoration within the space, saying that the restoration of the cemetery space is a monument in itself by "the large numbers of Blacks and whites who have worked side by side to restore it. It would be an affront to the entire community and a betrayal of America's quest for human dignity for all people to permit the disinterment of this cemetery when there is a possibility of its restoration and maintenance in perpetuity as a monument to Black traditions in our Nation's capital" (IV. The Court's Decision).

The history of African American Burial Grounds is one of movement and social inequality. When land became scarce within city limits, those individuals buried within historically African American cemeteries were preferentially chosen for movement to outlying areas. It is also clear from Richardson's (1989) work that when cemeteries were "moved" it didn't always include all the bodies. In the case of the second burial area established by the Colored Union Beneficial Society, when the grounds were closed in 1898, 500 bodies were to be moved to Woodlawn Cemetery, but only 390 openings were made at the new cemetery and only 129 of the deceased were actually moved. During the period studied (1880–1919) African Americans experienced a time of "economic despair, social anonymity, and losses of political power" (Richardson, 1989, p. 310). The social status in life is reflected in the loss of autonomy over the dead. Average mortality for African Americans in this time was roughly double that of whites as reported for the years between 1906 and 1910, average age at death was significantly lower for African Americas as well, with average age at death for whites of 44 and 30 for African Americans. Infant mortality amongst whites was approximately 90 out of every 1000 babies born and 208 out of every 1000 African American babies born. The District of Columbia also notes that African Americans had higher rates of tuberculosis, likely impacting both lived experience as well as higher mortality and lowering the average age at death. All told, the mortality and morbidity of African Americans during this time is indicative of a large degree of social inequality that is perpetuated in the restricted access to mortuary spaces. But these mortuary spaces were also seen as spaces of power and identity formation. The history of the Mount Zion cemetery is illustrative of this history, by 1975 the social climate was significantly changed to see value in these burials and in the Space they inhabited. The Space and place they inhabited was transformed, as is the experience of visiting the cemetery. It is now not only a place of burial for the dead, but also an expression of culture and an acknowledgement of history and the development of American identity.

While separated physically from white cemeteries, African American mortuary spaces were also places of social safety and places where identity could be performed. They operated in this way *because* they were separated from white social structures, and therefore didn't depend on them economically. As Cann (2020, p. 6) notes, mortuary ritual "allowed African Americans to reclaim their narrative in a space wholly undefined and not controlled by whites. This allowed for funeral homes to express their identity in ways that could not be expressed in other areas of shared social space." Effectively, mortuary space was one where "black identity mattered" (Cann, 2020, p. 7). Often framed as "homegoings", the social space created for and by mortuary performance has acted as a preservation mechanism for African American tradition while serving as a place for community building and protest. As Stanley (2016, n.p.) comments, "From their earliest incarnations, black funerals were political, subversive—a talking back to the powers that be." Burial Grounds "represented the domain of the departed, and a place to express continuing developments of cultural identities within African diasporas" (King, 2010, p. 125). Cemetery spaces were purchased by burial societies' weekly dues from members (DuBois, 1899). Burial societies built social and monetary capitol and ultimately generated a burial insurance industry. This, in turn, helped provide social support not only for the deceased but also for the bereaved and larger African-American community in general (Cann, 2020). The mortuary performances were uniquely African-American, incorporating some elements of African identity (discussed above), as well as some inclusions that occur only in the third space of African-American experience, such as the incorporation of shoes into the grave shafts (Davidson, 2010).

African American undertakers and later funeral directors were responsible for treatment of African American dead, and so the fees generated through burial expenses directly impacted the African American community, creating wealth. African American funeral directors took on roles as culture keepers; the spaces used for funeral services were also used for community meetings, counselling, and ultimately protests during the civil rights movement (Washington, 1992). Elijah Cook, for example, was a funeral director in Montgomery, Alabama. In the basement of his carpentry shop (where he made coffins), he opened a training school to teach formerly enslaved individuals carpentry skills as well as how to read and write (Johnson, 2004). Johnson (2004, p. 13) draws the relationship between the economic stability generated by the stable nature of the funerary industry and the social capitol that this brings about, noting that funeral directors acted as "bridges and conduits to the powers-that-be." The segregation of death industries allowed African American funeral directors to develop their own business and social networks (building bonding capitol along the lines of Putnam (2020)), ultimately leading to increased autonomy and social standing. African American funeral directors gained status and generational wealth that allowed them to take socially prominent roles (Johnson, 2004). For Williams (2008), African American funeral directors took on important roles as culture keepers performing several needs of the community: (1) they acted as centers for community information; (2) they served as counselors and advisors; (3) they became symbols of success within the community; and (4) they engaged in philanthropy. In particular, within the Civil Rights movement, Johnson (2004) notes that funeral directors could acts as conduits between the community and powers that be and could hold prominent roles without fear of retaliation in the business community. Essentially, they didn't have to worry about losing White business and so could be vocal and prominent in the Civil Rights Movement. Ultimately, Funeral Directors took advantage of this larger social platform to run for Congress and the Senate. The first African American senator from Michigan, Charles Diggs Sr, was a funeral director. His son, Charles Diggs Jr later became the first African American Congressman from Michigan where he helped to organize the Congressional Black Caucus (Johnson, 2004).

The death and burial of Emmett Till is further addressed in Chap. 9, specifically the role that Mamie Till-Mobley's decision to have an open casket funeral had as a catalyst in the Civil Rights Movement. Emmett Till's lynching and funeral served as rallying points for the development of African American identity that centered on demanding full citizenship in ways that would not have been possible without an individual to identify. We can see this thread from Emmett Till through modern deaths in police custody (George Floyd, Trayvon Martin, and many others), a link that is often made explicit. As Cann (2020, p. 13 of 15) concludes:

> The grandiosity of the funeral in Floyd's funeral is thus not only intentional but viewed as an act of resistance that demonstrates how much black lives do, in fact, matter. The open casket of Emmett Till, the closed caskets of the victims of the Birmingham bombing, and the gold casket of George Floyd all signal that black deaths matter and will continue to matter until Black Lives Matter more. Just as Emmett Till's body was the burden of proof to convict Jim Crow, the bodies of Trayvon Martin, Tamir Rice, George Floyd, and the others unjustly slain prove that Jim Crow's descendants—more covert forms of racism that plague our justice system are alive and well in the twenty-first century. The African American Funeral home continues to play an important role in American culture, and through its very existence, argues against racist policies and practices evident in the high mortality rates of blacks in the United States.

What began as a way to form group identity in the worst of circumstances becomes a way to perform and promote resiliency on a grand scale when the mortuary performance is examined in this light.

5.3 Summary

This chapter examines how the processing of the body communicates social information and fosters community building. Drawing on body-centered theoretical models, it shows that the unique qualities of the body become powerful amplification mechanisms when manipulated by the living. Whether in spaces that evolve gradually or those shaped by migration, mortuary rituals and body processing serve as key tools for forming and strengthening communities. The chapter focuses particularly on how African American mortuary practices have shaped social identity and resistance, acting as vital instruments for cultural preservation and empowerment. The bodies of the dead serve as significant sites for cultural expression and social negotiation. Through the lens of bioarchaeology, postcolonial theory, and social history, it explores how these mortuary rituals reflect broader social structures, power dynamics, and cultural hybridity. African American burial customs, which blend African traditions with new practices created in response to the unique circumstances of enslavement and oppression, become acts of resistance and empowerment. The chapter also highlights the pivotal role of African American funeral directors and mortuary specialists, who functioned as culture keepers and played an instrumental role in organizing and leading efforts during the Civil Rights Movement. Their influence continues today, with mortuary rituals remaining a significant part of contemporary movements such as Black Lives Matter. Ultimately, the chapter underscores that African American mortuary practices were not only essential for cultural survival but also for social and political change, reflecting the resilience and strength of African American communities in the face of racial oppression.

References

Anwar, N. H. (2014). The postcolonial city in South Asia. *Singapore Journal of Tropical Geography, 35*(1), 22–38. https://doi.org/10.1111/sjtg.12048

Blakely, R. L., & Harrington, J. M. (1997). Grave consequences: The opportunistic procurement of Cadavers at the Medical College of Georgia. In R. L. Blakely & J. M. Harrington (Eds.), *Bones in the basement: Postmortem racism in nineteenth-century medical training* (pp. 162–183). Smithsonian Institution Press.

Blakey, M. L. (1998). The New York African burial ground project: An examination of enslaved lives, a construction of ancestral ties. *Transforming Anthropology, 7*, 53–58.

Blakey, M. L. (2001). Bioarchaeology of the African Diaspora in the Americas: Its origin and scope. *Annual Review of Anthropology, 30*, 387–422.

Blakey, M. L., & Rankin-Hill, L. M. (2004). *The New York African burial ground skeletal biology final report*. Howard University.

Blanks, J., Abuabara, A., Roberts, A., & Semien, J. (2021). Preservation at the intersections: Patterns of disproportionate multihazard risk and vulnerability in Louisiana's Historic African American Cemeteries. *Environmental Justice, 14*(1), 1–13.

Blassingame, J. W. (Ed.). (1977). *Slave testimony: Two centuries of letters, speeches, interviews, and autobiographies*. LSU Press.

Bongers, J. L. (2019). Mortuary practice, imperial conquest, and sociopolitical change in the Middle Chicha Valley, Peru (ca. AD 1200–1650). (PhD). University of California Los Angeles, Los Angeles.

Cann, C. K. (2020). Black deaths matter earning the right to live: Death and the African-American Funeral Home. *Religions, 11*.

Christensen, A. M., Passalacqua, N. V., & Bartelink, E. J. (2014). *Forensic anthropology: Current methods and practice*. Elsevier.

Davidson, J. M. (2010). Keeping the Devil at Bay: The shoe in the coffin lid and other grave charms in nineteenth- and early twentieth-century America. *International Journal of Historical Archaeology, 14*, 614–649.

DuBois, W. E. B. (1899). *The Philadelphia Negro: A social study*. Shocken.

Duncan, W. N., & Schwarz, K. R. (2014). Partiable, permeable, and relational bodies in a maya mass grave. In A. J. Osterholtz, K. M. Baustian, & D. Martin (Eds.), *Commingled and disarticulated human remains: Working toward improved, theory, method, and data* (pp. 149–172). Springer.

Female Union Band Association et al., Plaintiffs, v. Unknown Heirs at Law, Devisees and Alienees of Joseph Mason, Deceased, et al., Defendants, and Neville Waters and Afro-American Bicentennial Corporation, Intervenors. Charles L. Norris et al., Plaintiffs, v. Unknown Heirs at Law et al., Defendants, and Neville Waters and Afro-American Bicentennial Corporation, Intervenors, No. Civ. A. Nos. 2591-63, 1705-65, 403 (United States District Court, District of Columbia 1975).

Geller, P. L. (2012). Parting (with) the dead: Body partibility as evidence of ancestor veneration. *Ancient Mesoamerica, 23*(1), 115–130.

Geller, P. L., & Suri, M. S. (2014). Relationality, corporeality and bioarchaeology: Bodies qua bodies, bodies in context. *Cambridge Archaeological Journal, 24*, 499–512.

Goody, J. (1962). *Death, property and the ancestors: A study of the mortuary customs of the LoDagaa of West Africa*. Stanford University Press.

Handler, J. S. (1995). An African-type Burial, Newton Plantation Barbados. *African Diaspora Archaeology Newsletter, 2*(3).

Handler, J. S. (1996). A Prone Burial from a Plantation Slave Cemetery in Barbados, West Indies: Possible evidence for an African-type witch or other negatively viewed person. *Historical Archaeology, 30*, 76–86.

Handler, J. S. (1997). An African-type Helaer/Diviner and his grave goods: A Burial from a Plantation Slave Cemetery in Barbados, West Indies. *International Journal of Historical Archaeology, 1*, 91–130.
Handler, J. S., & Corruccini, R. S. (1983). Plantation slave life in Barbados: A physical anthropological analysis. *Journal of Interdisciplinary History, XIV*, 65–90.
Hernberg, S. (2000). Lead poisoning in a historical perspective. *American Journal of Industrial Medicine, 38*(3), 244–254.
Jamieson, R. W. (1995). Material culture and social death: African-American burial practices. *Historical Archaeology, 29*(4), 39–58. Retrieved from http://www.jstor.org/stable/25616423
Johnson, C. L. (2004). *Undertakings: The politics of African-American funeral directing*. (PhD). University of South Carolina, (UMI Number 3130454).
King, C. (2010). Separated by death and color: The African American Cemetery of New Philadelphia, Illinois. *Historical Archaeology, 44*(1), 125–137. Retrieved from http://www.jstor.org/stable/27820824
Kohrs, S. (2021). *Down by the Riverside: Burial practices of the enslaved.* Retrieved from fnfsr.org: https://fnfsr.org/app/uploads/2021/02/Down-by-the-Riverside-Script-by-Kohrs-Transcript.pdf
LaRoche, C. J. (1994). Beads from the African Burial Ground, New York City: A preliminary assessment. *Beads: Journal of the Society of Bead Researchers, 6*, 3–20.
Locard, E. (1920). *L'enquête criminelle et les méthodes scientifiques: E.* Flammarion.
Mack, M. E., & Blakey, M. L. (2004). The New York African Burial Ground Project: Past biases, current dilemmas, and future research opportunities. *Historical Archaeology, 38*(1), 10–17. Retrieved from http://www.jstor.org/stable/25617128
Mains, S. P., Gilmartin, M., Cullen, D., Mohammad, R., Tolia-Kelly, D. P., Raghuram, P., & Winders, J. (2013). Postcolonial migrations. *Social & Cultural Geography, 14*(2), 131–144. https://doi.org/10.1080/14649365.2012.753468
Merleau-Ponty, M. (2005 [1962]). *Phenomenology of perception* (C. Smith, Trans.). New York: Routledge.
Meskell, L. (2000). Writing the body in archaeology. In A. E. Rautman (Ed.), *Reading the body: Representations and remains in the archaeological record* (pp. 13–21). University of Pennsylvania Press.
Mitchell, P. G. (1984). *The history of Mt. Zion United Methodist Church and Mt. Zion Cemetery* (Vol. 51, pp. 103–118). Records of the Columbia Historical Society. Retrieved from http://www.jstor.org/stable/40067847
Osterholtz, A. J. (2015). *Bodies in MOTION: A bioarchaeological analysis of migration and identity in bronze age cyprus (2400–1100 BC).* (Ph.D. Dissertation). University of Nevada Las Vegas, Las Vegas, NV.
Osterholtz, A. J. (2018). Bioarchaeological approaches to Southwestern violence. *Kiva, 84*(4), 399–402.
Parker Pearson, M. (1999). *The archaeology of death and burial*. Texas A&M University Press.
Patterson, O. (1982). *Slavery and social death*. Harvard University Press.
Putnam, R. D. (2020). *Bowling alone: The collapse and revival of American community (Revised and Updated).* Simon & Schuster Paperbacks.
Rainville, L. (2009). Protecting our shared Heritage in African-American Cemeteries. *Journal of Field Archaeology, 34*, 196–206.
Richardson, S. J. (1989). *The Burial Grounds of Black Washington: 1880–1919* (Vol. 52, pp. 304–326). Records of the Columbia Historical Society. Retrieved from http://www.jstor.org/stable/40067870
Rogers, E. M. (1995). *Diffusion of innovations* (4th ed.). The Free Press.
Rogers, K. (2005). Integrating the City of the Dead: The integration of cemeteries and the evolution of property law, 1900–1969. *Alabama Law Review, 56*, 1153–1166.
Said, E. (1978). *Orientalism*. Vintage.

Saul, F. P. (1972). The human skeletal remains of Altar de Sacrificios: An osteobiographic analysis. *Papers of the Peabody Museum of American Archaeology and Ethnology, Harvard University, 63*(2).
Saul, F. P. (1976). Osteobiography: Life history recorded in bone. In E. Giles & J. S. Friedlander (Eds.), *The measures of man: Methodologies in biological anthropology* (pp. 372–382). Peabody Museum Press.
Saxe, A. A. (1970). *Social dimensions of mortuary practices*. (Doctor of Philosophy). The University of Michigan, Ann Arbor, MI.
Scheper-Hughes, N., & Lock, M. M. (1987). The mindful body: A prolegomenon to future work in medical anthropology. *Medical Anthropology Quarterly, 1*(1), 6–41. Retrieved from http://www.jstor.org/stable/648769
Schrader, S. (2019). *Activity, diet, and social practice: Addressing everyday life in human skeletal remains*. Springer Cham.
Sofaer, J. R. (2006). *The body as material culture: A theoretical osteoarchaeology*. Cambridge University Press.
Stanley, T. (2016). *The disappearance of a distinctly black way to mourn*. The Atlantic.
Tilley, C. (2016). Phenomenological approaches to landscape archaeology. In B. David & J. Thomas (Eds.), *Handbook of landscape archaeology* (pp. 271–276). Routledge.
van Gennep, A. (1960 [1909]). *Les Rites de Passage*. Chicago: University of Chicago Press.
Vlach, J. M. (1991). *By the work of their hands: Studies in Afro-American Folklife*. University Press of Virginia.
Voskos, I., & Knapp, A. B. (2008). Cyprus at the end of the late bronze age: Crisis and colonization or continuity and hybridization? *American Journal of Archaeology, 112*(4), 659–684.
Walker, W. H. (1996). Ritual deposits: Another perspective. In E. C. Adams (Ed.), *River of change: Prehistory of the middle Little Colorado River Valley, Arizona* (pp. 75–91). Arizona State Museum, University of Arizona.
Washington, B. T. (1992). *The Negro in business*. DeVore & Sons.
Williams, C. T. (2008). *Paradigm shift in African American funeral customs: Looking through the lens of oral history and consumer culture*. (Doctor of Education). Georgia Southern University, Statesboro, GA.

Part II
Applying the Model

Part II focuses on the application of the model of Poetics within Bioarchaeology to historical processes, examining the role of the body in the creation of social spaces and as a medium for the creation/negotiation/renegotiation of social relationships. This section provides a multitude of scales

Chapter 6 traces the development of body preservation in mortuary practices, both in terms of embalming and mummification. Particularly in terms of embalming, the chapter traces the transition from home-based ritual to the medicalization of processing that required specialized training to perform this. The chapter illustrates how these processes are active, evolving activities that reflect social identity and cultural values. Chapter 7 examines inhumation (burial) practices with discussions of different types of inhumations that reflect different aspects of identity. Included here are historical discussions of churchyards, military cemeteries, Potter's Fields, and Green cemeteries. These spaces are active places for the living and reflect social concerns and values. Chapter 8 examines the history and development of four famous cemeteries (Arlington National Cemetery, St. Louis No. 1, Okunoin Cemetery, and Père Lachaise Cemetery) as examples of how this space can reflect and be reflective of social processes. Each cemetery offers a unique lens through which to examine the ways death rituals and memorialization practices interact with larger social and historical forces. Finally, Chapter 9 examines the death rituals surrounding some specific individuals and how these can be used to understand broader social contexts. The burial and memorialization of the individuals discussed (Marie Laveau, Frank Little, Marie Curie, Richard McKinley, John Byres, Richard Legg, Emmett Till, and the Presley family) underscore how memorialization practices of individuals are deeply tied to deeper social movements, historical narratives, and cultural values.

Chapter 6
Preservation as Processing

Abstract This chapter focuses on the preservation of the body in historical and cultural contexts. The first part of the chapter focuses on colonial and antebellum mortuary practices in the U.S., particularly within White Colonial contexts. It traces the shift from familial, home-based death rituals to the emergence of specialized mortuary professions, such as coffin makers and embalmers, reflecting broader changes in American society, including urbanization, religious diversity, and social stratification. The chapter further examines the profound impact of the Civil War (1861—1865) on U.S. mortuary practices, particularly the rise of embalming as a response to the mass deaths of soldiers. The war's scale necessitated the preservation and transportation of bodies, leading to the widespread adoption of embalming and the professionalization of the funeral industry. The chapter traces the development of embalming techniques in the US and the rise of the 'traditional' American funeral, with a focus on the rise of medicalization and the death industry.

The second part of the chapter focuses on mummification in two contexts. First, mummification traditions in Bronze Age Britain are detailed, particularly at Cladh Hallan, where intentional body preservation was linked to shifts in social and religious structures. It also examines Buddhist mummification practices, including the self-mummification of monks in Japan, and their spiritual significance as "living relics" that continued to exert power after death. Through these case studies, the chapter highlights how mummification and mortuary rituals have been used across cultures to navigate identity, power, and community formation.

This chapter addresses two disparate forms of processing, both revolving around preservation of tissues. First, I will give a brief history of mortuary performance within the US, centered on the processing of the body that allows for visitation and interaction with the dead body and how those interactions changed throughout time. Second, I will briefly discuss mummification and the intentional creation of mummies. Both the history of mortuary performance in the US and mummification in Bronze Age Britain, though, share a desire to interact with the dead by the living and those processes of preservation are reflective of social needs and the cultural contexts in which they developed and changed.

A. Osterholtz, *The Bodies Poetic*, Bioarchaeology and Social Theory,
https://doi.org/10.1007/978-3-031-85998-4_6

Mortuary performance can be viewed as part of a reciprocal loop, both influencing and being influenced by social factors. Here, I present a woefully short history of mortuary processes within the US as a way to illustrate how these practices reflect society and its changes. I focus particularly on the material culture correlates and social understandings of death. Mortuary rituals involve the processing of the body in identifiable ways, and the material culture that develops surrounding these performances has specific meanings. The changes to material culture come about through changes in the relationship between the living and the dead. For instance, the development of mortuary specialists such as coffin makers means that cabinet makers or family members no longer made boxes for the dead. The history presented here is biased toward a White Colonial way of death within the US. This bias is the result of numerous factors, including the prevalence of sources related to this particular way of death partly due to structural factors both in the study of death being primarily conducted until recent years with this focus and the underrepresentation of different perspectives in the material record. As noted in Chap. 5, the Collective for Radical Death Studies seeks to highlight the important role of different views and approaches to the study of death, highlighting the experiences of African Americans and immigrant groups to the United States. In the past 20 years, numerous studies of African American mortuary customs and the unique role of African American mortuary performances in the development of identity and community (e.g., Burg, 2008; Cann, 2020; Fletcher, 2021; Foster & Eckert, 2003; Frohne, 2002; Johnson, 2004; Jones, 2011; King, 2010; Murphy & Stephens, 2020; Williams, 2008). These studies are important steps towards presenting the important contributions to American mortuary performance, but are typically viewed as an alternative to the Colonial, White way of death.

In this volume, I have chosen to approach African American funerary customs in a different section specifically because those mortuary performances performed by African Americans highlight the resiliency of those groups brought to the US in different mechanisms from the Colonial, White approach to death. They are not lesser than but highlight different elements of identity and community construction. Whereas in African American mortuary rituals we see the development of a third space and the profound resilience of a newly formed community, the White Colonial way of death sought to maintain mortuary rituals with European origins, especially in the early days of Colonial settlement in the US. These rituals still form communities, and we see those rituals and meanings diverging from the European origins beginning in the late 18th and early nineteenth century (with the greatest changes coming out of the Civil War), but they are fundamentally different from African American ethnogenesis evident in new mortuary programs.

6.1 The Antebellum Way of Death

What is a *good* death? Going all the way back to the seventeenth and eighteenth century within the United States, the concept of a *good* death was one surrounded by friends and family. Death was expected to take place in the home, and the attitude of the dying person was thought to be a reflection of their ultimate fate, whether they would be welcomed into heaven or would be consigned to hell (Laderman, 1995). There was a general acceptance of death, and the bereaved had significant contact with the dead body: being present for the death, washing and laying out the body, and being actively involved with the burial process (particularly in rural areas). This is essentially a continuation of the *ars moriendi*, the medieval approach to dying well and one deeply rooted in a Christian framework that focused on the afterlife (Forcén & Forcén, 2016; Valentine, 2021).

With the increasing prevalence of a protestant population in North America, dying well took on an increased importance since the concept of purgatory was no longer present. Dying well was seen as an indication that the dying individual was granted admittance to heaven. Along with an appropriate method of dying, burial in a church cemetery was seen as the appropriate venue for burial. Tombstones were archives that told stories for the community in general, and the iconography of these tombstones can be read as social lessons. Deetz and Dethlesen (Deetz, 1996; Deetz & Dethlefsen, 1971; Dethlefsen & Deetz, 1966) tie changes in the iconography to changing social movements including the second great awakening and class distinctions.

By the antebellum nineteenth century, a *good* death still involved death at home and was replete with social lessons. In this era, we see the rise of the cemetery as an important place for the display of social status and identity. The rise of the rural cemetery focused on privately owned cemeteries that allowed for elaborate monumentation and the creation of family plots within more urban environments. Mount Auburn Cemetery in Cambridge, Massachusetts is one early example. Founded in 1831, it is essentially a park-like atmosphere. This was organized as a voluntary association of families and individuals. The creation of these idyllic landscapes created a distinction between a space for the living (in cities) and space for the dead (cemeteries). Owners of family plots were given license to create their own place for memorialization of their family members and were expected to maintain the graves themselves (Sachs, 2010). Individual plots could also be purchased, but these had restrictions on the amount of monumentation permitted and individual owners lacked voting rights in the overall management of the cemetery. The space was designed to be visited, and to encapsulate contradictions. Sachs (2010, p. 210) writes that the "ultimate magic of the cemetery lay in its ability to weave together seemingly opposing elements, to preserve wildness in the midst of the artificial city, to blend life and death, time and space, female emotiveness and male restraint… an arcadian realm whose fundamental hybridity suggests permeable borders and impure categories, a liminal landscape that punctures definitions of modernity." Along with more of an individual focus in the layout and ownership of the new cemetery plots, monumentality shifted as well, with more images consistent with hope, immortality and life. The symbolism on tombstones and monuments becomes more varied and more complex.

Death itself was still expected to occur within the home, surrounded by friends and family. This was a time of increasing urbanization, though, and those rituals had to change to accommodate that. Religious and ethnic diversity also led to increasing diversity in the mortuary rituals performed, as Laderman (1995, p. 30) notes, "with the lack of any one dominant understanding of death, the dead were thrown into the competitive marketplace of ideas that mirrored the increasingly diversified socio-economic spheres." The dead became materials by which social status could be performed by the living. Daily life was saturated with reminders of death, including the use of highly formulaic mourning dress and iconography. The dead themselves also became more ornately dressed, as opposed to the winding sheets typically used in earlier periods. Increasingly ornate material culture related to burial becomes available as well, including coffins that are mass produced (LeeDecker, 2001, 2009).

The burial of a woman in Lexington, Missouri can be used to illustrate the ornate nature of burial in this time. The burial was analyzed by Wescott and colleagues (2010) using a multidisciplinary approach that combined archaeological, historical, bioarchaeological, and chemical analyses after the coffin was unearthed in 2006. The woman, whose name is unknown, was buried wearing a corset, chemise, dress and shoes. She was buried in an ornate casket within a family plot. Wescott and colleagues traced historical records and found the coffin came from the John Shotwell Stewart plot. The plot was purchased in 1850 for $20. The casket contained the remains of an adult woman whose name is unknown between the ages of 20 and 30 years at death. Light colored tulle near the cranium was likely a face veil and shoe fragments found as well. The morphology of her ribs suggested that she wore a corset during life. The remains of a corset were also found within the coffin. Based on prevailing understandings at the time, she was likely laid out for burial as though she were resting (Stearns, 2021). Altogether, this analysis was holistic and her burial can be seen as a typical burial for a middle-class individual living in the nineteenth century. She was buried within a family plot wearing clothing (as opposed to a winding sheet or gown), and her burial would have been an important mechanism for her family's performance of status within the Lexington community. There may have been a postmortem photo taken of her.

The earliest postmortem photography dates to the 1840s and can be seen to be integral to the development of photography as a profession, providing lucrative income soon after the invention of photographic techniques (Borgo, Licata, & Iorio, 2016, p. 109). In postmortem photos, the dead are often portrayed as sleeping: "Death is negated in a warm sleep of a body that seems to rest" (Fig. 6.1). There is also a functional element to the use of postmortem photography. Early photography needed long exposures. As children are not known for the ability to keep still for long periods of time, often postmortem photos may have been the only photos of small children. The body is still being used to tell a social story, but in this case, it is not one of ultimate redemption, but one of memory. Significant material culture was created with antebellum nineteenth century mortuary performances, including postmortem photography, jewelry made from hair from the deceased, and a plethora of new types of mortuary monumentation. Aries (2004) describes this as the domestication of the corpse. Death was transformed to a time of hope and promise.

Fig. 6.1 Post-mortem of man with Legion d'Honneur ribbon, c. 1850. France. Daguerreotype, sixth plate; overall: 8 × 7 cm (3 1/8 × 2 3/4 in.). The Cleveland Museum of Art, Gift of Michael Mattis and Judith Hochberg 2021.81

6.2 Death and the Civil War

The Civil War (1861–1865), of course, changes everything. With the Civil War, we see the first generation of individuals dying far from home and while more died from disease than battle wounds, the nature of their deaths was fundamentally different from the generations before. Essentially, where people died and who was dying in larger numbers fundamentally altered the way of death in the US and therefore had significant impacts on the mortuary rituals performed. Death far from home changed how the living and the dead interacted, including the shipping of remains (if families could afford this and/or it could be arranged). Shipping of remains itself caused changes in the preference towards embalming. Prior to the Civil War, embalming was known but relatively uncommon in the US. After the Civil War, this became the norm, often codified in state law as a public health measure.

6.2.1 The Rise of Embalming and Preservation

What many of us think of as 'traditional' funerals are a product of the twentieth century commodification and professionalization of the death industry. It can be seen as a central shift in who has access to and responsibility for caring for the dead, shifting from the family and friends of the deceased to funeral homes who coordinate and provide services to the family. This can be viewed as a medicalization of death, a shifting of responsibility from the family and friends of the deceased to an unknown provider with an emphasis on the public good (public health). Foucault (2003 [1963]) argues that medical paradigms provide systems in which the body is understood and experienced. Underlying these paradigms is the possibility of understanding and mitigating death, of placing it into an identifiable box that can be understood and thereby controlled. Foucault (2003 [1963]) argues that medical encounters are disciplinary in that medicine provides guidelines about who the patient and who the doctor is, and provides a way to regulate or control the body. In the case of caring for the dead, this is the control over the dead body by the living. Death specialists (such as undertakers, embalmers, or other mortuary specialists) can be seen as links in power relations (Foucault, 1984 [1967]). Power is relational, a strategy bought into by living participants in caring for the dead, and employs elements of both the Social Body and the Body Politic (Scheper-Hughes & Lock, 1987). The Social Body is represented by the willingness of the living to participate in the medicalization process and the Body Politic by mandating and providing guidelines for the funeral industry.

This shift really began in the nineteenth century and accelerated with the Civil War in the US. Faust (2008, p. 62) argues that for nineteenth century Americans (both North and South), the "body was the repository of human identity in two senses: it represented the intrinsic selfhood and individuality of a particular human, and at the same time it incarnated the very humanness of that identity—the promise of eternal life that differentiates human remains from the carcasses of animals, who possess neither consciousness of death nor promise of either physical or spiritual immortality." Here she is arguing that the physical remains embodied an understanding of the relationship with the living and the dead and must be treated as separate from any other dead thing. In the US, prior to the large-scale mortality event of the Civil War, death was a personal process, dying away from friends and family alienated the living from the dead and vice versa. Leaving the dead unburied was seen as an insult not only to the dead but also to the cultural and religious ideals shared by many.

During the Civil War, military officials were forced to develop burial procedures quickly. This wasn't well planned, and records weren't always kept well, particularly with quickly changing fronts. This was particularly true with battles with high numbers of casualties or for the losing side of the conflict. The non-victorious side was rarely allowed to bury their dead with ceremony or in accordance with religious or cultural beliefs. Faust (2008) notes numerous instances where bodies from the opposing force were buried in shallow communal graves later disturbed by scavengers or simply left to decompose on the surface. The lack of preparation for burial of the dead became obvious very early on with the First Battle of Bull Run in July of 1861. The Union Army issued General Order No. 75 which made commanding officers responsible for the burial of soldiers who died under their command or in their jurisdiction and to report the deaths to the adjutant general. Merrifield (n.d.) details some of challenges to this order, namely the death of large numbers of soldiers during battles, including the commanding officers responsible for organizing the burial details. The survivors of battles were also often ill-equipped to handle burial details, and they lacked materials for grave markers. Fallen soldiers also often lacked any kind of identification, making it difficult to fully document who was buried where. This was expanded 6 months later with General Order No. 33 which ordered Commanding Generals to set aside Burial Grounds close to every battlefield and to bury the dead as soon as is possible with the graves marked and records of burials kept. Faust (2008, p. 65) refers to this as "utopian" measure, noting that the Union Army had no regular burial details, no graves registration units, and no ambulance service until 1864. The burials occurred in a haphazard fashion, but these Burial Grounds ultimately were transitioned into first National Cemeteries. By 1865, there were approximately 30 National Cemeteries (Merrifield, n.d.). All of these things lead to a large number of unidentified individuals from the war, possibly making national mourning through the establishment of National Cemeteries more socially pertinent.

The general picture drawn by Civil War accounts of death and burial are one of chaos and general social breakdown. Carefully constructed mortuary rituals of *good death* surrounded by friends and family were shattered, no one was present to care for the dead in the way that was expected. Those duties fell to strangers, names and identities were lost. In fact, the total number of dead continues to be counted for this war. Hacker (2011) now estimates total deaths to be between approximately 752,000, an increase of approximately 20% over the oft-cited total of 620,000+ deaths based on a 1% sample of individuals from the 1850—1880 censuses. Barceló and colleagues (2024) recently provided an updated analysis utilizing the full count and a sample of linked records across multiple census to provide an estimate of 689,000 civil war deaths. They also provide state-by-state accounting of excess male mortality (mortality rates above the expected number of deaths). These were generally higher for Confederate states, with Louisiana having the highest rate of excess male deaths (19.15%). They conclude by stating that it "seems reasonable to expect that enormous mortality in the US South shaped state demographics, impacting families through the death of bread-winners and triggering economic, political, and social legacies for generations" (p. 3 of 3).

Wealthy families had more resources to attempt to find and bring fallen sons and husbands home for burial. The problem created by transportation over perhaps hundreds of miles became important. Prior to the Civil War, when bodies were transported for burial, they were packed in ice, but this was not feasible for the majority of those dying in the Civil War. Faust (2008) recounts stories of families tracking down the bodies of their kin and being forced to bribe railroad officials to allow shipment of the bodies despite advanced states of decomposition. Decomposition,

Fig. 6.2 Lyford's Embalming Office, Gettysburg 1863. Image credit: Tiburon Audubon Center (originally: U.S. Army), Public domain, via Wikimedia Commons. https://upload.wikimedia.org/wikipedia/commons/4/4c/Lyford%27s_Embalming_Office%2C_Gettysburg_1863.jpg

with its cultural understanding of miasma leading to disease, needed to be retarded to allow for transport on the scale necessitated by the mass mortality of the Civil War. Decomposition also altered the appearance of the dead, something that the family members transporting the dead wished to stop or at least slowed (Brenner, 2014). These two issues, decomposition and the importance of maintaining a life-like appearance of the dead, led to the rise in embalming. Mortuary surgeons commonly set up shop near the battlefield hospitals and offered a sliding scale of cost for the procedure, with embalming for common soldiers running $5 but increasing with rank (Faust, 2008) (Fig. 6.2). Approximately 30,000—40,000 soldiers were embalmed during the Civil War (Colman, 1997). This increase in embalming allowed for the dead to be buried in community cemeteries more in line with accepted cultural norms and practices.

The preservation of the body was important as a selling point and a way to normalize embalming within American society. Civil War embalming is characterized by the use of incredibly caustic chemicals, including arsenic, creosote, mercury, turpentine and wood alcohol (Brenner, 2014). Dr. Thomas Holmes is seen as the father of modern embalming and received his start in the Civil War, where he set up an embalming practice in Washington DC, treating up to three bodies per day for a cost of $100 per body (Bentley & Chasteen, 2002). His methods were quickly adapted by others along with the general normalization of body preservation as part of the American mortuary program.[1]

6.3 Post-war Death in the US

By the 1880s, embalming was advertised as a way to prevent disease caused by the dead (Colman, 1997). Arsenic-based solutions were generally used for embalming by the end of the nineteenth-early twentieth centuries (Brenner, 2014), possibly due to arsnic's ability to keep a corpse more supple or movable (Bentley & Chasteen, 2002). Dr. Holmes' techniques were primarily arsenic based. The preservation of bodies using his techniques is excellent, and examples of his craft can still be found preserved at the Armed Forces Institute of Pathology. Bentley and Chasteen (2002) detail one such instance of a well-preserved head and torso prepared by Holmes dating to the 1850s that contains 28,000 ppm of arsenic. Embalmers typically used anywhere from 4 ounces to 12 pounds of arsenic in their work. Even after the introduction of formaldehyde, some embalmers preferred arsenic due to its lack of fumes.

The use of arsenic, however, does have significant drawbacks. Arsenic can contaminate ground water requiring remediation. Leaching of arsenic from bodies embalmed with arsenic may be a significant source of contamination of ground water supplies (Bentley & Chasteen, 2002). Arsenic is also a poison as well as additive in multiple medicines during the nineteenth and twentieth centuries, and so the use of arsenic as a conservation agent may mask death by arsenic poisoning, either intentional or accidental (Bentley & Chasteen, 2002). This latter point became important during a reexamination of the body of Zachary Taylor, who died on the ninth of July 1850. More than 140 years after his death, a new theory surfaced that he had been poisoned due to his stance on slavery. The crypt was opened in June 1991 and fingernail, hair, and tissue samples were taken by forensic anthropologist William Maples and run by three independent laboratories. Arsenic levels within the body were found to be between 3 and 9 times the modern-day rate expected (Parenti, 1996). William Maples, recounting this in his memoir *Dead Men Do Tell Tales* (1994) argues these levels are simply consistent with exposure to the nineteenth century, and would not have produced death or even illness. Leaching of embalming fluids from the body into soil and groundwater may have been one reason for the introduction of the vault in addition to the use of caskets or coffins.

Today, the Federal Trade Commission (FTC) is tasked with regulating the funeral industry pursuant to 5 U.S.C. 57a(a), 15 U.S.C. 46(g), and 5 U.S.C. 552. This sets forth definitions of what a "casket" is "a rigid container which is designed for the encasement of human remains and which is usually constructed of wood, metal,

[1] Here I want to make a distinction between funerary embalming and anatomization or preservation of human tissues for the purposes of medical research. Anatomization and medical embalming will be discussed later in this volume.

fiberglass, plastic, or like material, and ornamented and lined with fabric" (Part 452(c)), different types of funeral services, and so on. That this is regulated by a government agency (Commerce) suggests the separated nature of this process. The FTC sets out what providers of funeral services must provide and what information must be furnished to the consumer (i.e., the family and friends of the deceased). Violation of the rules provided by the FTC can result in penalties of up to $51,744 per violation as of January 2024 (FTC, 2024). The modern funeral process is formulaic by design. Funeral service providers are required to provide a General Price List (GPL) to anyone who begins a discussion about funeral arrangements, either directly after death or in "pre-need" planning. This GPL "empowers you, the consumer, to make the choices during the funeral arrangement conference which are right for your family"(Wyoming Funeral Directors Association 2013). It also sets out what the state requirements are regarding embalming, the use of alternative containers (such as cardboard coffins, and so on), and any charges for services offered by the funeral director. These may include embalming charges, fees for preparation of the body, cost of removal and transfer of the body to the funeral home, charges for use f facilities for viewing, funeral, or memorial service, charges for funerary vehicles, costs of cremation, and costs of equipment for graveside services. While this definitely simplifies a process the bereaved are ill-equipped to deal with at a time of heightened stress, it still removes them from the process of preparing the body and laying the dead to rest.

The modern vault is generally attributed to Wilbert Haase who assumed leadership of the L.G. Haase Manufacturing company in 1913 (this was reorganized in 1997 as Wilbert Funeral Services Inc.). The thinking behind the use of a vault was to separate the dead and contain fluids that could contaminate the groundwater and surrounding soil. Burial vaults are lined and sealed outer housings for a coffin, protecting the casket from the weight of the earth over it and preventing sinking graves. Haase developed the modern water-proof vault after a visit to the tomb of Tutankhamun in 1930 with the invention of an asphalt-lined concrete vault (Wilbert, 2020a). Funeral vaults are common, sometimes mandated by state law for certain cemeteries. Vaults are constructed of 5000 psi concrete with various lining materials including copper, bronze, stainless steel, and high impact plastics for "strength, water resistance and elegance"(Wilbert, 2020b). The use of the word 'elegance' here is interesting, given that the lining of a vault is designed never to be seen. By the time of a cemetery service, the vault is sealed, so this portion of the mortuary program is known only to those who contracted the service and those who provide the service. The Wilbert website lists notables buried in their vaults, including Jimmy Dorsey, Dwight D. Eisenhower, Lyndon Johnson, John F. Kennedy, Jaqueline Kennedy Onassis, Elvis Presley, Gilda Radner, and John Wayne (Wilbert, 2020a).

6.4 Twentieth Century Death and Mourning in the US: the Birth of a 'Traditional' Funeral

By the beginning of the twentieth century, changes in mortality trends brought about a new relationship with death. Walter (1993) identifies a few general trends: medicalization, secularization, privatization, individualism, and expressivism. For Ariès (1974), these can be summed up as a denial of death and a denial of mourning. The rise of hospitals in the late 19th and early twentieth century removed death from the home and into a medicalized setting. This medicalization had the added outcome of redefining death as a medical as opposed to a social process. Death in hospitals was hidden from the family and gradually, family and friends had less and less exposure to death itself. If death is no longer on display at home and hidden in hospitals, mourning must also become a hidden process, something done in private and lacking formalized rituals as had been common in earlier periods. As Walter (1993, p. 128) notes, "the lack of a social code of mourning means that many people keep out of the way of the grieving for fear of saying or doing the wrong thing." Mourning is to be done in private and to be experienced by the individual instead of the community (Walter, 1993). The hiding of death from the public sphere are indicative of increasing secularization, something we see reflected in the increasing role of the funeral home as a neutral location, open to people of all faiths or no faith at all.

The rise of the mortuary industry further removed the family from caring for the dead, making this a realm of specialists. By the end of the nineteenth century, licensing requirements were present in most states for funeral directors, and embalming was a routine element of mortuary preparation. Special training was offered by from mortuary training institutions. Funeral homes emerged in the early twentieth century, becoming the "epicenter of the business of death" (Beard & Burger, 2017, p. 51). A good number of these new activities took place behind the scenes, and new traditions for viewing developed. The funeral home was effectively a one-stop-shop for the bereaved. At a funeral home, the bereaved would meet with the funeral director who would arrange for the body to be picked up from the place of death, arrange for embalming. The funeral home became a neutral location for ceremony in the form of a funeral service: it is not a hospital, not home, not overtly religiously aligned with any one religion. New rituals included viewing the body in a coffin with a hinged lid with the lower body covered. Funerals can be officiated by clergy from the deceased or bereaved's religion, these may take place within a church or within the funeral home itself. The ritual is split, with a funeral (that may or may not involve a eulogy, depending on the wishes of the deceased and/or bereaved and religious tradition) and a burial service which takes place at the site of burial or dispersal (in the case of cremation).

This commodification of the death industry was criticized in the twentieth century, particularly by Jessica Mitford (Mitford, 2000). Throughout various editions of her book *The American Way of Death* (first published in 1963), Mitford questions the myth that twentieth century funeral traditions were indeed tradition. Mitford highlights the performativity of the funeral, stating "If the undertaker is the stage manager of the fabulous production that is the modern American funeral, the stellar role is reserved for the occupant of the open casket. The décor, the stagehands, the supporting cast are all arranged for the advantageous display of the deceased, without which the rest of the paraphernalia would lose its point" (Mitford, 2000, p. 50). She highlighted the role of profit motive on the part of funeral directors, and questioned the ethics of such an approach, noting that funeral services are rarely something that people comparison shop for, saying "choice doesn't enter the picture for average individuals faced, generally for the first time, with the necessity of buying a product of which they are totally ignorant, at a moment when they are least in a position to quibble"(p. 49). This highlights the fact that the average bereaved individual had less and less contact with death and mortuary processes and relied on specialists to inform and help them make decisions.

The American funeral industry held significant power, and backlash to Mitford's book was loud. West and McKerns (2009) detail the efforts of the funeral industry to paint Mitford as a communist, published pamphlets about her decorated with pink skulls, and so on. Mitford was even questioned by the House Unamerican Activities Committee, where she asserted her fifth amendment rights against self-incrimination when asked if she was a communist, showing that her critics had high ranging powers to throw shade on her motivations for writing. Partly in response to Mitford's critique of the American funeral industry, the FTC began to investigate the industry in 1973, with new regulations proposed in 1975. These new regulations prohibited a variety of practices including unnecessary embalming and bait-and-switch sales techniques. In 1982, the Funeral Rule was initiated by the FTC (2012). The Funeral Rule called for the full disclosure of funeral prices, legal requirements and incidental charges. It also eliminated the bundling of services and tie-in pricing, instead mandating an àla carte approach where the consumer picks individual services. According to the National Funeral Directors association (NFDA, nfda.org), the average funeral cost in 1960 was $708 (roughly $7600 in 2024 buying power[2]). By 1980, the average cost was $1809 (roughly $7319 in 2024 buying power). By 2000, the average cost was $5180 (roughly $9660 in 2024 buying power). The average cost of a funeral in 2024 was down a bit to $7848 according to the NFDA (coinciding with a rise in the interest in more cost-effective green funerals and cremation, topics discussed more in Chap. 11). In general, the trend for funeral cost has been relatively consistent with inflation, and represents a major expense for most families.

One byproduct of the Funeral Rule was an increase in pre-planning, discussed in greater detail in Chap. 11 and the rise in burial insurance to cover the cost. By the end of the twentieth century, a wide variety of options were open to individuals in terms of preplanning, body disposition, and memorialization. Many of these would fall under the heading of "new rituals". Roberts (2011, p. 17) describes the creation of new mortuary rituals as positive, noting that "those who had created their own rituals were proud that they had honored their loved ones in unique ways; each detail… indicated that their tribute, like their loved one was special." In general,

[2] Using the Bureau of Labor Statistics inflation Calculator, https://data.bls.gov/cgi-bin/cpicalc.pl

funerary performances become less uniform towards the end of the twentieth century, more reflective of individual identity than collective identity (explored more in Chap. 11). It may just be that the focus on individual identity, though, *is* reflective of a society that is focused on individual identity and accomplishment over that of the collective.

6.5 Making Mummies

Mummification can be the result of both natural and intentional processes. Despite the method by which this initially occurs, mummies must be viewed and interacted with in order to maintain social importance. Mummies have occurred in a variety of times and places, and so the approach here is to examine a few case studies rather than to completely investigate the process of mummification as a whole.

How do we identify mummified individuals? In some cases, it is obvious, the body has been recovered in a mummified state. But not all individuals who were mummified will be found in such a state of preservation. Booth et al. (2015) present one method that may be useful in interpreting the process of mummification in dry bone material. They examined bacterial intrusions into Haversian systems as indicative of putrefaction-related bioerosion. Intentional retarding of soft-tissue decay through mummification may be reflected in a lack of this bioerosion.

6.5.1 Ancestor Veneration to Dedication? The Changing Role of the Body in Bronze Age Britain

Parker Pearson and colleagues (2005, 2007) describe mummification in Bronze Age Britain (c. 2200–700 B.C.), an area and time frame not well known for mummification of the dead. At the site of Cladh Hallan on the island of South Uist on the Outer Hebrides excavations uncovered a well-preserved group of Late Bronze Age to Iron Age Roundhouses within which multiple burials were found. One round house contained the burial of two adults and a child, the central house contained the burial of a child and two dogs, and the southern house had the burial of one child. The burials of the two adults and the children in the central house gave evidence for mummification and curation of their bodies. Parker Pearson and colleagues (2005) argue that these sets of remains were placed at the time of house construction. The north house contained remains of two adults 'individuals', burials 2613 and 2638.

Burial 2613 was initially thought to be that of a single adult female (Parker Pearson et al., 2005). The skeleton's lateral maxillary incisors had been removed after death and paced in the hands before burial, the left tooth in the left hand and right tooth in right hand. Further ancient DNA (aDNA) analysis of this 'individual' revealed that is a composite consisting of elements from a minimum of 6

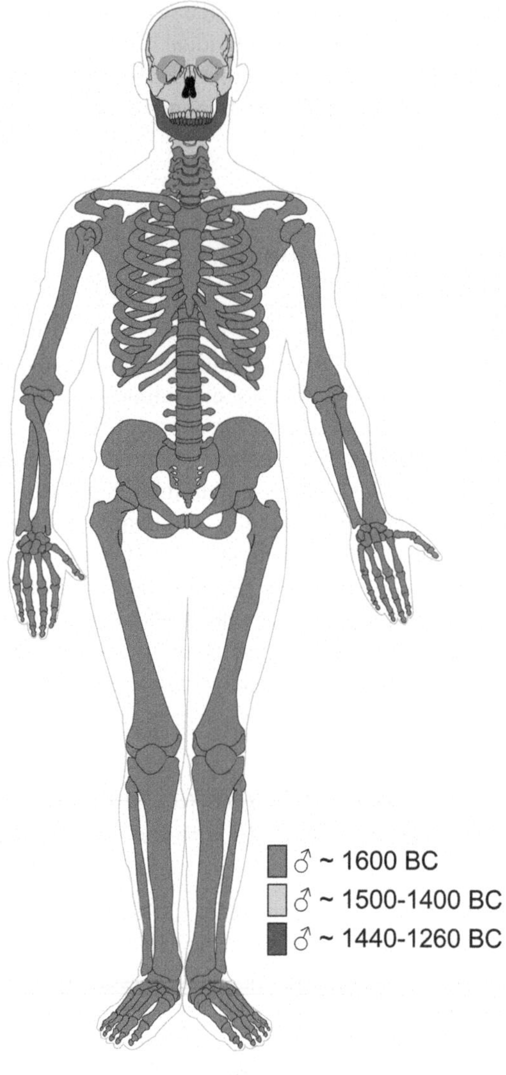

Fig. 6.3 Composition of the bones of an adult male mummy from Cladh Hallan. Originally published in *Journal of Archaeological Sciences,* 2012. Image attribution: Bullenwächter, CC BY-SA 3.0 <https://creativecommons.org/licenses/by-sa/3.0>, via Wikimedia Commons. https://upload.wikimedia.org/wikipedia/commons/c/cf/Cladh-Hallan_Man.svg

individuals, with mandible, femur, and humerus contributed by different individuals than that which provided the vertebrae, pelvis, and cranium. Burial 2638 contained the remains of at least three different individuals, compiled together to form one body shape (Fig. 6.3). The postcranial skeleton belonged to one male, the head and cervical vertebrae to another male, and the mandible from a third male individual. Based on the secure context of the burial, Parker Pearson and coworkers identify this as a composited individual whose body was intentionally constructed out of remains from three males.

One child (Burial 2792) was entirely disarticulated except for the pelvis and vertebrae. To understand whether the bodies had been altered, three approaches were taken: 1) establishing the date of death as significantly before date of deposition; 2) determine from the degree of microbial impact on the bones whether soft tissue decay had been arrested after death; and 3) to ascertain any trace of pre-depositional modification of the bones indicating soft tissue preservation.

In the case of the adult male (Burial 2638), the radiocarbon dates derived from the burial are earlier than those from the earliest levels of roundhouse construction. The radiocarbon dates from the different parts of the 'individual' also varied, with the skull dating to between 1500 and 1260 cal BC, 1500–1210 cal BC for the mandible, and 1620–1410 cal BC for the tibia. The femur from Burial 2613 dates to between 1370 and 1050 cal BC. The disarticulated child skeleton dates to between 1440 and 1130 cal BC. Roundhouse construction dates to between 1330 and 1100 cal BC. The image created by the different radiocarbon dates in conjunction with the composite nature of the adult remains suggests that the mummies were repaired over a long period of use by adding elements from different bodies. They were continually modified and adapted for continued use prior to their ultimate deposition.

The articulated nature of the remains in conjunction with the dating of the individuals possibly over a century before deposition implies that soft tissue preservation must have been performed. Parker Pearson and colleagues argue that "The soft tissues—at least the ligaments and perhaps the skin—could have been reserved in a number of ways… the body must be eviscerated; this entails removal of the internal organs which contain most of the bacteria that initiate decay" (Parker Pearson et al., 2005, p. 541). Microbial degradation is caused by collagenolytic gut bacteria that enters the bone after death via the blood supply. For Burial 2638, an unusual pattern ws detected using light microscopy, with dense microbial attack at the junction between lamellar and Haversian bone on the periosteal surface and more diffuse microbial infiltration to the endosteal surface. "The pattern is both intense and restricted, indicating that there was some initial decay which was then interrupted." The method of preservation was inferred using infrared spectrometry. Changes to the bone caused demineralization of some of the surface layers of the bone as well as the recrystallization of some parts of the bone. This pattern of recrystallization within the postcranial skeleton suggest that the body was exposed to an acidic environment for a short period of time. Parker Pearson et al. (2005) attribute this to burial within an acidic peat bog.

Parker Pearson and colleagues bring together a suite of analytical methods that present an intriguing picture. Between 1600 and 1000 BC, Britain's landscape transitioned from one dominated by burial locations to one filled with settlements and field systems. This change "is particularly evident at Cladh Hallan around 1100 BC when these mummified, ancestral dead were deliberately buried within the solid and imposing roundhouses which marked a significant change from the small and ephemeral houses of the Earlier Bronze Age." Changes to the social structure were reflected in changes to the mortuary program. "The care with which the two adult skeletons were interred under the north house is suggestive of a formal and respectful removal from the world of the living" (Parker Pearson et al., 2005, p. 543). Parker Parker Pearson and colleagues attribute this to a shift from ancestor veneration, and so the burial of the individuals may mark the end of the need to maintain these mummies. "[T]his moment of burial of preserved bodies as foundation deposits beneath at least one of the houses may represent a fundamental religious transformation in island life when the old beliefs gave way to the new, not through their total rejection but their incorporation into the very foundations of the new" (Parker Pearson et al., 2005, pp. 543–544). In effect, the preserved and reconstituted bodies of the ancestors were still important and still carried power into both the new structure and new social system.

6.5.2 Buddhist Mummification and Veneration

Buddhist practices of mummification are seen by Sharf (1999) as akin to creating a relic (relics are discussed in depth in Chap. 16). In Buddhist mummification, the body of an eminent master is desiccated via the use of charcoal, lime, or salt in low-air environments (Gildow, 2005), wrapped in layers of lacquer-impregnated cloth, dressed and adorned, and installed in on an alter. They can also be covered with a protective layer of lacquer. Due to the process used to create the mummy, the result can be difficult to distinguish from an icon. "In the case of a Buddhist mummy the identity of relic and image—substance and form—has been fully realized" (Sharf, 1999, p. 83). These mummies are seen as granting requests and sending dreams (Gildow, 2005). Methods used in the mummification were typically seen to merely assist in the process of preservation initiated by the virtue of the deceased (Gildow, 2005). In northeastern Japan, mummies are seen as 'living buddhas', and are capable of transmitting telepathic messages, curing disease, and saving lives during natural disasters (Dahl, 2020). The whole-body relic created through the process of mummification transforms the deceased into a posthumous Buddha, or more rarely, to gain more power within the preserved body.

The concept of the *sokushinbutsu* should be examined in more depth. These are the "self-mummified" monks enshrined and venerated in temples around Mount Yudono (Dahl, 2020, p. 293). These mummies that serve as full-body relics attract pilgrims, ascetics, and tourists. This process of self-mummification was practiced from the eighth century AD up into the twentieth century. Monks who wished to undergo this process gradually reduced their food intake over a long period of time as well as vomiting to expel bodily fluids and repel parasites. The final years of life, the individual became a walking skeleton and was then buried alive. He would die due to dehydration and starvation. This was seen as a kind of sacrifice to alleviate the suffering of people who would ultimately visit the shrine. After 3 years, the body would be exhumed from the underground stone chamber to see if the body had mummified. If the body was found to have become a natural mummy, the deceased had achieved enlightenment. If the body was not mummified, it would be immediately reburied (Rosso, 2014), sometimes after some alterations such as removal of organs and packing with material (Lobetti, 2011).

These mummies were seen as "resting in a perfect state between life and death, untouched by impurity and producing power and benefit from their perfect stillness (Lobetti, 2011, p. 241). The preservation of the body also involved the preservation of the identity of the deceased. Lobetti (2011, p. 242) describes one such body, that of Honmyōkai, whose mummy dates to 1683. Honmyōkai was a retainer of a feudal lord of Tsuruoka who left his family to pursue asceticism and a life at the Chūren-ji temple to pray for the recovery of his feudal lord from serious illness. He abstained from consumption of cereal grains and ate a diet consisting of pine needles, tree bark, pinecones, chestnuts and occasionally stones and crystals. This diet may have been designed for those seeking immortality (Lobetti, 2011). After 10 years of this diet, Honmyōkai entered a small stone chamber in the ground. His body was soon exhumed, dried by incense and charcoal fumes and then reinterred. After 3 years, the body was completely mummified and placed on an alter at the temple he founded as a living relic. In some ways, the stories of the lives of the *sokushinbutsu*, whether they be based on fact or myth, serve as lessons for those who visit their venerated bodies. Other *sokushinbutsu* include individuals who were atoning for acts such as the killing of a samurai or neglecting their duties to safeguard the community. Their acts of contrition, including years of fasting and slow suicide, acted not only as a sacrifice for the community but also as an example that duty to the community and social standards should come before personal comfort or safety.

6.6 Summary

Though different approaches, both the history of mortuary processes in the US and the process of mummification both show the importance of preservation for mortuary performance. Once preservation was an option in the US, the focus of mortuary performance tended to revolve around the preserved body and our relationship with death changed, with new social structures developing to separate the living from the dead. The process of creating mummies, embalming bodies, and further modifying them through continued care demonstrates the power of the body for communicating social messages about group identity. The concept of the 'good death' was indicative of the view of life and death in colonial and antebellum America. The modification of bodies in the Orkneys functioned for hundreds of years, across many generations, tying communities together through their continual adding to and modification of the mummy. When they no longer served that social function, the bodies were again utilized as social messaging that denoted the change in society. Buddhist mummies were and are touchstones between the living and the dead, and their continual care and monitoring is an important element of daily life. The preservation of the body, then, can be seen as everyday Poetics ritual that strengthens the ties between the living and the dead.

References

Ariès, P. (1974). The reversal of death: Changes in attitudes toward death in western societies. In D. E. Stannard (Ed.), *Death in America* (pp. 134–158).

Ariès, P. (2004). The hour of our death. In A. C. G. M. Robben (Ed.), *Death, mourning, and burial: A cross-cultural reader* (pp. 40–48). Blackwell Publishing.

Barceló, J., Jensen, J. L., Peisakhin, L., & Zhai, H. (2024). New estimates of US Civil War mortality from full-census records.*Proceedings of the National Academy of Sciences, 121*(48), e2414919121.

Beard, V. R., & Burger, W. C. (2017). Change and innovation in the funeral industry: A typology of motivations. *Omega, 75*(1), 47–68.

Bentley, R., & Chasteen, T. G. (2002). Arsenic curiosa and humanity. *Chemical Educator, 7*, 51–60.

Booth, T. J., Chamberlain, A. T., & Parker Pearson, M. (2015). Mummification in Bronze Age Britain. *Antiquity, 89*(347), 1155–1173. https://doi.org/10.15184/aqy.2015.111

Borgo, M., Licata, M., & Iorio, S. (2016). Post-mortem photography. The edge where life meets death? *Human and Social Studies, 5*, 103–115.

Brenner, E. (2014). Human body preservation – Old and new techniques. *Journal of Anatomy, 224*(3), 316–344. https://doi.org/10.1111/joa.12160

Burg, S. B. (2008). "From troubled ground to common ground": The Locust Grove African-American Cemetery Restoration Project: A case study of service-learning and community history. *The Public Historian, 30*(2), 51–82.

Cann, C. K. (2020). Black deaths matter earning the right to live: Death and the African-American funeral home. *Religions, 11*, 390.

Colman, P. (1997). *Corpses, coffins, and crypts: A history of burial*. Henry Holt and Company.

Dahl, S. A. P. (2020). Buddhist mummy or 'living buddha'? The politics of immortality in Japanese Buddhism. *Anthropological Forum, 30*(3), 292–312. https://doi.org/10.1080/00664677.2020.1786804

Deetz, J. (1996). *In small things forgotten: An archaeology of Early American Life* (Rev Exp Su ed.). Anchor.

Deetz, J., & Dethlefsen, E. N. (1971). Some social aspects of New England Colonia Mortuary Art. *Memoirs of the Society for American Archaeology, 1971*(25), 30–38.

Dethlefsen, E. N., & Deetz, J. (1966). Death's heads, cherubs, and willow trees: Experimental archaeology in Colonial Cemeteries. *American Antiquity, 31*(4), 502–510.

Faust, D. G. (2008). *This republic of suffering: Death and the American Civil War* (1st ed.). Alfred A. Knopf.

Federal Trade Commission. (2012). The FTC Funeral Rule. Retrieved from https://consumer.ftc.gov/articles/ftc-funeral-rule.

Federal Trade Commission. (2024). FTC sends warning letters to funeral homes after first undercover phone sweep. Staff found that some funeral homes provided inconsistent pricing information, while others failed to provide pricing entirely. https://www.ftc.gov/news-events/news/press-releases/2024/01/ftc-sends-warning-letters-funeral-homes-after-first-undercover-phone-sweep#:~:text=The%20letters%20reiterate%20that%20the,up%20to%20$51%2C744%20per%20violation

Fletcher, K. (2021). African American funeral directors as culture keepers: An interview with Karla Holloway. *Mortality, 27*, 1–7. https://doi.org/10.1080/13576275.2021.2010346

Forcén, F. E., & Forcén, C. E. (2016). Ars moriendi: Coping with death in the late middle ages. *Palliative & Supportive Care, 14*(5), 553–560.

Foster, G. S., & Eckert, C. M. (2003). Up from the grave: A sociohistorical reconstruction of an African American community from cemetery data in the rural midwest. *Journal of Black Studies, 33*(4), 468–489. Retrieved from http://www.jstor.org/stable/3180875

Foucault, M. (1984 [1967]). Of other spaces: Utopias and heterotopias. *Architecture /Mouvement/ Continuité*, 1–9.

Foucault, M. (2003 [1963]). The birth of the clinnic: An archaeology of medical perception (A. M. Sheridan, Trans.). Taylor & Francis.

Frohne, A. E. (2002). *The African Burial Ground in New York City: Manifesting and representing spirituality of space*. (Ph.D.). State University of New York at Binghamton, Ann Arbor. Retrieved from https://login.proxy.library.msstate.edu/login?url=https://www.proquest.com/dissertations-theses/african-burial-ground-new-york-city-manifesting/docview/305471634/se-2?accountid=34815. ProQuest Dissertations & Theses Global database. (3051877)

Gildow, D. M. (2005). Flesh bodies, stiff corpses, and gathered gold: Mummy worship, corpse processing, and mortuary ritual in Contemporary Taiwan. *Journal of Chinese Religions, 33*, 1–37.

Hacker, J. D. (2011). A census-based count of the civil war dead. *Civil War History, 57*, 307–348.

Johnson, C. L. (2004). *Undertakings: The politics of African-American Funeral Directing*. (PhD). University of South Carolina. (UMI Number 3130454).

Jones, D. (2011). The city of the dead: The place of cultural identity and environmental sustainability in the African-American Cemetery. *Landscape Journal, 30*, 2–11.

King, C. (2010). Separated by death and color: The African American Cemetery of New Philadelphia, Illinois. *Historical Archaeology, 44*(1), 125–137. Retrieved from http://www.jstor.org/stable/27820824

Laderman, G. (1995). Locating the dead: A cultural history of death in the Antebellum, Anglo-Protestant Communities of the Northeast. *Journal of the American Academy of Religion, 63*(1), 27–52.

LeeDecker, C. H. (2001). The coffin maker's craft: Treatment of the dead in rural eighteenth-century delaware. *Journal of Middle Atlantic Archaeology, 17*, 1–14.

LeeDecker, C. H. (2009). Preparing for an afterlife on earth: The transformation of mortuary behavior in nineteenth-century North America. In T. Majewski & D. R. M. Gaimster (Eds.), *International handbook of historical archaeology* (pp. 141–157).

Lobetti, T. F. (2011). *Faith in the flesh: Body and ascetic practices in a Contemporary Japanese Religious Context*. (PhD Dissertation). University of London.

Maples, W. R., & Browning, M. (1994). *Dead men do tell tales: The strange and fascinating cases of a forensic anthropologist*. Doubleday.

Merrifield, K. (n.d.). From necessity to honor: The evolution of National Cemeteries in the United States. Retrieved from https://www.nps.gov/nr/travel/national_cemeteries/Development.html

Mitford, J. (2000). *The American way of death revisited*. Vintage.

Murphy, R., & Stephens, T. (2020). *Section 27 and Freedman's village in Arlington National Cemetery: The African American History of america's Most Hallowed Ground*. McFarland and Company Inc.

Parenti, M. (1996). The strange death of President Zachary Taylor: A case study in the manufacturing of mainstream history. *Nature, Society, and Thought, 9*, 5–29.

Parker Pearson, M., Chamberlain, A., Collins, M. J., Cox, C., Craig, G., Craig, O. E., et al. (2007). *Further evidence for mummification in Bronze Age Britain*. Antiquity.

Parker Pearson, M., Chamberlain, A., Craig, O. E., Marshall, P., Mulville, J., Smith, H., et al. (2005). Evidence for mummification in Bronze Age Britain. *Antiquity, 79*, 529–546.

Roberts, P. (2011). What now? Cremation without tradition. *OMEGA-Journal of Death and Dying, 62*(1), 1–30.

Rosso, A. M. (2014). Mummification in the ancient and new world. *Acta Med-Hist Adriat, 12*, 329–370.

Sachs, A. (2010). American arcadia: Mount Auburn Cemetery and the nineteenth-century landscape tradition. *Environmental History, 15*(2), 206–235.

Scheper-Hughes, N., & Lock, M. M. (1987). The Mindful Body: A prolegomenon to future work in medical anthropology. *Medical Anthropology Quarterly, 1*(1), 6–41. Retrieved from http://www.jstor.org/stable/648769

Sharf, R. H. (1999). On the allure of Buddhist relics. *Representations, 66*, 75–99.

Stearns, P. N. (2021). Death in Modern North American History. In P. N. Stearns (Ed.), *The Routledge history of death since 1800* (pp. 213–234). Routledge.

Valentine, D. (2021). *Ars Moriendi as social reality: Articulation of the good death in the United States*. (PhD). University of Missouri-Columbia.

Walter, T. (1993). Death in the New Age. *Religion, 23*(2), 127–145.

Wescott, D. J., Brinsko, K., Feaerman, M., Golda, S. D., Nichols, J., Spigelman, M., et al. (2010). A fisk patent metallic burial case from Western Missouri: An interdisciplinary and comprehensive effort to reconstruct the history of an early settler of Lexington, Missouri. *Archaeological and Anthropological Sciences, 2*, 283–305.

West, S. C., & McKerns, J. P. (2009). Death and communists: The funeral industry's attack on Jessica Mitford's The American Way of Death. *American Journalism, 26*(1), 31–53. Retrieved from https://search.ebscohost.com/login.aspx?direct=true&AuthType=ip,shib&db=lfh&AN=40208773&site=ehost-live&custid=magn1307

Wilbert. (2020a). History. https://www.wilbert.com/about/history/. Accessed December 2, 2020.

Wilbert. (2020b). What is a Burial Vault? https://www.wilbert.com/burial/. Accessed December 2, 2020.

Williams, C. T. (2008). *Paradigm shift in African American funeral customs: Looking through the lens of oral history and consumer culture*. (Doctor of Education). Georgia Southern University.

Wyoming Funeral Directors Association. (2013). Understanding the General Price List. http://www.wyomingfda.org/Understanding-the-General-Price-List.html. Accessed December 2, 2020.

Chapter 7
Inhumation and Cremation

Making Places Matter with Bodies

Abstract This chapter explores the evolving significance of burial and cremation practices in shaping social identity, cultural norms, and historical transformations. It examines inhumation practices, focusing on burial contexts such as churchyards, military cemeteries, and marginalized burial grounds like Potter's Fields. These sites are analyzed as material and symbolic expressions of community values, reflecting social status, religious affiliations, and collective memory. The chapter highlights how burial spaces are designed to demarcate social boundaries, such as the hierarchical arrangement of graves in churchyards and the exclusion of marginalized communities in segregated burial grounds. Furthermore, the chapter explores the historical development of cremation as a key practice in Eurasian societies, beginning in the Neolithic period. It distinguishes between primary cremation (involving fresh bodies) and secondary cremation (the processing of bones), considering how these rituals evolved in response to major societal changes such as the transition from foraging to agro-pastoralism, population growth, and shifts in kinship and social hierarchies. Case studies from regions like Britain, Ireland, the Aegean, and Germany illustrate how cremation practices were used to express variations in social status, gender roles, and community bonds. By examining both burial and cremation practices, this chapter highlights the dynamic role of death rituals in shaping social memory, identity, and belonging across time, reflecting both historical and contemporary societal shifts.

This chapter will center on two forms of body processing: inhumation and cremation. Inhumation, the burial of the body in a place set aside for that explicit purpose. The location of a Burial Ground (whether it is a pauper's field, churchyard, cemetery, church interior, or modern-day green cemetery, discussed in Chap. 11) with relation to a community has changed through time but always has meaning and gives a clue into the social identity of the deceased individual as well as the relationship between the living and the dead. Burials of individuals within church structures

A. Osterholtz, *The Bodies Poetic*, Bioarchaeology and Social Theory,
https://doi.org/10.1007/978-3-031-85998-4_7

suggest high status relative to the rest of the community and indicate affiliation with that religious worldview. Burial within a mass grave associated with epidemic disease suggests that death occurred during a time where the normal 'rules' for burial were suspended due to massive social upheaval. Military cemeteries (such as Arlington National Cemetery covered in Chap. 8 in greater detail) amplify one aspect of identity (soldier) while crosscutting class, race, and gender boundaries.

Cremation is examined historically, with a focus on the power of cremation to shape social and cultural identities from the Neolithic period onward. Beginning in the Neolithic, cremation can be seen as a transformation of the body that reflects a shift in social structure and a shift from agro-pastoralism to agriculture. Also discussed is the role of cremation within the Roman world, examining how shifts between cremation and inhumation were driven by changing religious beliefs, social dynamics and the importance of status display.

Bioarchaeology is focused on understanding the body as a dataset to interpret past lived experience. Part of that within the biocultural model is understanding the context in which the body is found. This includes the material remains found with the body, interpreted as the remnants of the activities surrounding burial (in the case of interments) or disposal. This approach necessitates the examination of different life-histories of objects found in mortuary contexts as indicative of mortuary ritual. This mortuary ritual, in turn, is indicative of the relationship between the living and the dead. These mortuary rituals have different foci, and are conducted for different purposes. Some, such as covering pictures and mirrors, stopping clocks to avoid bad luck, and careful preparation of the body, objects left in or on the coffin, and sometimes altering the body itself were designed to control the dead and limit their influence over the lives of the living (Blakely & Harrington, 1997).

Methods and modes of body disposal can help us to understand the nature of how death is viewed within society, and the relationship between the living and the dead and the living to each other. In highly individualistic societies, such as the United States and western Europe, death rituals and methods of disposal have become incredibly diverse, ranging from a 'traditional' burial in a community or faith-based cemetery to shooting remains into space to burial in mushroom suits to hasten decomposition in a minimally environmentally destructive manner.

To understand the life history of the body as an artifact, we must first identify what typical processing should look like. Understanding the typical mortuary program allows us to identify how specific individuals may diverge from that pattern. Differences in mortuary programs may indicate group affiliation, gender, age at death-based identities, locations or modes of death, or other subaltern identities. Mortuary programs tend to be very conservative and so changes to them are usually seen as indicative of large-scale cultural changes.

Defining what is atypical is contingent on an understanding of what typical means, and as with much of the topics covered in this volume, will vary based on space and time. Numerous classification systems for the disposal of human remains exist based on a variety of elements. For inhumations, one distinction made is between primary and secondary burials. Primary burials are those that refer to the interment of an individual that is not later recovered and processed (Sprague, 2006). The designation as a primary burial does not mean that the body was not modified prior to burial, simply that it was not altered or disturbed after the initial deposition. Martin et al. also note that primary burials tend to be mostly articulated. Secondary burial denotes the final deposition of remains after (potentially significant) processing and modification of the remains from their primary context. Common examples of secondary burial can be the gathering of remains after cremation for disposal, gathering of remains for deposition in bundle burials, scaffold burials, and tree burials. Processing of the body is implicit in the identification of a burial as secondary. Both primary and secondary burials may consist of single individuals or multiple burial.

An additional element to consider is the position of the body, whether extended or flexed, and the position of the arms. There will always be some flexibility and variability within what can be identified as the typical mortuary program, but some changes should be viewed as reflective of an atypical burial. One example of normal variation is the position of the arms in medieval cemeteries. At the cemetery at Đurđevac-Sošice in eastern Croatia (where I excavate), individuals buried within the church had hand and arm positions that varied from extended along the sides of the body, crossed over the chest, and crossed over the pelvis, with the crossed over the pelvis being the most common. All individuals, however, are buried in an extended, supine manner.

7.1 Identifying Cemetery Spaces, Churchyards, Burial Grounds and the Differences Between Them

Inhumation, or burial, of individuals within designated spaces are used to perform different elements of identity, whether that is socioeconomic status, illness and public fear of contagion, religious affiliation, or environmental concerns. All of these activities take place within spaces designated for burials. Chapter 8 will focus on some specific cemeteries as case studies for how these spaces can be reflective of social processes and identity formation; this chapter will provide some context for processes themselves. Before examining specific types of cemetery spaces, we must first understand exactly what a cemetery is. Many writers discussed here focus on elements that make up the specific space of a cemetery: (1) a space specifically designed for the burial of the dead; (2) a space that is bounded and sacred, a place for ritual; and (3) a space that is designed for a community, not a specific religious or cultural community. Härke (2001, p. 18) argues that cemeteries are unique liminal spaces, "islands of subjective (as against scientific, industrial) time—islands where monuments evoke the past, and where rituals link past, present and future." This may include spaces under occupation room floors, caves, or within communal spaces; what is important is the socially-accepted nature of the location for burial. The location of the burial space conveys meaning and can be used to understand the relationship between the living and the dead.

In a Poetics sense, the cemetery presents a stage upon which mortuary and memorial performances can be conducted and so in developing the definition of such a space, it is a functional definition that has the most utility for us. In writing an entry for the Oxford Dictionary of Architecture, Curl (2021, n.p.) identifies a cemetery as a space intentionally created for the burial of the dead "not being a churchyard attached to a place of worship." Kolbuszewski (1995) builds on the intentionality of a cemetery space by layering on an understanding of ritualization with respect to the disposition of individuals within the prepared space. For Kolbuszewski, this refinement was necessary to distinguish cemetery spaces from mass graves resulting from ethnic cleansing. The ritualized nature is also recognized by Francis (2003, p. 223), who defines a cemetery as "the appropriate sacred space where the living and the dead are separated and symbolically joined as one people through the performance of transition and memorial rites." This ritualized nature is described as decisions for location based on "nonfunctional considerations (such as the desire to create a focus for ritual and remembrance…)" by Härke (2001, p. 11). Rugg (2000) argues that the internal features of a cemetery such as the layout and boundary of the physical site as well as the 'sacredness' (p. 261) of the site may also be important for identifying a space as a cemetery. These physical characteristics are important for Rugg, as "the boundary and the entrance gate define the cemetery as a separate place with a special purpose"(Rugg, 2000, p. 262). This sacred aspect of the space is related to the role of ritual, not a specific religious or spiritual outlook. It is instead, a celebration of community and a focus for remembrance and memory.

Cemeteries can be distinguished from Churchyards based on ownership and location. Churchyards are generally associated with places of worship, and house only the remains of those who were members of the church body. Cemeteries, on the other hand, tend towards the more secular, though they may retain faith-based sections. The prerequisite of religious affiliation, though, is not necessary for the identification of a cemetery (Rugg, 2022, p. 20), as Rugg states, "cemeteries can either accommodate or marginalise differences." Sections within cemeteries, discussed below and in Chap. 8, may be reflective of different elements of identity, such as ethnic identity, or may reflect military service and/or other status markers (such as cemetery sections or spaces set aside for individuals who died while incarcerated). Churchyards, on the other hand, tend to be smaller and governed by ecclesiastic regulations.

Burial Grounds are more difficult to define; Rugg (2000) uses the term 'Burial Ground' to refer to burial spaces centered on the identity of the bereaved, specifically "minority ethnic groups" (p. 266) denied access to cemetery or churchyard burial. "In many instances these minorities either chose or are compelled through exclusionary practices to lay out their own sites for interment. In many instances the purpose of the site extends beyond the need to accommodate the dead. The Burial Ground may be an essential component of that minority's expression of identity, and its difference from mainstream provision becomes imperative" (J. Rugg, 2000, p. 267). These communities can use these specific Burial Grounds to conduct distinctive funerary practices. In physical layout, Burial Grounds can resemble cemeteries, having boundaries, monumentation, and the like. The monuments may be more uniform in burial grounds than might be expected in a community cemetery, especially in multi-ethnic communities.

7.2 Making Places Matter with Bodies

The location of a Burial Ground can indicate aspects about identity, often amplifying one specific element of identity over others. The sections below will examine some specific types of burial locations and what they tell us about social structure and group identity.

7.2.1 Potter's Fields

Potter's Fields can be defined as indigent cemetery spaces, places where individuals whose relatives lack the resources (social or financial) for church burial can be buried at municipal expense. The term itself is biblical, referring to the field near Jerusalem bought with the 30 pieces of silver that Judas returned to the chief priests of the temple. The land was worthless for farming and would instead be used to bury strangers (Matthew 27:7). The term *strangers* is important here, since strangers are those lacking social networks within the community. A 1953 section of Ohio Code (§ 1713.38) uses the term "stranger" in their codes relating to the use of unclaimed bodies for medical education (see discussion in Chap. 15 for more about the use of unclaimed bodies), an allusion to the biblical importance of social ties for proper burial. These can be viewed as a buffering mechanism, with the State providing care when individuals cannot. It is also a reflection of the social power, with the State deciding where these cemetery spaces should be placed, deciding on the mortuary rituals performed, and performing those actions. According to Willis (2018, p. 539), "it is a location where the everyday techniques of governance, and their byproducts are concentrated and made material."

Where cemeteries and Burial Grounds were placed was often a reflection of society. Blakely and Harrington (1997) note that the creation of Potter's Fields separate from churchyard and municipal Burial Grounds was a reflection of the lower status of individuals within those Potter's Fields. African American Burial Grounds were commonly located near the Potter's Fields. The separation of people of color and the poor from the rest of the community "served to emphasize the powerlessness" of these groups (Blakely & Harrington, 1997, p. 168). Again, the imagery of land useless for farming and the origin of the term "Potter's Field" is important for thinking about landscape. Potter's Fields are located in places where the land is deemed to be worthless for any other purposes.

One such example of a Potter's Field is Hart Island in New York City. Individuals who die unclaimed or as a ward of the state are subject to burial on Hart Island as their bodies are technically then the property of the city. As property of the city, they were subject to being used in medical schools for dissection if needed (bringing to mind the issues of anatomization and medical education of indigent individuals discussed elsewhere in this volume) until 2016, when New York overturned a 162-year old law that allowed for the use of unclaimed bodies to be used in medical education after being unclaimed for 48 h after death (Bernstein, 2016a). On Hart Island, bodies are buried in mass graves dug by heavy machinery with burials conducted by prison inmates (Fig. 7.1) paid $.50/h (Bernstein, 2016b). Long trenches are dug with excavators, and burials are stacked in pine coffins within the trenches until full. In the over 150 year history of Hart Island, over one million individuals have been buried there. Twice a week, bodies are taken to the island via ferry for interment. Near the dock, a visitors' gazebo was installed in 2007. The City of New York maintains a website for Hart Island (https://www.nyc.gov/site/hartisland/index.page). On this website, they term Hart Island as the city's public cemetery and list links for finding loved ones, how to go about requesting disinterment, and guidelines for visiting the island. Individual burial data is only available for individuals buried since 1977; those data are available in the Hart Island database.

Visitation is limited to family members, close friends, and partners of the deceased and must be scheduled ahead of time. The website states that visits are available twice a month on weekends. The right to visit the graves on Hart Island was granted in 2015 after a class-action lawsuit filed by the American Civil Liberties Union (ACLU) of New York (NYCLU) resulted in a settlement. The settlement allowed for family members and close friends to visit the graves and leave

Fig. 7.1 Inmates placing coffins in the Potter's Field on Hart Island, NY (circa 1890). Image attribution: Jacob Riis, Public domain, via Wikimedia Commons. https://upload.wikimedia.org/wikipedia/commons/b/be/A_trench_at_the_potter%27s_field_on_Hart_Island%2C_circa_1890_by_Jacob_Riis.jpg

mementos such as flowers, stuffed animals, prayer cards, flags, and blankets at the graves (NYCLU, 2015). The island has historically been used for other purposes such as a jail, halfway house, military training camp, sanitarium, drug treatment center, and a missile base, all while still being used as a cemetery. The Department of Corrections, which currently owns the island, allowed visitation to the island prior to the settlement, but visitors were not allowed to visit individual graves, only to go to a visitor's gazebo that was constructed in 2007 (Bernstein, 2016b; Kilgannon, 2015). Here, a monument was placed that served as a proxy tombstone for all those buried on Hart Island (Fig. 7.2), similar to how the Tombs of the Unknown Soldier (see Chapter 13) acts as a proxy marker for any unidentified solder.

Buelow (2010, p. 3) describes the landscape of Hart Island as one of "negated identity" and notes that the physical relationship between New York City (visible in the distance) and the Island highlights the socioeconomic disparities visible within the city itself. During the nineteenth century, when Hart Island was initiated as a Potter's Field, cemetery development in the US tended toward monumentality and

Fig. 7.2 Stone monument at the Potter's Field on Hart Island, NY. Image attribution: RoySmith, CC BY-SA 4.0 <https://creativecommons.org/licenses/by-sa/4.0>, via Wikimedia Commons. https://upload.wikimedia.org/wikipedia/commons/e/ee/Hart_Island_Potter%27s_Field_Monument.jpg

ornate burials, displays of conspicuous wealth and community identity (e.g., Tarlow, 2000). So, the spartan nature of Hart Island, while being visible from the City of New York, presented a physical reminder of the cost of dying poor. Writing from an architectural point of view Buelow focuses on the role of landscape and the function of a cemetery, to bury the dead, but socially cemeteries fulfill the need to reconnect with the dead through the physicality of the cemetery landscape. Because Hart Island lacks grave markings, landscaping, and open visitation, it cannot provide the physical link between the living and the dead. As Buelow (2010, p. 38) states when "a landscape meant to serve the function of preserving memory is completely removed from public perception, identity is effectively eliminated." Rugg (2000, p. 269) argues that the use of Potter's Fields can be seen as a method that the state puts into place to deter the poor from seeking public assistance, noting that in the UK "during the nineteenth century, paupers dying 'on the parish' were often buried in mass graves in unconsecrated ground, in coffins so badly constructed that the body could be seen through the cracks."

7.2.2 *Plague pits and Bone Beds: Epidemic and Mass Burials*

The burial of individuals who have died from epidemic disease are typically treated differently from a typical mortuary program. Fear of contamination as well as social breakdown altered what was socially acceptable and propitious. The best-known episodes of Black Death occurred during the fourteenth century. In the fourteenth-century, the Black Death was responsible for 30–60% death rates in Europe, amounting to tens of millions of people (DeWitte & Kowaleski, 2017; Green, 2014). Recurrences of the Black Death occurred through the seventeenth century. Harding (1993) examines the burial of epidemic victims in London in the seventeenth century known as the Great Plague. Parishes were primarily in charge of burial duties. Local inhabitants were entitled to burial within parish cemeteries, and vestry and churchwardens collected burial fees, directed grave layouts and acquired new space for burial as needed. Parishes were the local units of government, but the government of the city of London was responsible for issues of public hygiene and issued plague orders (including some guidelines for burials). A sudden and dramatic increase in the number of individuals requiring burial placed strains on all aspects of infrastructure, both governmental and parish-based. Disease burdens were not spread equally throughout the city, with some areas (particularly those with higher population density) more heavily impacted. Households had responsibilities for organizing and paying for burials within the parishes, so the practice of shutting up an entire household after the infection of a single household member impacted the overall order of the Parish. Individual desires conflicted with plague orders and the orders of the Privy council who sought to restrict public assemblies and processions. Parishes, on the other hand, were the entities usually responsible for burial processions, and were faced with enormous financial burdens in the wake of large increases in the number of dead needing burial. Those responsible for gathering the dead or in the burial process were physically and socially separated from the rest of the community.

As the numbers of the dying began to rise in London in 1665, traditional mortuary processes began to break down (Fig. 7.3). The elderly and those of low socioeconomic status may have been more heavily impacted by the Plague (DeWitte & Kowaleski, 2017). The first common grave was dug in late August, and likely filled up over a space of days as another was opened at the beginning of September. By mid-September, over 30 people per day required burial, and individual rites were replaced at least partially by mass burial in a common grave. The first day with no recorded burials was the 11th of November and not until December was the monthly total of individuals needing burial back to pre-epidemic levels (Harding, 1993). This history reflects two major differences to the typical mortuary program, organized collection of corpses and the use of mass graves. Individuals found in these contexts can be understood as victims of the epidemic through the incorporation of textual sources. Their differential burial can be understood and explained as a divergence from the typical burial program due to social changes. It is a temporary change, but one that is reflected in different processing of the dead (coffin burials were less likely, the shift from single to multiple burial) and therefore different life history of the bodies as artifacts. These mass graves lack individual identity and may lack monumentation generally associated with Burial Grounds, Cemeteries, or Churchyards. Additionally, they may lack boundaries around the mass grave or the boundaries may be hastily erected (Rugg, 2000).

7.2.3 Mass Graves and Conflict

Mass Graves may also be associated with war or conflict, and in those cases, mass graves may be intentionally hidden. There is, at present, no universally accepted definition of what constitutes a mass grave associated with warfare atrocities or ethnic cleansing. For Skinner (1987), a mass grave must contain at least 6 individuals, but the United Nations proposed a definition of a mass grave based on the circumstances that created it. This definition argues that a grave of three or more victims of extra-judicial, summary or arbitrary executions who did not die in combat can be identified as a mass grave. The Bournemouth protocol (Klinker & Smith, 2020, p. 4) identify a mass grave as "a site or defined area containing a multitude (more than one) of buried, submerged or surface scattered human remains (including skeletonized, commingled and fragmented remains), where the circumstances surrounding the death and/or the body disposal method warrant an investigation as to their lawfulness." For Haglund et al. (2001), the number of individuals interred is only one element of identifying a mass grave. Another important element for them is the internal configuration, whether the bodies are adjacent or separate. This will have an impact on how the dead decompose, and these differences in taphonomy may relate to reconstructing the actions that created the mass grave (a single incident versus an open grave into which individuals are added over time).

Fig. 7.3 Illustration by George Cruikshank of Men burying the bodies of plague victims used in an 1876 edition of Daniel Defoe's "A Journal of the Plague Year" (drawing circa 1835), CC BY 4.0 <https://creativecommons.org/licenses/by/4.0>, via Wikimedia Commons. https://upload.wikimedia.org/wikipedia/commons/3/3f/Men_burying_the_bodies_of_plague_victims_in_a_pit._Engraving_Wellcome_V0010606ER.jpg

Mass graves associated with ethnic cleansing or war may be identified or excavated for numerous reasons, generally revolving around creating documentation for human rights work and forensic medico-legal investigation (Haglund et al., 2001). Klinker and Smith (2020, p. 6) argue for a 'do no harm' approach to determining whether to disturb an identified mass grave, noting that "Communities may be undergoing rapid social change which could include a degree of fluidity in social and power structures and the re-establishment of social norms. A 'do no harm' approach in these circumstances will actively seek to avoid undermining existing structures and relationships that are essential to the maintenance of peace and community coexistence. It should also endeavor to avoid creating inequalities, including in relation to gender. It will include a clear respect for and, where possible, adherence to cultural sensitivities and norms, and the known religious beliefs of victims and/or their families should be taken into consideration as far as possible and to the extent that they do not adversely affect the achievement of an effective investigation." This highlights the potential conflict that is evident in the excavation of mass graves. Whose voices should be listened to? The living or the dead?

Kovras (2008) discusses the complicated politics of looking for mass graves on Cyprus. Cyprus gained its independence from the United Kingdom in 1960; conflict soon erupted as to the relationship between Cyprus, Turkey, and Greece. At the time of independence, the population of Cyprus consisted of roughly 80% Greek Cypriots, 18% Turkish Cypriots, with the remaining 2% identifying as some other ethnic minority. Greek Cypriot and Turkish Cypriot groups regarded the Zurich agreements as unsatisfactory, leading to ethnically centered conflict and violence. Between 1963 sand 1964, there were disappearances of approximately 230 Turkish Cypriots and 42 Greek Cypriots (Mikellide, 2014). A larger second wave of violence erupted in 1974. Between July and August of 1974, 1464 Greek Cypriots and 270 Turkish Cypriots went missing (Mikellide, 2014). It was also in 1974 that the island became divided, with a UN-established green zone in the city of Nicosia and a guarded buffer zone along the width of the island. Contact between the Turkish Republic of Northern Cyprus and Cyprus was not permitted until 2003 when 9 checkpoints opened, allowing connection between the North and South for the first time since 1974 (United Nations Secruity Council, 2003). The lack of contact between North and South on the island allowed for the conflict to be remembered and memorialized very differently depending on which size of the buffer zone one's family currently (and previously) lived on. For Bryant (2012, p. 337), this partition itself became part of the issue with memory and closure, stating "the fissure in the Body Politic is mapped onto bodies, especially through analogizing partition to a wound."

In 1981, the Committee on Missing Persons in Cyprus (CMP, cmp-cyprus.org) was established by agreement with both Greek Cypriot and Turkish Cypriot communities. The CMP was founded to establish the fate of the missing persons from both the 1963–1964 and 1974 disappearances. The CMP began exhumations of identified graves in 2005. Mikellide (2014, Table 1) identifies different types of locations for these graves: burial in open fields (n = 64); superficial burial and surface depositions on mountains (n = 34); burial in wells (n = 24); burial in streambeds (n = 16); burial in symbolic locations (n = 16), and burials in preexisting features (n = 3). Mikellide (2014) makes a basic distinction between public burials, those burials that occur in relatively public places not far from the location of death and without any attempt to conceal the bodies and/or evidence (burials in open fields, burials in house yards, surface deposition and superficial burials on mountaintops, and burials in symbolic locations) and clandestine burials those burials defined by the desire of the perpetrators to conceal the bodies and/or evidence (burial in wells, burial in preexisting structures, burial in streambeds). Burials contained different numbers of individuals, ranging from individual burials to mass graves containing up to 39 individuals. The identification of such graves is based more on the functional definition than the strict numerical definition, given the relatively small numbers of individuals interred within some these graves. But when viewed as a whole, the pattern is consistent with mass graves resulting from war or conflict.

Burials in open fields tended to have more multiple burials, while burials in house yards tended to contain fewer individuals (at most 4). Conflicts in the 1960s tended to have lower visibility and a greater emphasis on concealment of the bodies, while those from 1974 tended to have greater visibility. The CMP has a mandate to investigate, identify, excavate (in association with the International Red Cross), and repatriate individuals to their communities. Of interest is that their mandate specifically prohibits collecting evidence as to the perpetrators of the mass deaths. As their mandate states, "The committee will not attempt to attribute responsibility for the deaths of any missing persons or make findings as to the cause of such deaths" (CMP, 2023).

This focus on identification and repatriation while specifically not seeking evidence of responsibility is unique, and Ktori and Baranhan (2018) cite this program as a model for a successful humanitarian programs in post-conflict environments. Part of its success has been the decision not to attribute blame or responsibility for the mass graves, allowing both Greek Cypriots and Turkish Cypriots to work towards a common goal of repatriation and reburial in community cemeteries (Fig. 7.4). The return of individuals who simply disappeared has an impact on what Hadjigeorgiou (2022) refers to as ambiguous loss. For individuals whose family members disappeared, the lack of proper mortuary ritual can lead to a lack of closure. The lack of mortuary ritual impacts the individual family members as well as the community. This unknowing can lead to an implied threat, where hostilities are only seen as paused. As Bryant (2012, p. 347) notes "institutions of memory are built upon particular temporalities understood through the figure of the wound—either one that will ultimately be healed through a return to wholeness, or a wound as rupture that represents sacrifice for a future in which there can be no return."

The case study of Cyprus is illustrative from a Poetics point of view for several reasons. First, it highlights the importance of mortuary ritual. The lack of proper mortuary ritual inhibits the culturally meaningful mourning process. The development of the CMP, while it has sustained significant criticism (see Hadjigeorgiou, 2022) relating to the length of time it has taken to receive resolution or identify relatives, was an effort by those living on Cyprus in conjunction with the international

Fig. 7.4 An OSCE PA delegation visits the CMP Anthropology lab where the work of identifying the missing was done (June 19, 2019). Image attribution: OSCE Parliamentary Assembly from Copenhagen, Denmark, CC BY-SA 2.0 <https://creativecommons.org/licenses/by-sa/2.0>, via Wikimedia Commons. https://upload.wikimedia.org/wikipedia/commons/3/32/OSCE_PA_Delegation_visits_the_Committee_on_Missing_Persons_in_Cyprus_%28CMP%29%2C_UN_Buffer_Zone%2C_19_June_2019_%2848101955016%29.jpg

community to provide a new form of mortuary ritual (i.e., the excavation, identification, repatriation). It is effectively the development of new ritual in a newly created third space. Second, the CMP approach highlights the importance of landscape with its distinction between public and clandestine burial locations and the motivations behind those. In the 1960s, it was socially important to keep those locations more hidden, while in 1974, the public nature of mass burials sent a strong cultural message. The landscape used was dictated by the nature of the burial activities, and the landscape is reciprocally changed by its usage. Third, the case study of Cyprus highlights the importance of memory as well as its historicity.

Bryant has highlighted the importance of memory formation and how Turkish Cypriots and Greek Cypriots interact differently with their past, while both groups have felt a disconnect with each other and with how to move forward. This is, at its heart, a disconnection about memory. As time passes, and the number of those who had first hand memories of the conflicts decreases, newer generations will continue to grapple with the conflicts, the disappeared, and the mass graves. The archive of first-hand accounts curated by the CMP are an effort to create a stable repository for that memory. But it is important to keep in mind that every generation must interact with history in their own way. For many, this comes down to reconciliation. For Hadjipavlou (2007, pp. 58–59), reconciliation can be seen as a "positive cyclical relationship… a cycle in which progress in one realm fosters progress in another… Reconciliation is more than co-existence of formerly hostile groups living near each other. It is also more than formerly hostile groups interacting and working together… I view reconciliation as the capacity to reach each other, feel empathy for their suffering and engage in shared social activities. This means mutual acceptance and trust-building preceded by the task of forgiveness whereby victims and perpetrators come to acknowledge past hurts. They come to see the humanity of one another, and see the possibility of a new constructive relationship."

Essentially, this involves the active forgetting of individual hurts and the intentional creation of a new third space. The generational effect must not be denied here, with younger generations seeing the separation of north and south as their 'normal'. Webster (2005) notes that individuals aged 55—65 (81.7%) were more in favor of reunification between the north and the south than individuals aged 18—24 (55.3%). Webster attributes this to the fact that these individuals have no personal memories of a unified Cyprus but may have heard family stories of fighting or the disappearance of family members. Sitas and colleagues (2007) note that young people (as of 2007) were less likely to have visited across the buffer zone, so it is more likely that their experiences are based solely on the exposure to family stories and photos. Memory, and the mutability of memory, in this instance is important for the creation of identity. The identification of the missing and their repatriation (along with the new mortuary rituals described above) may have an important role to play in upcoming years for Cypriots on both sides of the buffer zone.

7.2.4 Burial Within Churches in Medieval Europe

Burials within and around medieval churches are physical representations of social standing and power (Osterholtz & Valent, 2025). As Härke (2001) notes, burial rites were essentially a monopoly of the Church in medieval Europe and location of burial was dictated based on donation of burial offerings on behalf of the dead to the Church. This is a very visible reflection of the Body Politic (with the Church hierarchy having control over the burial space and the burial program within that space). Specifically, here, burial within churches refers to the placement of burials under the floors of churches, often resulting in a pattern of overlapping burials that contain the disturbed remains of previous graves. From an archaeological perspective, this leads to a complicated depositional history.

Burial within church spaces has a very long history, beginning with the founding of the church based on the burial location of St. Peter in Rome. The burial of popes and other notables within St. Peter's Basilica in Rome provides a good case study for the role of church burial and the performance of power and identity. Numerous good case studies could be the focus here, but I will focus on the founding burial, the burial that marked the initial location of the founding of St. Peter's Basilica. The Apostle Peter was believed to be killed on the orders of the Emperor Nero sometime after the fire that destroyed Rome in A.D. 64. He was possibly crucified and then his body would have been denied typical burial rites. But as Curran (1996) notes, early Christians were careful to recover and provide burial rites to martyrs. The body would likely have been recovered and given burial within a necropolis. With the legalization of Christianity by the Roman emperor Constantine, a church was built over the location of St. Peter's tomb in 326. By the medieval period, the basilica was in need of significant repair which began in 1506 and would be completed by 1626.

During excavations ordered by Pope Pius XIII in 1940s to reorganize the cemetery space used by Popes, a monument was discovered directly under the center of the dome of the original basilica. Beneath the monument, a cache of bones was uncovered, with many believing this was the original grave of St. Peter. This was the original interpretation by the archaeologists who conducted the excavations as well as later scholars who worked from the report produced by the archaeologists. O'Callaghan (1953) notes that while it is impossible to identify the bones as those of St. Peter, the archaeological context (including the tomb architecture and numerous coins found in the deposit) date to the right time frame. Toynbee (1953) notes that the monument was not a tomb in itself, but was a small shrine constructed after the date of the Apostle's death (likely by the first Christian Roman Emperor Constantine), likely placed to commemorate the original location of the burial. Generally, this is presented to visitors as the location of the tomb of St. Peter, and it has been largely accepted as an important place of power for Catholic identity and the basilica in general. This became a preferred location of burial for popes. Space for the Papal Tombs was created by raising the basilica floor directly above this necropolis. Burial of those popes who have been beatified (recognized as saints) has occurred in different spaces, including chapels within the basilica itself (such as Pius X, fig.) or directly below the basilica floor, as in John Paul II.

After the excavations in the 1940s, 1950, Pope Pius XII noted in his Christmas radio address that the tomb was that of Peter's, but that identification of the bones as Peter's could not be made. The human bones were analyzed by Venerando Correnti, a professor of Medical Anthropology at the University of Palermo in the 1950s. Correnti identified the remains of at least three people, at least one of whom was skeletally female, as well as animal bones (Curran, 1996). Additional remains from the surrounding area (the loculus) were then incorporated into the analysis. Amongst these remains, Correnti was able to identify the remains of an elderly male consistent with the historical understanding of the Peter's age at death. In June 1968, Pope Paul VI reported to an audience in the basilica that the remains found were indeed those of Peter (Curran, 1996). In November 2013, Pope Francis put these remains on public display for the first time. The bronze display case held nine bone fragments and sat on the altar during mass.

This founding history is one that was celebrated and recreated by later Catholics within their own church spaces. For the remainder of this section, I want to focus on the use on church space for burials within medieval and early modern Europe. Medieval and early modern Europe is a particularly fruitful place to seek to understand how burial location can reflective of and influence group identity due to the rich textual sources and long history of historical research that has been conducted.

First is the issue of parish versus cathedral spaces. Forrest (2010, p. 1110) examined the role of politics in late medieval Hereford, and sees "collective ideas about the afterlife and more worldly concerns for commemoration and 'memorialisation'... as a realization of socially conditioned values and beliefs." Social relationships were contested and negotiated using burial location. Disputes "were full of political potential... and the potential for social relations, particularly communal identities and leadership, to be reshaped"(Forrest, 2010, p. 1111). Some parishes, as discussed above, established parish cemeteries and collected fees from inhabitants of the parish for burial. Other parishes, such as the outlying parishes of St. Peter, All Saints, St. Martin, St. Nicholas, and St. Owen, were required to pay fees and bury their dead in the Hereford cathedral cemetery. This created a "burial monopoly" (Forrest, 2010, p. 1111), where families were required to bring their dead to the cathedral, incurring increased costs. This highlights the growing divide between rich and poor and between rural (parish) and urban (cathedral).

Burial could take place in cemetery space or within the church itself (e.g., Fig. 7.5). Burials of high-profile (relative to the rest of the community) individuals often took place within the church itself, these individuals became part of the cathedral architecture through the burial and so this space was highly sought after but also represents restricted access. While the sources used for this section span a great deal of time and space, they all relate centrally to the role of burials within the Catholic church. Central to the beginning of burial within churches was a change during the Medieval period with respect to understandings of the soul and the decreasing importance of keeping the body whole after death. Binski (1996) notes that relics made of Saints' bones were seen to be emblematic of the whole individual, and argues that this change also allowed for the reuse of burial space in medieval cemeteries and churches, since a single element was representative of the whole individual. Space within the churches, particularly close to the relics in the altar, became politicized (Stewart, 2017) and burial near them an opportunity to perform status and identity. Stewart (2017) summarizes the work of Chavarría Arnau (2009, published in Italian) who derived a list of common assumptions regarding preferential locations for burials within medieval Italian churches:

- Locations as close as possible to the altar, where mass is celebrated and martyrs and bishops are buried
- The area around the baptistry
- Areas near the main entrance to the church "due to the constant movements and attention of the faithful visitors and clerics as they enter and exit the church"(Stewart, 2017, p. 42).
- Areas protected by the façade of the church
- For those individuals outside the church, as close as possible to the church where their bodies will come into contact with rainwater running off the church roof.

These should be seen as a general pattern, with individual churches exhibiting burial patterns consistent with their own particular historical context and reflecting the combination of folk and Catholic dogma. James (1989, p. 30), stated "one might be able to detect aristocrats and others within an individual cemetery, because the

Fig. 7.5 A burial placed in the floor of a late medieval/early Modern period church in Trogir (Croatia). Photo by the author, 2018

community which used it would devise their own methods of distinguishing particular graves by means of grave goods or other signals, but that one could never be sure that the thinking behind burial practices was necessary the same in a neighbouring community, let alone in one on the other side of the kingdom." This highlights the need for contextualization. While we can look at medieval cemeteries across Europe to understand general patterns, any deep understanding must be based on understanding a single cemetery in its own context. In their work, Stewart (2017) noted that while status differences could be identified based on burial location for one cemetery in their sample, the other site had no such identifiable patterns.

As a social construct, the Catholic church varied between locations with the introduction of local tradition layered on dogma provided by the papacy. This central dogma, though, provided baseline traditions that were present regardless of where the church structure was located. The size and location of church Burial Grounds was defined by the radius of 60 yards from the altar in main churches and 30 yards for chapels in 1059 by Pope Nicholas II. (Schmitz-Esser, 2016, p. 33 as cited by Kolnberger, 2018). This was related the efficacy of relics to sanctify space and place. This ensured that the church cemeteries could not be expanded indefinitely, and may have been instrumental in the development of status-based locations for burials. Those closer to the relics in the alter would have benefitted more from their power, those buried within the church itself would become part of the architecture holding the relic. In their dissertation, Stewart (2017) examined burial location along with demographic factors in two medieval churches in Italy. They found that for the site of Trino Vercellese, there is a clear difference between those buried inside the church and outside the church in terms of bone length (as a measure of overall health potentially an indicator of status) and diet (measured through isotopic ratios).

As with other parts of Europe, in Sweden the tradition of burial under church floors began during the Medieval period and extended through to the Early Modern period. At first, this practice was limited to clergymen, but became less restrictive in the seventheenth and eighteenth centuries (Kallio-Seppä & Tranberg, 2020). Between 1630 and 1670, approximately 50 individuals were buried under the church floor. By the time the practice ended in 1769, between 2040 and 2050 individuals were buried under the church floor. This was based on the idea of restricted use of space, it was not until the seventeenth and eighteenth centuries that restrictions allowed groups other than the clergy to be buried beneath the church.

Burial within the church can be seen as a marker of status, with a clear hierarchy of placements within the church, with those closer to the altar (and the relics within it) being of higher status than other places (Kenzler, 2019). Death was the great equalizer, taking the Great and not-Great alike, but the burial locations were distinctly not equal. The equalizing elements of the use of the *memento mori* allowed for the creation of group or community identity, the stratified locations of burial highlighted the class distinctions and inequalities inherent in the system. Thus, both class-based inequality became structural (both figuratively and literally) for the Medieval church, with the bodies of the high-status members forming part of the literal church structure. Though still in the process of excavation, the church at Đurđevac-Sošice in eastern Croatia has so far yielded at least 200 burials dating to between the twelfth and seventeenth centuries. These burials overlap and disturb earlier burials, but most are oriented in a rough east-west orientation and have similar body positioning. During the placement of a new burial, when an older burial was disturbed, the remains were added to the burial fill. The effect of this was that the fill layer of the burials contains numerous isolated human remains. These individuals, once placed in considerate, high status burials under the floors of the churches, to have lost their individual identity when disturbed. The personhood of these individuals was no longer of central importance, they had become part of the architecture of the church, another element that had to be moved to make room for an additional burial (Osterholtz and Valent, 2025). The important element was the inclusion of the dead within the "divine service," with the church forming a "roofed cemetery" (Kenzler, 2019, p. 127).

In their work, Kallio-Seppä and Tranberg (2020) look into the sensory experience that parishioners would have experienced with burials below the church. They note that scents were tied to early Christian theology, with bad smells associated with satanic power. Good smells, such as incense, were used in churches to combat bad odors. This addition of another sense to the phenomenological experience of attending a ritual performance in a church (mass) heightened the experience of the individual. As Kallio-Seppä and Tranberg (2020, p. 4 of 17) argue, sensory experiences create memories, and a "sensorial perception is never a prereflective, pristine experience. Instead the work of memory is based on senses, and the senses rely on the materiality and physicality of the world." Essentially, the more senses involved with an experience, the more intense it can be for memory association and creation. Specifically, the smell of decomposition would have been prominent in medieval churches, particularly in the warm summer months. As Kallio-Seppä and Tranberg (2020, p. 8 of 17) note, the "smell of death created an olfactory image of the place that reminded churchgoers of their mortality and the brevity of life. The notion of 'memento mori'—remember that you must die—was materialized by the sense of smell." Parishioners would use 'nose herbs' or 'church bouquets' of marjoram, oregano, wormwood, thyme, sage, lemon balm, lavender, hyssop, valerian, and mint to try to mask the odors. In addition to having a strong scent, many of these herbs also had an association with healing. Norway spruce was also common placed on coffins themselves, but this would not cover an odor significantly. Spruce was believed to be a disinfectant against airborne disease. Putrid odors from under-floor church burials were a common source of complaint beginning in the seventeenth century. This was also the period of slight democratization of the church space, when commoners could then be buried under the church floors.

By the eighteenth century, general complaint of the odor was a common argument in parish and provost meetings for cutting back or stopping the number of church burials, "based on the idea of harmful toxic evaporations and the fear of the spread of contagious diseases, such as the plague"(Kallio-Seppä & Tranberg, 2020, p. 9 of 17). This burial practice only changed when the church ran out of space. Burials moved to spaces outside the church. The rituals involving scents during church services remained but took on different meanings. No longer was incense burned to cover the smell of putrefaction under the floorboards, but it was still used. And it provided a sense memory for parishioners that continues to this day.

The inter-cutting of graves within churches was an important aspect of identity formation for elite social classes within medieval Europe. They were quite literally closer to the source of power within the church when they were buried there, became part of the church architecture and foundations. The individual identity was less important than the show of status and piety that burial within the church denoted for the survivors. The dead were the building materials of faith. As Ariès (1975, p. 22) notes, "the body was entrusted to the Church. It made little difference what the Church saw fit to do with those bodies so long as they remained within its holy precincts." James (1989) highlights the importance of individual context and situating analysis of burial patterns into a local setting. The general patterns of relatively higher status burials within churches can be established, but additional patterns can also be visible archaeologically. These additional patterns may give clues about the community engagement with the Church, illustrating the interaction between the Social Body (local folk beliefs and practices) and the Body Politic (the Church ritual and hierarchy).

7.2.5 *Faith-Based Cemeteries: the Materiality of Belief*

With the reformation, the monopoly held by the Catholic church over burial form began to break down, at least in some areas. Ultimately, this helped to lead to the formulation of more individualistic, faith-based spaces were developed where inclusion within the cemetery space was predicated on one aspect of identity over anything else, religious affiliation. In multi-faith communities, the rituals surrounding faith-based burial practices and the space they occur in can be instrumental in creating and maintaining community identity. For Yazidis in Turkey, these rituals have a role in "keeping the past alive"(Sağır & Tayanç, 2023, p. 1). In their ethnographic study of Yazidis, Sağır and Tayanç (2023, p. 4) wanted to identify how "cemeteries, with their public appearances, reflect the memory of a minority/religious community—Yazidis' belief system contributes to the repro-duction and construction of physical space and how the cultural significance of that space is maintained and transmitted using social memory."

Essentially, some rituals can only take place in cemetery spaces and these rituals are integral to how this group maintains their unique identity in a country where they are a minority group. The cemetery is a space to make Yazidi beliefs manifest. Even while maintaining group identity, the Yazidis have adapted to larger Turkish society in that they have truncated the length of the funeral rites from 7 to 3 days to comply with Turkish cultural traditions. This might be interpreted as an adaptation creating a third space, and in that third space, the ritual and its truncation become even more amplified. The change in mortuary performance further separates this group from Yazidis outside of Turkey while bringing them closer in line with the rest of Turkish society. In effect, within Turkish Yazidi groups, the new rituals provide new forms of bonding capital, while also providing forms of bridging capital to the larger Turkish society. Yazidi cemeteries are located outside of the settlements, on promontories where they would be viewed on a daily basis, emphasizing their importance in everyday life (Fig. 7.6).

7.2.6 *Cemeteries for the Executed*

The mode of death can result in differential burial practices as well, using US servicemen who are executed as the primary way to understand how judicial execution can impact burial practice. US servicemen convicted of capital crimes and who have been executed are not buried in regular military cemeteries. Under the Uniform Code of Military Justice, 14 offences are publishable by death in both war and peacetime: Mutiny or Sedition (article 94); Misbehavior before the enemy (article 99); subordinate compel-ling surrender (article 100); improper use of countersign (article 101); forcing a safeguard (article 102); aiding the enemy (article 104); espionage (article 106a); improper hazarding of a vessel (article 110); murder (article 118); and rape (article 120). Additionally, in times of war, desertion (article 85),

Fig. 7.6 Yazidi village Koçan in Turkey, with village on the left side of image and cemetery on the promontory (right side of image). Image attribution: Êzîdîxan, CC BY-SA 4.0 <https://creativecommons.org/licenses/by-sa/4.0>, via Wikimedia Commons. https://upload.wikimedia.org/wikipedia/commons/0/04/Ko%C3%A7an1.jpeg

assaulting or disobey-ing a superior commissioned officer (article 90), lurking as a spy (article 106), or misbehavior of a sentinel or lookout (article 113) can be punishable by death. Cur-rent guidelines for the use of capital punishment (Department of the Army, 1959) allow for the next of kin to claim remains of those servicemembers who have been executed. If the remains are not claimed, they will be buried in the United States Disciplinary Barracks inmate cemetery located at Fort Leavenworth, KS. This cemetery contains the bodies of 298 military personnel who die while imprisoned or are executed whose bodies were not claimed by their families as well as the re-mains of 14 German Prisoners of War from World War II who were executed for the murder of other Prisoners of War while in custody. The graves are marked with military headstones inscribed with the names of the dead. In contrast to gravestones in National cemeteries, markers of the US servicemembers do not bear the rank of the deceased, just name, date of death and age at death. The German Prisoners of War, however, do bear their ranks.

The Oise-Aisne American cemetery holds the remains of 6012 American war dead from World War I within Plots A, B, C, and D. Plot E, however, holds the remains of 98 American servicemembers executed during the European theater of operations during World War II (Fig. 7.7). These individuals were executed for the crimes of murder and/or rape and were initially buried near the sites of their executions in England, France, Belgium, Germany, Italy, and Algeria. In 1949, these remains were gathered from their primary burial locations and reinterred in Plot E, a small, hidden section located 100 meters from the other cemetery plots. All individuals had been dishonorably discharged the day before their execution. The plot is not designed to be visible or to attract visitor, as it is only accessible through the office of the cemetery superintendent's office.

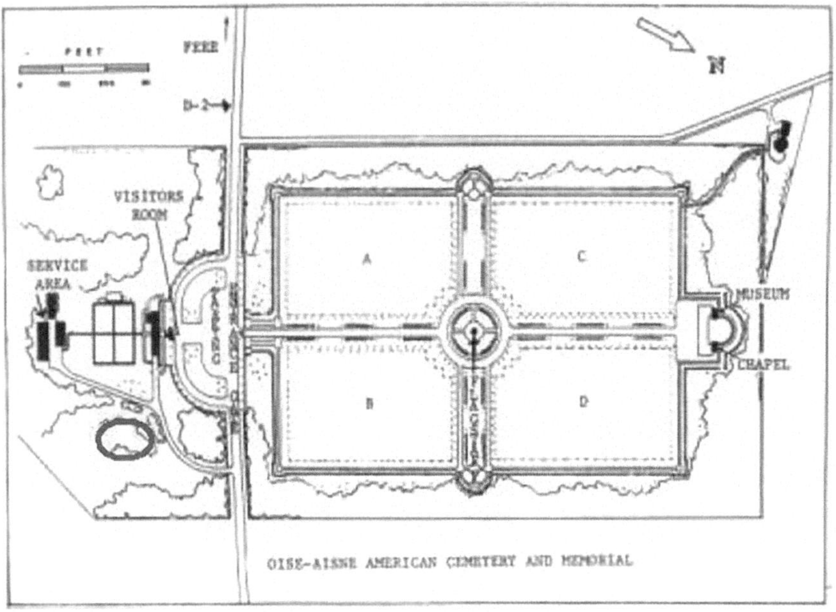

Location of Cemetery Features

Fig. 7.7 Modified google map showing the layout of the Oise-Ainse American Cemetery Memorial. Plot E, the burial location for executed servicemen is visible in the lower right side. https://commons.wikimedia.org/wiki/File:Location_of_Plot_E_in_Oise-Aisne_War_Cemetery.jpg

Of the personnel executed during WWII between March 1943 and June 1945, 56% were African American, 17% were Hispanic, and 28% were White (Lilly, 1995). Of the over eight million army inductees during July 1, 1944—December 31, 1945, 87.3% were designated as White, with 11% African American, and 3% Hispanic (Selective Service System, 1948), so servicemembers of color are over-represented in the numbers of individuals who were executed. US servicemembers who were executed were also much more likely to be from lower military rank (94% held the rank of private or private first class), with no one over the rank of Corporal executed. Lilly and Thomson (1997) argue that the military selectively punished lower ranks as well as those who were more socially disadvantaged. Military justice was swift, with some cases allowing only 10 days to prepare for trial and trails routinely lasting under 2 days. In the majority of cases, the defense called only one witness (the defendant), offered no exhibits and made no motions to the court (Lilly & Thomson, 1997). One of these individuals was Louis Till, an African American soldier who was executed for the murder of one Italian woman and the rape of two others, in July of 1945. Till was the father of Emmett Till and the husband of Mamie Till-Mobley (Emmett Till is further discussed in Chap. 9).

7.3 Cremation as Social Process in the Past (Eurasia)

Dry (also known as flame-based) cremation has a long history, being used as a method for disposal of the dead since the Neolithic (e.g., Bocquentin et al., 2020; Kutterer, 2012; Lloyd-Smith, 2012). Cremation has long been seen as transformative, as Oestigaard (1999, p. 346) notes, cremation is "a medium of change and transmutation" and can be "analysed as a result of three but interdependent processes; first as a technological transformation, secondly as a social transformation and finally as a ritual transformation." It effectively amplifies the liminal period of Hertz's (1960 [1907]) model, necessitating a different set of rituals including the creation of a pyre, the tending the pyre, and the interval of time necessary for allowing the body to cool before collecting bones and ashes for disposal. Oestigaard (1999, p. 351) examines transformations relating to cremation both in terms of time and Space, focusing on the use of landscape both in terms of location of the processing site (pyre) and disposal site as well as the differential uselife of objects used in cremation including grave goods and containers for ashes (e.g., an urn). In his view, the urn is an "ideologically transformed multi-vocal container with several functions, characteristics, and properties. Functionally, it is transportable and thus represents alliances, relationships and circles of acquaintances. Socially, it is an object of economic status which may reflect social differentiation and hierarchy. Ritually, it is a charnel-house for the burnt bones, through which cosmological and religious ideas are expressed." This builds on the concept of use-life of ritually charged objects (Walker & Lucero, 2000; Walker & Schiffer, 2006) and the differential disposal necessary for those objects.

The inclusion of ash and bone into an urn also turns that object into a portable cemetery of sorts, with the possibility of splitting those remains into multiple containers for distribution or circulation amongst groups as a way to reinforce social bonds (see also Booth & Brück, 2020; Brück, 2004; Brück, 2014). If mortuary processes can be seen to revitalize those elements most central to the reproduction of social order (as argued by Bloch & Parry, 1982), rituals surrounding cremation and ash dispersal/disposal establish a ritual bond between generations based on what Knipe (1977) sees as a kind of gift ex-change between the living and the dead.

Here I want to focus on how mortuary programs changed with the introduction of cremation or the removal of cremation. Typically, these shifts coincide with large-scale social changes such as the transition from foraging to agro-pastoralism or the intensification of agriculture along with population growth. Cremation has been used in many different locations, this particular mortuary program reflected different social processes. The history presented here is not meant to be comprehensive but to provide some case studies as to how mortuary programs involving cremation can be used to reconstruct some elements of lived experience, social structure, and the relationship between the living and the dead.

A distinction between primary (of fleshed corpses) and secondary (of dry bone bone) cremations should be made, as these represent more or less extended periods and types of processing (Bocquentin et al., 2020). Secondary cremations involve the extensive processing of the body to reduce it to a state of dry bone or allowing the body to decompose naturally prior to cremation, a far more intensive period of processing than that seen in primary cremations. Differentiation between these two types may be difficult if the cremation is extensive, but Bocquentin et al. (2020) suggest that the presence of a pyre pit, the representation of the elements (particularly the presence of multiple elements from the same limb close to each other and the presence of small bones such and hand and foot phalanges in association with long bones from the same limb) within that pit, and examination of the context may provide a way to distinguish between the two types.

Neolithicization, or the process of shifting from an economy based on hunting and gathering to agro-pastoralism, caused shifts in lived experience in the past, from how and where individuals lived, to how individuals died and were processed. As noted elsewhere in this volume, mortuary performances tend to be conservative, and so changes tend to be indicative of large-scale social changes. In the case of the introduction of cremation in the Neolithic, changes to the mortuary program may indicate a shift in belief (Bocquentin et al., 2020). Lloyd-Smith (2012) focused on examining cultural change and burial practice at Niah Cave, Sarawak, and noted a transition between pre-Neolithic burials in flexed, seated, or highly processed burials to multiple and cremation burials in the Neolithic. Bocquentin et al. (2020) examined this change for the Upper Jordan Valley in Israel (7031–6700 cal BC)and also noted that cremation does not appear to have been a major component to pre-Neolithic burial programs. Iriarte et al. (2020) examined the earliest cremated bones from the site of Kharaysin, Jordan (ca. 9000 year. cal BP), and come to the conclusion that they involved the burning of dry bone, not fresh corpses.

In Ireland and Britain, the transition from the Neolithic was marked by a shift from communal burial to individual burial, either inhumation or cremation, with cremation taking a central role from approximately 1700–1100 BC (Brück, 2006; Ellison, 1980). Cremains were either placed in a container of some kind (such as pottery vessels or leather bags) and placed in individual graves or placed directly into a collective grave. The co-occurrence of multiple mortuary programs has been interpreted as a marker of social status, with those receiving cremation seen as having lower social status (Braithwaite, 1984; Pierpoint, 1980) due to their secondary locations in relation to a primary inhumation. It has further been suggested that cremation had a negative connotation, destroying the individuality of the deceased, while inhumation was seen as maintaining that identity. Males were more likely to be interred while females were more likely to be cremated, so all together, some have argued that women held lower social positions in the Early Bronze Age in Britain (e.g., Mizoguchi, 1993; Pierpoint, 1980).

Brück (2014) argues, however, that the role of cremation in the Early Bronze Age was far more complex. She notes that cremated burials can be seen as a form of fragmentation as they contain only a portion of an individual, a pattern not explained by differential preservation of the remains. She draws an association with the concept of fragmentation, or splitting, of other types of material culture such as beads, noting that bead assemblages contain both older and newer beads suggesting a recombination or circulation of artifacts that "gave material form to interpersonal relationships" (Brück, 2014, p. 127), and argues that cremains could have been similarly subjected to processes of curation and circulation amongst community members. She also pushes back on the concept of cremation as destruction, arguing that this is a "product of modern, Western ideologies of the body: the suggestion that cremation acted as a means of 'destroying' the body and 'negating' the self works only in a context where an idealized model of the person as a bounded and homogenous individual is prioritized over other ways of constructing personal and social identity" (Brück, 2014, p. 130). The association with gender and mortuary program is also in need of more nuance, as Brück (2006, 2014) notes, inhumation was more common for both males and females at the be-ginning of the Early Bronze Age.

Cremation was more common for both males and females toward the end of the Bronze Age. Females were still more likely to be cremated, though, something that Brück (2014) argues may be due to gendered systems of kinship and marriage. Parker Pearson (1999b) argues that patrilineal descent was most common during this time and place, with women relocating after marriage. Those marriages and the transfer of women would have been invaluable to strategic alliances of kin groups. "In such contexts, women's locations at the interstices means that they may come to be seen as partiable persons—sometimes with a distinctly fracture and dividual sense of self"(Brück, 2014, p. 134). Their cremains may have been circulated as a means of maintaining intergroup links, and this interstitial social role informed their mortuary processing. Thus, the splitting of the body in Bronze Age Britain may have been a perfect reflection of identity as partiable and community based (along the lines of Oestigaard, 1999).

As opposed to the British Bronze Age, cremation was rare throughout the Aegean Bronze Age, particularly in the western Mediterranean. Cremation technique likely consisted of wood pyres, and are portrayed on at least one vase painting (Musgrave, 1990). As with Neolithic and Bronze Age cremations elsewhere, these do not involve the further processing of the burned bone into ash, but rather leave large pieces of bone (e.g., Fig. 7.8). The period known as the Late Helladic IIIC (LH IIIC) on the Greek mainland and Late Minoan IIIC (LM IIIC) dates to the period after the destruction of the Mycenaean palaces in the twelfth century BC that was accompanied by a breakdown of the palatial political and economic system (Ruppenstein, 2010). Though still not the most common form of mortuary performance, the number of cemeteries with cremation burials increased dramatically after the fall of the palatial system. These can be broken into two groups according to the cemetery structure. The first consists of the cremation burials in Mycenaean chamber tombs; these small number of cremations were laced in the same tombs as inhumations. Ruppenstein (2010) attributes the mixed cremation/inhumation nature of these tombs as indicative of full community incorporation of individuals who were cremated. The Greek Austrian excavations of the cemetery at Elateia, for example, documented 61 Mycenean chamber tombs, 49 of which contained human remains. Approximately 2000 individuals were represented, 23 of which were cremated (Deger-Jalkotsky, 2013).

The second type of cremation cemeteries are those consisting primarily of cremation burials. These have been found at Argos and Mycenae-Chania on the

Fig. 7.8 Example of Cretan cremation, photographed by Tom Oates in an archaeological museum on the island of Crete. Image attribution: Nabokov at English Wikipedia, CC BY-SA 3.0 <https://creativecommons.org/licenses/by-sa/3.0>, via Wikimedia Commons. https://upload.wikimedia.org/wikipedia/commons/c/c1/CrematedHumanRemains.JPG

main-land and Atsipades on Crete and exhibit a very different pattern. As opposed to incorporation of cremated remains into the community, cremated individuals in these cemeteries are separated from the rest of the community who follow the traditional pattern of chamber tombs. These cremated individuals were set aside, with cremains placed in tumuli set apart from the rest of the burials, suggesting a social difference between these individuals and those within the inhumation cemeteries (Ruppenstein, 2010). This suggests increasing variability in mortuary pro-grams after significant social change and reorganization. Without the consistent political control of the palatial system that likely included proscribed mortuary performances, local variants became more common.

As with the Bronze Age, the Early Iron Age mortuary program consisted primarily of inhumation graves, though Ruppenstein (2010) notes that regional variation was common. In general, cremations increased with the introduction of iron weapons, consistent with the beginning of the Iron Age. Cremation becomes more common during the Iron Age, Lloyd (2015) notes that for Athens and Knossos it was particularly important and was the type of mortuary program most likely to be associated with the inclusion of weapons. In particular, 'killed' swords are included with cremated individuals. 'Killed' swords are those that are bent out of shape, usually in a U-shape or a circle prior to inclusion as grave goods. This is a deliberate reshaping of the sword that would have required a smith. This additional processing would have been a part of the mortuary ritual, and should probably be seen as a modification as opposed to a destruction of the sword, changing its life history as an object.

'Killed' weapons appear in other Iron Age contexts much later in time (as the Iron Age is defined by material used to create implements and weapons, timing varies greatly), such as Iron Age Europe and the Baltic region (e.g., Fig. 7.9). This

Fig. 7.9 Finds from the Prince's Tomb at Oss (Vorstengraf) dating to around 700 BCE (Netherlands). The bronze urn contained cremated remains, a bent sword inlaid with gold, remnants of clothing, metal rings, knives, a clothing pin, and other goods. After cremation, the bone fragments were likely bundled together, wrapped in fabric, and placed with the grave goods (unburned) in the urn. Image attribution: Bron: Rijksmuseum van Oudheden, Leiden, CC BY 3.0 <https://creativecommons.org/licenses/by/3.0>, via Wikimedia Commons. https://upload.wikimedia.org/wikipedia/commons/4/40/Vorstengraf_Oss_-_bronzen_situla_met_inhoud_en_zwaard_--_700_v_Chr_-_RMO.png

should not necessarily be viewed as cultural contact or the transmission of ideas or rituals from one group to another. Instead this should be seen as a sacrifice by the living of valuable objects as part of the mortuary process. Karvonen (1998, p. 5) gives some possible explanation for 'killing' grave goods based on Finnish folk beliefs: (1) breaking the objects was a way to kill them and liberate their 'souls' to follow the deceased to the other world; (2) marking the possession of the object by the dead; (3) breaking the grave goods reflected a belief that the otherworld was a mirror image of this world. He also gives some functional explanations, including a desire to fit large objects into small spaces, but notes that a functional explanation does not negate the symbolic meaning that may have also accompanied it. Nor should these objects be viewed as biographical, indicating a warrior occupation for the dead due to the inclusion of a sword (particularly a 'killed' sword) in a burial (Whitley, 2002). Keeping in mind that the dead don't bury themselves (Parker Pearson, 1999a), the burials and their included goods are a creation of the living and may be used to reinforce gender ideals or to recreate heroic myths (Whitley, 2002).

In Germany during the Iron Age, particularly the Hallstatt (750–450 BC) and the LaTene (450–15 BC) periods, approximately 1/3 of the identified graves contain cremations. Iron weapons have the benefit of being non-organic and therefore more likely to preserve in the archaeological record. Karvonen (1998) examined Finnish Iron Age mortuary processes with special focus on the grave goods accompanying different types of cremation cemeteries as well as inhumation cemeteries that co-occurred. 'Killed' objects occurred in both cremation cemeteries and inhumation cemeteries, either cut or bent so as to make them useless. This suggests a varied mortuary program, with some common elements between cremation and inhumation.

7.3.1 Marking Social change with cremation and Inhumation in the Roman World

In general, Roman social structures were formalized, and there is a level of consistency that is visible across the Roman world. For adults within the Roman Republic, inhumation was much more common and considered the original form of burial. It was widely practiced throughout the Roman republic from the eighth through sixth century BCE (Toynbee, 1996), when a transition occurs toward cremation. Rebay-Salisbury (2012) links the use of cremation to ideas about pollution with cremation as a purifying act. By around 400 BCE, cremation was the predominant practice; this remained the case until the second century CE when inhumation again became more common throughout the empire. In some places, cremation and inhumation were practiced along side each other, and the relative prominence of one over the other likely reflected local traditions at play. The Roman world stretched from Britain through the middle east and was composed of the ethnic groups the Romans conquered during their expansion. This necessitated the use of social systems that provided a sense of Roman-ness or Roman identity.

Roman regulation of mortuary performance would have provided one such mechanism. While the overall regulation of the mortuary program is consistent across the Roman Empire, local tradition would have varied somewhat. This can be seen as contested space, where the loyalties of the local inhabitants were both to their new identity as Romans but also held onto local customs and traditions, incorporating them into the larger Roman patterns. In Roman Gaul, for example, two types of cremation burial were practiced from the first to third centuries CE. The first consisted of a small rectangular pit containing a secondary cremation deposit. Cremations were carried out at the *ustrina* located in the cemetery but apart from the gravesite itself. The second type (*bustum*) is a larger rectangular pit containing remains of a funeral pyre that has been constructed over the grave itself, leading to a primary cremation. Bustum graves generally also contain grave goods, and both types may include animal offerings.

Some remnants of earlier inhumation practices may have been retained for some time. Cicero mentions a practice in which a small portion of the body (a finger joint) was cut from the body and buried separately (Graham, 2011; Rebay-Salisbury, 2012). According to Rebay-Salisbury (2012), this practice may have helped to legitimize cremation during the transition from inhumation to cremation in the Republic. Even when cremation was most common form of burial for Roman citizens, some portions of the community still received inhumation. Infants who died toothless (roughly under 1 year of age at death) (e.g., Fig. 7.10) and individuals struck by

Fig. 7.10 Grave 72 from the Trogir Communal Necropolis (Trogir, Croatia), an infant buried in an amphora within a section of the city necropolis set aside for the very young). Photo by Kaukal d.o.o. Arheologija I geodetski mjerenje, used with permission of the Trogir Town Museum

lightning were given inhumations, while the bodies of the poor were sometimes included in trash heaps (Nock, 1932). Differential mortuary processes are likely to reflect the social identities of these groups. With high infant mortality rates (some estimates of over 30% (e.g., Carroll, 2011)), it isn't unexpected for infants to have a different social identity than the rest of the community. This shouldn't be read to mean that infants were not loved or cared for. Infants in Italy are often buried with other children or adults and found with grave goods of an apotropaic (protective) nature and feeding bottles. The inclusion of child-specific grave goods indicates that these individuals had a specific social identity, but it was not the same as the adults and so the burial performances are necessarily different.

By the second century CE, inhumation was again the predominant mortuary performance. Why did these changes occur? The timing around the second century CE was once linked to a higher number of individuals within the Roman empire that identified with Jewish and then later Christian religions, but the widespread nature of the switch to inhumation and the timing is simply too tight for Christian and too wide for Jewish traditions (Nock, 1932). A rise in Christianity in provinces in the fourth century may be responsible for the final and permanent change to inhumation, but it is unlikely to have been a cause for the switch to inhumation in Italy or in areas closer to Rome or with fewer Christian and Jewish groups present. Nock (1932) ties the change to changing views within Roman society, as well as a desire to distinguish between ranks in cases of mixed inhumation and cremation. Graham (2015) also notes the rising influence of mystery cults such as Orphism and Mithraism which encouraged inhumation within the eastern half of the Empire. The creation of mausolea associated with inhumations allowed for a greater display of wealth than would be possible in a Roman columbarium. Graham (2015) acknowledges the conspicuous display elements of inhumation but also notes that cremation and inhumation are fundamentally different approaches centered on an intent to modify the body through fire or to leave it whole.

In particular, Graham (2015, p. 42) focuses on the phenomenological aspects of the body itself, and the power of that experience to create collective identity. In particular, she states "the materiality of the corpse itself might act upon the senses, emotions and embodied memories of the living in such a way as to contribute towards the creation or maintenance of particular senses of personal or collective identity." Changing from cremation to inhumation fundamentally changes the amount and type of interaction with the corpse. Cremation involved a complex set of rituals including the construction of the pyre, the placing of the body, the maintenance of the fire, collection of cremains, and placement of the cremains within an urn for burial (Fig. 7.11). As part of the cremation ceremony, the body of the deceased could lie in state for extended periods of time (for the elite, up to a week), meaning that by the time the cremation occurred, the body would show significant signs of decomposition. For Graham, this decomposition is important from a phenomenological perspective, symbolizing the change in relationship between the living and the dead. "Cremation rites, with the materiality of the corpse at their heart, allowed the living to reconstitute both their own sense of personhood and that of the deceased" (Graham, 2015, p. 51).

Fig. 7.11 Grave 35 (a cremation urn containing the remains of an adult and grave goods) from the Trogir Communal Necropolis (Trogir, Croatia), showing the stages of excavation (left-lid intact, center with lid removed and unburned grave goods including unguentaria revealed, right with unburned grave goods removed. Photos by Kaukal d.o.o. Arheologija I geodetski mjerenje, used with permission of the Trogir Town Museum

With inhumation, an entirely different set of behaviors is needed. Decisions about how to dress the body, the form of the grave (stone-lined, earthen pit, amphora, stone sarcophagus?), arrangement of the body within the space, placement of goods to be interred with the deceased, and so on, would have needed to be answered. As Graham (2015, p. 46) notes, all of these actions "require that those responsible for the burial come into contact with and experience the materiality of the corpse in its entirety. As a result, they can also impact upon the nature of the relationship between the living and the dead." In contrast to the performance of decomposition being important for symbolizing changing relationships, inhumation (particularly the use of shrouds) suggests an increasing denial of decomposition, and should be seen as an attempt to preserve the integrity of the body. The presence of shrouds can be inferred through the placement of the limbs within burials close to the body when excavated and potentially through the discovery of shroud pins. Shrouds and coffins conceal the body, a major change from cremation where the body may have been on display. This distancing of the body from mourners is consistent with the changes we see in the US (see Chap. 5), where decreasing proximity to the corpse or the process of preparing the body decreases the intimacy of the average citizen to the dead. This likely was the case for the Roman empire, particularly for the wealthy who were able to use funeral specialists. The coffin or other burial container becomes the focus of contemplation and visitation, not the actual corpse. In other words, the built landscape of the cemetery begins to embody the mortuary process and the relationship between the living and the dead.

7.4 Summary

Inhumation has been with us as a species for a very long time. Inhumation is, to some degree, an easier venue through which to explore human mortuary behavior because cemeteries, graveyards, churchyards (etc.) are visible features on the landscape and are part of our every-day experience. They imply a permanence or a final place for the dead, one that is not ensured, and often disruption of the grave space is part of the mortuary ritual (such as we see with within-church burials). Inhumation is a particular form of land use, and one that tends to change the use of that land from then on. Burial sites serve as both material and symbolic expressions of community values, marking distinctions of social status, power, and belief systems. Ultimately the chapter demonstrates that burial practices are not passive acts of interment but dynamic, evolving spaces where social identity, cultural values, and environmental concerns intersect. By examining these sites and practices, we can shed light on how death rituals shape collective memory, identity, and social belonging over time.

References

Ariès, P. (1975). *Western attitudes toward death: From the middle ages to the present* (P. Ranum, Trans.). Johns Hopkins University Press.

Bernstein, N. (2016a). New York state bans use of unclaimed dead as cadavers without consent. *New York Times*. Retrieved from https://www.nytimes.com/2016/08/20/nyregion/law-bans-the-use-of-unclaimed-dead-as-cadavers-without-consent.html

Bernstein, N. (2016b, May 15). Unearthing the secrets of New York's mass graves. *New York Times*. Retrieved from http://www.nytimes.com/interactive/2016/05/15/nyregion/new-york-mass-graves-hart-island.html

Binski, P. (1996). *Medieval death: Ritual and representation*. Cornell University Press.

Blakely, R. L., & Harrington, J. M. (1997). Grave consequences: The opportunistic procurement of cadavers at the Medical College of Georgia. In R. L. Blakely & J. M. Harrington (Eds.), *Bones in the basement: Postmortem racism in nineteenth-century medical training* (pp. 162–183). Smithsonian Institution Press.

Bloch, M., & Parry, J. P. (1982). *Death and the regeneration of life*. Cambridge University Press.

Bocquentin, F., Anton, M., Berna, F., Rosen, A., Khalaily, H., Greenberg, H., et al. (2020). Emergence of corpse cremation during the Pre-Pottery Neolithic of the Southern Levant: A multidisciplinary study of a pyre-pit burial. *PLoS One, 15*(8), e0235386. https://doi.org/10.1371/journal.pone.0235386

Booth, T. J., & Brück, J. (2020). Death is not the end: Radiocarbon and histotaphonomic evidence for the curation and excarnation of human remains in Bronze Age Britain. *Antiquity, 94*(377), 1186–1203. https://doi.org/10.15184/aqy.2020.152

Braithwaite, M. (1984). Ritual and prestige in the prehistory of Wessex, c. 2000—1400 BC: A new dimension to the archaeological evidence. In D. Miller & C. Tilley (Eds.), *Ideology, power, and prehistory* (pp. 93–110). Cambridge University Press.

Brück, J. (2004). Material metaphors: The relational construction of identity in Early Bronze Age burials in Ireland and Britain. *Journal of Social Archaeology, 4*, 307–333.

Brück, J. (2006). Fragmentation, personhood and the social construction of technology in Middle and Late Bronze Age Britain. *Cambridge Archaeological Journal, 16*, 297–315.

Brück, J. (2014). Cremation, gender, and concepts of the self in the British Early Bronze Age. In I. Kuijt, C. P. Quinn, & G. Cooney (Eds.), *Transformed by fire: The archaeology of cremation in cultural context* (pp. 119–139). The University of Arizona Press.

Bryant, R. (2012). Partitions of memory: Wounds and witnessing in Cyprus. *Comparative Studies in Society and History, 54*(2), 332–360.

Buelow, D. A. (2010). *Peripheral memory: New York's forgotten landscape*. Massachusetts Institute of Technology.

Carroll, M. (2011). Infant death and burial in Roman Italy. *Journal of Roman Archaeology, 24*, 99–120.

Chavarría Arnau, A. (2009). *Archeologia delle chiese: Dalle origini all'anno Mille*. Carocci.

Committee on Missing Persons. (2023). Terms of Reference and Mandate. Retrieved from https://www.cmp-cyprus.org/terms-of-reference-and-mandate/

Curl, J. S. (2021). Cemetery. In *The Oxford dictionary of architecture* (4th ed.). Oxford University Press.

Curran, J. (1996). The bones of Saint Peter? *Classics Ireland, 3*, 18–46.

Deger-Jalkotsky, S. (2013). Cremation burials in the Mycenaean cemetery of Elateia-Alonaki in Central Greece. In B. Eder & U. Schuh (Eds.), *Proceedings of the international symposium held at the austrian academy of sciences* (pp. 221–229). Östereichische Akademie der Wissenschaften.

Denyer Willis, G. (2018). The Potter's field. *Comparative Studies in Society and History, 60*(3), 539–568. https://doi.org/10.1017/S001041751800018X

Department of the Army. (1959). *U.S. Army corrections system: Procedures for military executions*. Report No. AR 633-15. US Government Printing Office.

DeWitte, S. N., & Kowaleski, M. (2017). Black death bodies. *Fragments: Interdisciplinary Approaches to the Study of Ancient and Medieval Pasts, 6*, 1–37.
Ellison, A. (1980). Deverel-Rimbury Urn Cemeteries: The Evidence for Social Organization. Settlement and Society in the British Later Bronze Age. J. C. Barrett and R. J. Bradley. Oxford, British Archaeological Reports: 115–126.
Forrest, I. (2010). The politics of burial in late medieval Hereford. *English Historical Review, CXXC*, 1110–1138.
Francis, D. (2003). Cemeteries as cultural landscapes. *Mortality, 8*(2), 222–227.
Graham, E.-J. (2011). From fragments to ancestors: Re-defining the role of os resectum in rituals of purification and commemoration in Republican Rome. In M. Carroll & J. Rempel (Eds.), *Living through the dead: Burial and commemoration in the classical world* (pp. 91–109). Oxbow.
Graham, E.-J. (2015). Corporeal concerns: The role of the bod in the transformation of Roman mortuary practices. In Z. L. Devlin & E.-J. Graham (Eds.), *Studies in funerary archaeology. Death embodied: Archaeological approaches to the treatment of the corpse* (Vol. 9).
Green, M. H. (2014). Editor's introduction to 'Pandemic disease in the medieval world: Rethinking the Black Death'. *The Medieval Globe, 1*, 9–26.
Hadjigeorgiou, N. (2022). Truth and closure in Cyprus: An assessment of the committee on missing persons. *Israel Law Review, 55*(1), 3–24.
Hadjipavlou, M. (2007). Multiple stories. *Innovation: The European Journal of Social Sciences, 20*(1), 53–73. https://doi.org/10.1080/13511610701197866
Haglund, W. D., Connor, M., & Scott, D. D. (2001). The archaeology contemporary mass graves. *Historical Archaeology, 35*(1), 57–60.
Harding, V. (1993). Burial of the plague dead in early modern London. In J. A. I. Champion (Ed.), *Epidemic disease in London* (pp. 53–64). The Centre for Metropolitan History.
Härke, H. (2001). Cemeteries as places of power. In M. de Jong & F. Theus (Eds.), *Topographies of power in the Early Middle Ages* (pp. 9–30). Brill.
Hertz, R. A. (1960 [1907]). *Death and the right hand: A contribution to the study of the collective representation of death* (R. Needham & C. Needham, Trans.). The Free Press.
Iriarte, E., García-Tojal, J., Santana, J., Jorge-Villar, S. E., Teira, L., Muñiz, J., & Ibañez, J. J. (2020). Geochemical and spectroscopic approach to the characterization of earliest cremated human bones from the Levant (PPNB of Kharaysin, Jordan). *Journal of Archaeological Science: Reports, 30*, 102211. https://doi.org/10.1016/j.jasrep.2020.102211
James, E. (1989). Burial and status in the Early Medieval West. *Transactions of the Royal Historical Society, 39*, 23–40. https://doi.org/10.2307/3678976
Kallio-Seppä, T., & Tranberg, A. (2020). The materiality of odors: Experiencing church burials and the urban environment in Early Modern Northern Sweden. *Historical Archaeology, 55*, 65–81. https://doi.org/10.1007/s41636-020-00264-2
Karvonen, J. (1998). Deliberately damaged objects in iron age cremation cemeteries. *Fennoscandia Archaeologica, 15*, 3–13.
Kenzler, H. (2019). Post-medieval burial customs in Germany—An archaeological perspective on materiality and spatiality. *Mortality, 24*(2), 123–144. https://doi.org/10.1080/13576275.2019.1585781
Kilgannon, C. (2015, July 8). New York city to allow relatives to visit grave sites at Potter's field. *New York Times*. Retrieved from https://www.nytimes.com/2015/07/09/nyregion/new-york-city-to-allow-visits-to-grave-sites-on-hart-island.html
Klinker, M., & Smith, E. (2020). The Bournmouth protocol on mass grave protection and investigation. Retrieved from Bournemouth: https://www.icmp.int/wp-content/uploads/2022/02/mass_graves_project_english-4.pdf
Knipe, D. M. (1977). Sapindikarana: The hindu rite of entry into heaven. In F. E. Reynolds & E. H. Waugh (Eds.), *Religious encounters with death* (pp. 111–124). Pennsylvania State University Press.
Kolbuszewski, J. (1995). Cemeteries as a text for culture. In O. Czerner & I. Juszkiewicz (Eds.), *Cemetery art*. Wroclaw.

Kolnberger, T. (2018). Cemeteries and urban form: A historico-geographcial approach. *Urban Morphology, 22*(2).
Kovras, I. (2008). Unearthing the truth: The politics of exhumations in Cyprus and Spain. *History and Anthropology, 19*(4), 371–390.
Ktori, M., & Baranhan, G. (2018). Development and future perspectives of a humanitarian forensic programme: The committee on missing persons in Cyprus example. *Egyptian Journal of Forensic Sciences, 8*, 1–12.
Kutterer, A. (2012). Neolithic cremation in south-east Arabia: Archaeological and anthropological observations at FAY-NE10 in the Emirate of Sharjah (UAE). *Arabian Archaeology and Epigraphy, 23*, 125–144.
Lilly, J. R. (1995). Military executions during WWII: The case of David Cobb. *American Journal of Criminal Justice, 20*(1), 89–104.
Lilly, J. R., & Thomson, J. M. (1997). Executing US soldiers in England, World War II: Command influence and sexual racism. *The British Journal of Criminology, 37*(2), 262–288. Retrieved from http://www.jstor.org/stable/23638647
Lloyd, M. (2015). Death of a swordsman, death of a sword the killing of swords in the Early Iron Age Aegean ca. 1050 to 690 BCE. In G. Lee, H. Whittaker, & G. Wrightson (Eds.), *Ancient warfare: Introducing current research* (Vol. I, pp. 14–31). Cambridge Scholars Publishing.
Lloyd-Smith, L. (2012). Early Holocene burial practice at Niah Cave, Sarawak. *Journal of Indo-Pacific Archaeology, 21*, 54–69.
Mikellide, M. (2014). Burial patterns during times of armed conflict in Cyprus in the 1960s and 1970s. *Journal of Forensic Sciences, 59*(5), 1184–1190.
Mizoguchi, K. (1993). Time in the reproduction of mortuary practices. *World Archaeology, 25*, 223–235.
Musgrave, J. (1990). Dust and Damn'd Oblivion: A study of cremation in ancient Greece. *The Annual of the British School at Athens, 85*, 271–299.
Natural Death Centre. (2020a). The Natural Death Centre. http://www.naturaldeath.org.uk/index.php?page=home. Accessed December 2, 2020.
Nock, A. D. (1932). Cremation and burial in the Roman empire. *The Harvard Theological Review, 25*(4), 321–359. Retrieved from http://www.jstor.org/stable/1508378
NYCLU. (2015). Victory allows families to visit Hart Island Graves [Press release]. Retrieved from https://www.nyclu.org/press-release/victory-allows-families-visit-hart-island-graves
O'Callaghan, R. T. (1953). Vatican excavations and the Tomb of Peter. *The Biblical Archaeologist, 16*(4), 70–87.
Oestigaard, T. (1999). Cremations as transformations: When the dual cultural hypothesis was cremated and carried away in urns. *European Journal of Archaeology, 2*(3), 345–364.
Osterholtz, A. J., & Valent, I. (2025). The taphonomy of status: The creation of group identity and social inequality in Medieval Croatia. *Journal of Archaeological Method and Theory, 32*(1), 4.
Parker Pearson, M. (1999a). *The archaeology of death and burial*. Texas A&M University Press.
Parker Pearson, M. (1999b). The Earlier Bronze Age. In J. R. Hunter & I. B. M. Ralston (Eds.), *The archaeology of Britain: An introduction from the Upper Paleolithic to the industrial revolution* (pp. 77–94). Routledge.
Pierpoint, S. (1980). *Social patterns in Yorkshire prehistory, 3500-75 BC*. British Archaeological Reports.
Rebay-Salisbury, K. (2012). Inhumation and cremation: How burial practices are linked to beliefs. In M. L. S. Sørensen & K. Rebay-Salisbury (Eds.), *Embodied knowledge: Historical perspectives on technology and belief* (pp. 15–26). Oxbow Books.
Rugg, J. (2000). Defining the place of burial: What makes a cemetery a cemetery? *Mortality, 5*(3), 259–275. https://doi.org/10.1080/13576270050177225
Rugg, J. J. (2022). Further remarks on modern sepulture: Twenty years of cemetery studies and eight core questions defining cemetery research. *Eastern and Northern European Journal of Death Studies, 1*(1), 16–46.

Ruppenstein, F. (2010). Cremation burials in Greece from the Late Bronze Age to the Early Iron Age: Continuity or change? In M. Lochner & F. Ruppenstein (Eds.), *Brandbestattungen von der mittleren Donau bis zur ÄgÄis zwischen 1300 und 750 v. Chr.: Akten des internationalen Symposiums an der Österreichischen Akademie der Wissenschaften in Wien, 11.-12.Februar 2010* (pp. 185–196). Wien.

Sağır, A., & Tayanç, M. (2023). Imaginary views of death: Cemetery, identity, and memory (the case of Yazidis in Turkey). *Journal of Asian and African Studies, 00219096231188957.*

Schmitz-Esser, R. (2016). *Der Leichnam im Mittelalter. Einbalsamierung, Verbrennung und die kulturelle Konstruktion des toten Körpers*. Thorbecke.

Sitas, A., Latif, D., & Loizou, N. (2007). Prospects of reconciliation, co-existence and forgiveness in Cyprus in the post-referendum period. PRIO Cyprus Centre Report 4: 2007. Retrieved from https://d1wqtxts1xzle7.cloudfront.net/6748912/1740_report_4_07-libre.pdf?1390847942=&response-content-disposition=inline%3B+filename%3DProspects_of_reconciliation_co_existence.pdf&Expires=1740849908&Signature=atSvjkTd43tUr7rehENYU42yEQwriwBR8oE~dpQyvGVx3CKXCYjpNpFCOB6J9a-SSNoZpu0LUwF9QLDZuILMb~SELfawWy0Uz5ejVZ45gqbdEeifUxMMYnaMvnbDJfxH4kzprHe6WY9L9ITMgwBqQAVnBb8a2~IoBblWU~oGgnLNLSR1ffpN6d4FmqZVtc1ZZeIwsiHkzBGtrs6d9soUHkS1YCpZY4p-E-YDJcDHdAZvso29sRx5aQ67n2YhGB0PBQU-e2EkYzmCtz0N2tueD~al5zv269rebI67Ozpyz9Xfh~DhlT9sRr3fysGcR1FX82SNFKVI7RKaWFcQYdR9HhA__&Key-Pair-Id=APKAJLOHF5GGSLRBV4ZA

Skinner, M. (1987). Planning the archaeological recovery of evidence from recent mass graves. *Forensic Science International, 34*(4), 267–287.

Stewart, M. (2017). *Bioarchaeological and social implications of mortuary behavior in medieval Italy*. The Ohio State University.

System, S. S. (1948). *Selective service and victory: The 4th report of the Director of Selective Service 1944-1945 with a supplement for 1946-1947*. U.S. Government Printing Office.

Tarlow, S. (2000). Landscapes of memory: The nineteenth-century garden cemetery. *European Journal of Archaeology, 3*(2), 217–239. https://doi.org/10.1179/eja.2000.3.2.217

Toynbee, J. M. C. (1953). The shrine of St. Peter and its setting. *The Journal of Roman Studies, 43*(1-2), 1–26.

Toynbee, J. M. C. (1996). *Death and burial in the Roman world*. Thames and Hudson.

United Nations Security Council (2003). Security Council Extends Cyprus Force Until 15 December, Unanimously Adopting Resolution 1486 (2003) Press Release SC/7786. online, United Nations.

Walker, W. H., & Lucero, L. J. (2000). The depositional history of ritual and power. In M. A. Dobres & J. Robb (Eds.), *Agency in archaeology*. Routledge.

Walker, W. H., & Schiffer, M. B. (2006). The materiality of social power: The artifact-acquisition perspective. *Journal of Archaeological Method and Theory, 13*(67-88).

Webster, C. (2005). Division or unification in Cyprus? The role of demographics, attitudes and party inclination on Greek Cypriot preferences for a solution to the Cyprus problem. *Ethnopolitics, 4*(3), 299–309.

Whitley, J. (2002). Objects with attitude: Biographical facts and fallacies in the study of Late Bronze Age and Early Iron Age Warrior Graves. *Cambridge Archaeological Journal, 12*(2), 217–232.

Chapter 8
A Few Famous Cemeteries

Abstract This chapter examines four iconic cemeteries—Arlington National Cemetery (USA), Okunoin Cemetery (Japan), Père Lachaise Cemetery (France), and St. Louis No. 1 Cemetery (USA)—as spaces where culture, history, and memory converge. Each cemetery's development is explored in relation to its unique cultural and historical significance. Arlington National Cemetery, originally a Confederate plantation, evolved into a national symbol of military commemoration and reconciliation, reflecting shifts in American national identity, race, and memory. St. Louis No. 1 in New Orleans, established in 1789, stands as a testament to the city's diverse cultural heritage, with its above-ground tombs and rituals like the Second Line funeral offering a window into New Orleans' social and racial complexities. Okunoin Cemetery on Mount Koya, Japan, serves as a sacred pilgrimage site for Shingon Buddhists, with its intricate layout and spiritual significance underscoring the ongoing relationship between the living and the dead. Père Lachaise Cemetery in Paris, renowned for its cultural prominence and the interment of figures like Jim Morrison and Oscar Wilde, illustrates the intersection of personal and public memory, where individual legacies are entwined with broader social movements and cultural shifts. Each of these cemeteries, through their design, historical context, and role in public and private memory, exemplifies how spaces of death function as sites of identity formation, cultural performance, and ongoing negotiation between past and present. Together, they highlight the diverse ways in which cemeteries shape and reflect collective social and cultural experiences across time and place.

Cemetery spaces are places forever changed by the incorporation of graves. People interact differently with each other within a cemetery. There is a general code of conduct in cemeteries that varies by culture and the time of year. Like most aspects of human behavior, how we behave in a cemetery is a product of culture and history. Cemetery spaces may be quiet places of contemplation, or they may be places of celebration and communion between the living and the dead. In this chapter, I wanted to present a few case studies detailing the development and history of a few famous cemeteries that have taken on roles larger than a simple home for the dead. In each of

A. Osterholtz, *The Bodies Poetic*, Bioarchaeology and Social Theory,
https://doi.org/10.1007/978-3-031-85998-4_8

the cases presented here, the living make use of the space and the space has become central to how local groups define themselves. All have also become places of visitation for tourists, visiting famous graves or monuments. In choosing the case studies, these cemeteries were chosen for geographical breadth (2 US, 1 European, and 1 Japanese) to show that similar patterns of the importance of cemetery space exist within different cultural frameworks. The examples show the importance of local identity. These different local identities show a reciprocal relationship between the living and the dead, with the living interacting with the dead in ways that amplify communal identity.

Some cemeteries have become known for their occupants. They become cultural Spaces that allow for commemoration and celebration of the dead. Others, such as Okunion Cemetery have long histories of quiet contemplation and meditation. Some cemeteries, such as Arlington National Cemetery are spaces where inclusion within that space is restricted to those who performed national service (and sometimes their relatives) reflect the amplification of one aspect of an individual's identity in the service of a larger entity (the state). Burial in this space acts as an amplification of national identity, an act of the Body Politic. The Cimetière du Père Lachaise became not only a place of burial but a place of civil expression. The use of space during episodes of civil unrest are amplifications of the social body, as are the continued commemoration of those events. St. Louis No. 1 (New Orleans) is an excellent example of how cemetery space is used to create local identity shaped by migration, changes in governmental structure, and social change throughout its history. This space is one designed for the living utilizing the bodies and memorialization, and it continues to hold an important place in the cultural identity of the vibrant city of New Orleans.

The thread that combines these seemingly disparate cemeteries is their use as a Space that is larger than just a place where the dead are buried or placed. The use of space, how it has changed and morphed, has become important for group identity. Cemeteries are contested Spaces, constantly changing with architectural styles, or what Foucault (1984 [1967]) identifies as *hetertopias*. These spaces celebrate different intersectional identities and allow for cultural expression grounded by a common experience of visiting and interacting with the dead. The cemeteries presented here and discussed very briefly are an extremely small sample. All cemeteries provide community Space for remembrance of the dead and interaction with the living, these cemeteries are presented as highly public examples of the types of interactions that exist in every cemetery.

8.1 Arlington National Cemetery (Arlington, VA, USA)

8.1.1 Origins and Development

The land that makes up Arlington National Cemetery wasn't always set aside as a space for the dead. It began as a plantation owned by the Custis family, linking back to George Washington through the marriage of Martha Custis to George Washington.

It was linked to an economic system dependent upon the labor of enslaved individuals, with the mansion built by enslaved people, land cleared, planted, and harvested by enslaved people. In 1778, John Parke Custis purchased 1100 acres on the Potomac River that he decided to name Arlington. John Parke Custis was the son of Martha Custis, and when he died, his small children were sent to Mount Vernon to be raised by Martha and George Washington. The eldest son became the adopted son of George Washington and heir to Arlington. Arlington House was competed in 1818. In 1831, his daughter Mary Custis married Robert E. Lee at Arlington House. The plantation was inherited by Mary Custis Lee in 1857 along with 196 enslaved people, including 63 who lived at the plantation. Robert E. Lee was the executor of the estate of George Custis, and chose to circumvent the will's provisions for the freeing of those enslaved individuals as a way to maintain the plantation. Lee sold enslaved individuals or rented them out as laborers to other plantations in order to honor the cash bequests in George Custis' will (Murphy & Stephens, 2020).

Enslaved individuals had numerous intertwined relationship with the Custis family. George Washington Parke Custis acknowledged fathering a child with one enslaved woman. In 1826, he acknowledged Maria Carter Syphax as his child. He freed her mother and the rest of her siblings and gave them 17 acres of land on the western end of the Arlington plantation. A deed was not given to the Syphax family though, until granted the US congress in 1866 (Murphy & Stephens, 2020).

At the onset of the Civil War, Lee was offered commissions in both the Union and Confederate armies. He chose allegiance to the Confederacy and took a position as a major general. He urged his wife to vacate Arlington as it was located so close to Washington DC, and on May 13, 1861, she fled to Richmond, Virginia.

The plantation and its mansion would not remain empty for long. Once Virginia seceded from the Union, the land was seized for purported non-payment of taxes levied to support the war effort. In 1862, taxes were levied against the property under the Doolittle act (Labbe, 2018) Mary Custis Lee was informed about these taxes in 1864, and sent representatives to pay the taxes, but her representatives were informed that the land-owner needed to pay those taxes in person. This led to formal seizure of the land, auction of the land and purchase by the US government for $26,800 (Murphy & Stephens, 2020). Union soldiers who took over the property were supported by the enslaved individuals who remained when Mary Custis Lee fled. They cleared trees, built trenches and generally helped to prepare the site in case of confederate attack. The Custis Mansion became headquarters for General McDowell, and troops shared space with those fleeing Washington DC's outbreaks of whooping cough, measles and fever. A formal settlement of freed and escaped enslaved persons was founded as Freedman's Village on May 15, 1863 on the grounds of the plantation. It would ultimately house approximately 1000 people of color. This community was planned, contained a church, hospital, and school and existed until it was evicted to enlarge the cemetery in the 1890s. The placement of this settlement on land seized from the most lauded general in the confederacy was intentional, and the different use of this space could only be seen as a poke in the eye to Lee, his family, and the Confederacy. The view of enslaved people as property now used to promote the welfare and education of those same people of color would have been a performative use of the space, designed to be seen and known.

In 1864, the land was once again transformed. Lincoln was seeking a solution to filling the cemeteries of Washington DC with war dead, and looked to Union Quartermaster Montgomery Meigs for a solution. In 1862, Meigs had been placed in charge of purchasing land for national cemeteries for military dead. The grounds at Arlington were seen as a good solution to this problem. This decision to begin to use Arlington as a cemetery space for Union Soldiers can also been as an intentional transformation of that space away from productive plantation land to a place of memorial. It can also be seen as a punitive act to Lee specifically. Even though Mary Custis Lee was the actual inheritor of the land, it was generally associated with Robert E. Lee, and so changing his land to a memorial to those who died in the war where he chose sides against the Union would have been seen as fitting. Meigs viewed himself as a protégé of Robert E. Lee prior to the war, and so he took the betrayal personally, as Murphy and Stephens (2020, p. 14) note, the bodies of war dead and African Americans would poison the land for further use, "Meigs was determined to contaminate the land and underground drinking wells of the Arlington Estate with bodies."

Section 27 began as a Potter's Field where the enslaved population had built cabins and lived. Those cabins were torn down to make room for the Potter's Field, and the remained freedmen moved to Freedman's Village. The ground was smoothed and reshaped to serve as a cemetery. On May 3, 1864, William Christman was buried at Arlington in Sect. 27 (27-19). From Pennsylvania, Christman was a laborer who enlisted as a private into Company G, 67th Pennsylvania Volunteer Infantry. He served only a month before contracting rubella and was buried at Arlington because his family could not afford to have his body returned to Pennsylvania. The same day, William H. McKinney was also buried (Sect. 27–98). Private William B. Blatt of the 49th Pennsylvania Infantry became the first battle-related casualty to be buried in the cemetery on May 15 (Poole, 2009).

In June, 1864, Meigs ordered burials to be conducted in Mary Custis Lee's rose garden as a way to completely remake the landscape of the plantation and to make it unhabitable as a residence. "He took every step possible to bury more Union soldiers and many former enslaved on the property, making in unfarmable and thus politically and economically worthless due to the sheer number of the Union dead" (Murphy & Stephens, 2020, p. 186). This included the burial of his own son, John Rodgers Meigs (Labbe, 2018). Shortly after suggesting Arlington as a national cemetery, Meigs ordered the delivery of bodies from Washington and personally supervised the destruction of the rose garden close to the house and interment of those individuals (Murphy & Stephens, 2020). One interesting burial within the garden bears no name, just the number 5232. It contains three amputated legs from Union soldiers treated at Judiciary Square Hospital in May 1864. The legs are believed to belong to James Carey, a Pennsylvania Private who lived until 1913, Arthur McQuinn of the 14th US Infantry, and Sergeant Michael Crighton who survived his amputation but died 2 weeks later on June 9, 1864. His body (minus his leg) was interred in the lower cemetery. He is the only known person to have remains in two different graves at Arlington National Cemetery (Poole, 2009). Poole (2009, pp. 63–64) speculates that the burial of three limbs from non-commissioned personnel in a place of honor within the cemetery was intentional, remarking "Why was this trio singled out perhaps their symbolism mattered more than their individual identities, another rebuke for the Confederate general who had caused so much Union suffering. The gesture could not have been an afterthought—someone had to label the remains, transport them from the hospital across the river, record the particulars in a ledger, ready a marker for the grave, and bury the legs in the garden."

This transformation of the land from a place for the living to a place for the dead began in earnest after that decision. By 1865, almost 16,000 war dead were buried on the property, including approximately 4000 "unknowns". The monument atop one of the grave of unidentified soldiers from Bull Run reads "Beneath this stone repose the bones of two thousand one hundred and eleven unknown soldiers gathered after the war from the fields of Bull Run and the route to the Rappahannock. Their remains could not be identified but their names and deaths are recorded in the archives of their country, and its grateful citizens honor them as of their noble army of martyrs, may they rest in peace! September A.D. 1866" (Fig. 8.1). At least 1800 African American soldiers who had served as part of the United States Colored Troops were segregated into a separate section of the cemetery that would become Sect. 27. This area also would hold the graves of approximately 3800 formerly enslaved individuals who died in the Washington DC area (Murphy & Stephens, 2020). These African American citizens generally helped to construct and maintain

Fig. 8.1 Marker above collective burial of unidentified soldiers from the battle of Bull Run (Manassas). Photo by Tim from Washington, D.C., United States of America, CC BY-SA 2.0 <https://creativecommons.org/licenses/by-sa/2.0>, via Wikimedia Commons. https://upload.wikimedia.org/wikipedia/commons/f/fb/Civil_War_Unknowns_Memorial_-_E_side_-_Arlington_National_Cemetery_-_2011_%286,799,175,147%29.jpg

Washington DC's defenses during the civil war. "As former slave and segregated cemeteries all along the South today are disappearing due to neglect, abandonment and development, Section 27 holds a unique and rich history of the African American war dead. In life and death, the men of the United States Colored Troops faced discrimination, segregation and disrespect, even during the time of their passing and burial" (Murphy & Stephens, 2020, p. 5).

The continued segregation of African Americans in death is reflective of larger social structures in the nineteenth century in the US. Even though the Civil War was fought to end the institution of slavery, many northerners did not argue for equality or the elevation of formerly enslaved people to equal status or rank. The social structures evident in life continued in the treatment of the dead by the living, including those who made the decisions about where to place burials. Segregation of burials by social race would continue until President Harry Truman ordered integration within the Armed forces (Executive Order 9981: Desegregation of the Armed Forces (1948). The burial of formerly enslaved and free African Americans on the Lee property was in fact seen as a way to ensure that the family would never want to live there again so they didn't have to share space with those they would see as inferior to themselves. The Union was punishing Lee's treason with the loss of his land to be a Place and a Space for African Americans.

Burials were segregated by military class and rank as well as social race. Privates continued to be buried in Sect. 27, but Union captains and lieutenants and other offers began to be buried in more prized locations. All markers, however, were the same size and shape after 1902. All would be 36 inches tall, 10 inches wide, and 4 inches thick. The tops would have a rounded appearance. The layout would regimented, with long rows of burials in straight lines. After 1902, unidentified individuals would have "Unknown" carved on the face of the stone instead of small square blocks as markers.

The Civil war has been estimated to have caused approximately 620,000 lives (Poole, 2009). By the end of the Civil War, approximately 16,000 war dead were interred at Arlington National Cemetery, with 42.5% of those buried as "unknowns." The high rate of unidentified individuals was due to several factors, including the lack of infrastructure surrounding collection of remains after battle, long intervals between death and collection of the remains. These unidentified were buried in individual graves, though, to maintain their individual identities after death. If a family could travel to Arlington, they could visit any of those unidentified individuals and have a commemorative experience. There was always a possibility that they were visiting their fallen family member.

After the conclusion of the Civil War, a portion of Arlington was used as residential community (Freedman's village), a portion was set aside as a military cemetery, a fort (now known as Fort Myer), and subdivided into 10-acre farms. Until 1883, approximately 20% of the space was used as a military cemetery.

The Confederate dead were also buried in a separate section in Arlington. President McKinley, when preparing to repatriate the war dead from the Spanish-American war, promised to assume the care of Confederate graves in the north as an act of reconciliation. The grave markers would be the same size as the Union markers, but would come to a point at the top. The layout would also be different, with the graves placed in concentric rings around a central point. This central point would eventually be filed by a monument to the Confederacy would also come to this section in 1914. Ultimately, the Confederate burials were moved from other parts of the cemetery to form an "honor guard" around this monument. Murphy and Stephens (2020) tie to this to the rise of Jim Crow segregation. President Woodrow Wilson unveiled the monument on June 4, 1914 (Jefferson Davis' birthday). The Daughters of the Confederacy claimed ownership to the Arlington Mansion in the 1920s, and were ultimately successful in having Arlington House named as a memorial to Robert E. Lee to ensure "the correct interpretation" of its history, which Murphy and Stephens (2020, p. 7) call "a single framework, an updated version of the Lost Cause."

The Joint Resolution passed by Congress in 1955 making the Lee Mansion in Arlington National Cemetery a permanent memorial to Robert E. states that congress does "hereby pay honor and tribute to the everlasting memory of Robert E. Lee, whose name will ever be bright in our history as a great military leader, a great educator, a great American, and a truly great man through the simple heritage of his personal traits of high character, his grandeur of soul, his unfailing strength of heart"("Joint Resolution: Dedicating the Lee Mansion in Arlington National Cemetery as a permanent Memorial to Robert E. Lee," 1955, p. 190). This acted as an official memorial to Lee until 2017, when the National Parks Service quietly changed their website in the wake of increasing calls to remove Confederate monuments. According to Berman (2017), language on the NPS website still acknowledged Lee as a figure who hoped to see peace and unity within our nation (the original wording) but added additional language inviting contemplation of difficult aspects of American history, including citizenship, slavery, loyalty, and freedom (https://www.nps.gov/arho/index.htm). On December 15, 2020, a resolution was referred to the House Committees on Veteran's Affairs and Armed Services to redesignate the Lee Mansion from "the Robert E. Lee Memorial in Arlington National Cemetery" to simply "Arlington House"("H.J. Res. 105—Redesignating the Robert E. Lee Memorial in Arlington National Cemetery as "Arlington House"," 2020), an indication that the role that Lee has played in history, and the monumentation of Confederates in modern twenty-first century is still contentious.

In 1882, the Supreme Court determined that the land now containing the National Cemetery was taken from the Custis-Lee family without due process as the Doolittle act did not specify in-person payment by the land owner (Gaughan, 2012). Mary Custis Lee's son, George Washington Custis Lee then sold the property back to the government for $150,000 (Labbe, 2018). The use of the land as a cemetery as well as the creation of the Freedman's Village had indeed changed the land so much that they could no longer contemplate inhabiting the house or using the land. The concern had shifted from regaining the land to compensation for the land. By the 1890s, Freedman's village was gone, memorialized by a plaque in a park outside the National Cemetery and a bridge that spans Columbia Pike. Many of the inhabitants moved a bit south to form the neighborhood of Nauck.

With the Spanish-American War, a need for increased burial space would necessitate the opening of new sections. The military also learned from their experience with the Civil War, with increased structures in place to ensure a higher rate of identification for returning dead. After death on foreign soil, American soldiers were buried in local cemeteries. This "splendid little war", as described by US ambassador John Hay to Theodore Roosevelt in 1898, established the US as a world power and placed the US as a central player in international conflicts. During the Spanish-American war, 460 US soldiers died fighting, while 5200 were killed by malaria, typhoid, dysentery, and yellow fever. Another 4300 died in the Philippines between 1899 and 1901. President McKinley, a veteran of the Civil War ordered mortuary teams to Cuba and Puerto Rico as soon as fighting ceased. They were responsible for finding and marking battlefield graves and identifying individuals as much as possible. Funds were appropriated by congress to disinter and repatriate the remains of all Americans who died in the war, the first time that the government pledged to repatriate fallen soldiers from foreign soil if their next of kin requested repatriation.

The War Department sent D.H. Rhodes, a landscape gardener at Arlington and inspector of national cemeteries to provide accounting of the dead from the Spanish-American war. He and his crews surveyed grave sites, discovering 654 graves of US soldiers on the island of Cuba in August of 1898. He identified 141 of the dead from the markers on their temporary graves, interviews with fellow soldiers, uniform insignia and glass bottles containing slips of paper identifying the deceased, possibly left by burial crews (Poole, 2009). The repatriation program ended a few years later. The repatriation crews were able to identify all but 13.6% of the dead, with the percentage of unknown decreasing as the repatriation went on. For example, for the 1073 individuals repatriated in 1901, only 15 were unknowns. For the unknowns, maps of their graves and sketches of the burials accompanied notes in the hope of future identifications. Interestingly, the use of the bottle buried with the dead in their temporary graves became the inspiration for the dog tag still used by the armed forces. Charles C. Pierce, an Army chaplain who helped to recover the dead from the Philippines is usually credited with this idea (Poole, 2009).

The Spanish-American War was the first war fought on foreign soil, and so the dead began to arrive at Arlington for burial in waves. The first wave of 366 arrived on April 6, 1899. Government offices closed for the afternoon and flags were lowered to half-staff throughout Washington DC. The fourth Artillery Band welcomed the bodies, and President McKinley, his cabinet, and high-ranking military officers joined at graveside for the interments. Multiple religious faiths were represented and conducted services. The ceremonies lasted several hours, and at dusk the labor of placing the bodies into their graves commenced. It would take 2 days to complete the task.

A total of 116,516 Americans died in World War I, with the vast majority of these occurring overseas. This war led to social disarray in Europe, and into this difficult landscape, mortuary teams were sent into the European countryside to begin the task of locating, identifying, and interring Americans from temporary graves into permanent cemeteries. Charles G. Pierce, the chaplain responsible for the development of dog tags was recalled to service. He led over 900 men from the Army's Graves Registration service in the work. They surveyed approximately 2400 temporary European cemeteries after the 1918, identifying over 98% of the 79,351 American combat burials. The prevailing feeling of military leaders was to allow the soldiers to remain buried in Europe. Roosevelt, who lost his son to the war was noted for quoting Ecclesiastes 11:3 "Where the tree falls, there let it lie".

The logistics of returning war dead from Europe were immense, and the command argued would present challenges to the infrastructure in Europe at the time, impacting Europe's recovery and reconstruction. The scale of repatriation was simply so much larger than that for the Spanish-American War. Major General Frank McIntyre argued that there was a greater honor in leaving the fallen where they lay, that their sacrifice was greater than that of sacrificing one's life only for country, but as a service to civilization, saying that each soldier should be buried "where he fell, fighting the foes of civilization.... What can be a better testimonia to the valor and devotion of her dead sons... than that the Nation should secure and maintain in perpetuity vast cemeteries in France—cemeteries which, in themselves, with their thousands of graves, will arouse sentiments and emotions in pilgrims of future generations, which mere monuments can never do?" (Memorandum of Brig. Gen. Peter C. Harris, June 4, 1919, NARA RG 407, Box 566, File 293.8–293.9., quoted in Poole (2009, pp. 144–145)). This would have effectively made American cemeteries overseas into monuments not only to visitors from the US but also to those that would see them daily and act as a reminder of the sacrifice that Americans made to protect Europeans.

Letters began to arrive from grieving family members, though, and on October 29, 1919, the War Department announced that servicemembers would be returned from Europe if their family members requested repatriation. Burial would be conducted either at one of the national cemeteries or in private or religious cemeteries at the families' discretion. If repatriation was not requested, the bodies of servicemembers would be relocated to one of eight newly established American Cemeteries in England, Belgium, and France; those countries ceded the land for the cemeteries, so the individuals would be buried in American soil. Repatriation of remains back to the US began in November 120 and would continue actively through July 1922. Approximately 5800 were buried in national cemeteries, with 5241 occurring in Arlington. Another 30,000 remained in Europe in the American cemeteries created for this purpose.

Prior to the Korean Conflict,[1] American forces had temporarily buried soldiers in cemeteries within allied areas for future repatriation. In Korea, however, this was impossible as the lines shifted often. The pace of this war meant that allied dead were often abandoned where they fell in quick retreats. While American troops attempted to gather the dead where they could, it was not possible to have stable, temporary interments during the conflict. Instead, the US developed a policy of "concurrent return" in 1951, meaning that the war dead would be collected from the battlefield, transported to Japan, and shipped home for burial. This marks the first

[1] This may be referred to as either a "war" or a "conflict." The use of the term "conflict here" reflects that congress never technically declared war on Korea.

time that America's war dead were repatriated during wartime. At the time, this policy was appreciated by family members as well as the press. By the Korean Conflict, the armed forces were integrated, and the shipment of the soldiers side by side "without regard to race or rank" was seen as "proof that the nation had recovered its sense of democratic propriety" (Poole, 2009, p. 199). Though the bodies would be buried according to rank in Arlington, they travelled together, arrived in flag-draped coffins, and were afforded honors at Arlington. A truce was made between North and South Korea in 1953, and the recovery of the dead began in earnest. They ultimately recovered 30,450 sets of remains of which more than 29,500 were identified. Mortuary recovery teams had become sufficiently professionalized to bring the unidentified rate to less than 3%, likely due to increases in methods of body recovery, the use of dog tags, and scientific identification methods. It should be noted that this coincides with professionalizing and creation of standards in forensic anthropology that are still used today (Christensen et al., 2014). It is frustratingly difficult to determine how many dead from the Vietnam War are interred at Arlington. At least 300 helicopter pilots and crew members who died during the Vietnam Conflict are buried at Arlington and are memorialized by a monument within Section 35 (Arlington National Cemetery, 2025), but I have been unable to find statistics detailing the total number of US servicemen buried at Arlington from this conflict. The repatriation of fallen servicemen is ongoing for any conflict. While active repatriation of World War I remains was in its most active form through July 1922, for example, the repatriation of individuals who have been uncovered after that date continues through the present day. Soldiers who died during World War II, Vietnam, and Korea are also still being repatriated to US soil.

8.1.2 The Current State of Arlington

Arlington House has been restored to its 1857 appearance instead of a later time when it was occupied by Union troops, something that Murphy and Stephens (2020, p. 105) argue "is a *choice* [italics in original] made by preservationists, a choice that elevates the relationships between the slavers and the enslaved. The choice elevates one epoch or owner over another. Both of those times at Arlington have important, yet different lessons for current tourists."

The use of Space and creation of Place at Arlington has changed through time. What started out as a punitive act directed towards a single individual (Robert E. Lee) morphed into a place of reverence and honor, the burial place of presidents, war heroes, and a place where the cost of our democracy is on display for all to see. The history of Arlington, where the burials were placed, segregated, separated, moved, is *performative* and intentional. The placement of burials near the house is important. These actions change the nature of the Space significantly, moving it from a place of living to a place for the dead. The placement of three amputated limbs of non-commissioned soldiers in a place where only the higher-status burials are is important. The burial of these limbs, when most amputated limbs were buried in communal pits or burned as medical waste, is a testament not only to the death that occurred but the changes to the bodies of those that survived. It memorializes the sacrifice that soldiers made apart from their lives.

Who an individual was in life was reflected in various parts of their burial programs at Arlington. Confederate and Union soldiers are both buried at Arlington, but they are separated. Their grave markers are the same size, but different shapes, and the layout of their graves follows a different design. Even without the cultural context and historical knowledge presented here, it would be clear from a survey of the site that these groups are different, that different intersectional identities are expressed in their grave sites. It would also be clear, from an archaeological perspective, that the Confederate graves were reburials, further distinguishing them from the rest of the burials at Arlington.

Arlington is a constantly evolving Space. The modern plan for the cemetery can be traced back to the McMillan Plan published in 1902. This proposed a regional park system within the District of Columbia with an expanded national mall at the center. Arlington was seen as integral to this plan as a contemplative space. They argued against the erection of large monuments that would obstruct the visual field. Their recommendations would eventually be adopted at Arlington, but only after several notable burials. One of these was the reburial of L'Enfant. L'Enfant, responsible for the plan of Washington, DC. D.H. Rhodes was responsible for recovering the remains from the Digges Farm in Maryland on April 22, 1909. The remains were recovered from the initial grave, wrapped in a box in a silk American flag and transferred to a vault in a nearby cemetery until his reburial in front of the Arlington House.

His reburial at Arlington occurred on April 28, 1909. President Taft and Vice President Sherman attended, as did the French ambassador. The remains were pulled onto a caisson pulled by six horses down Pennsylvania Avenue, by the White House, through Georgetown and into Arlington. A Catholic service was conducted and Taps was played. Two years went by while L'Enfant's monument was finished. The reburial of L'Enfant in Arlington solidified the cemetery's status as the nation's most esteemed war memorial. With this burial, the revolutionary war was also celebrated at Arlington. Fourteen other war dead who perished in the War of 1812 were moved as well as soldiers from the Mexican-American War. These individuals were all mingled in Sect. 1, known as the Officer's Section. A section was designated for nurses from the Spanish-American war, an area where women still have more burial plots than men.

There is homogeneity in appearance of the markers, but individuals' status and some aspects of their identity can be gleaned from both location, orientation of the burials, and the different styles of their markers. The thing that links every person in Arlington is their service. With decreasing space available for burial at Arlington, individuals requesting interment at Arlington must meet specific requirements. Table 8.1 gives the eligibility requirements for interment or placement of ashes in above-ground structures for the 2018 Fiscal Year (adapted from Torreon (2019) Table 1). There are different requirements for in-ground burial due to a lack of space. As cremains can be displayed in a vertical orientation, the requirements for eligibility with Columbarium Court or wall niches is not as strict as those for in-ground burial. Torreon (2019) gives details of proposed changes to these eligibilities based on an anticipation of future needs for the cemetery. She notes that 95,000

Table 8.1 Eligibility Requirements for burial and placement of ashes at Arlington National Cemetery[a]

Servicemembers with Honorable Discharge	Eligibility in Fiscal Year 2018	
	Placement of ashes in Columbarium Court or Niche Wall	In-ground burial (casket or urn)
Dies on Federal Active Duty (non-training)	Yes	Yes
Dies on Federal Active Duty (training)	Yes	No
Veteran-retired from active duty, receiving retirement pay	Yes	Yes
Veteran-at least 1 day active duty other than training	Yes	No
Veteran-received Medal of Honor, Distinguished Service Cross (Air Force or Navy Cross), Distinguished Service Medal, Silver Star, or Purple Heart	Yes	Yes
Reservist in the Armed Forces, death occurs under honorable conditions while on active duty for training or performing full-time service	Yes	No
Former Prisoner of War who, while a POW served honorably in the military; and who died on or after November 30, 1993	Yes	Yes

[a]Adapted from Table 1 in Torreon (2019)

burial spaces remain in the cemetery if the current boundaries are maintained, and that this will not be sufficient for future needs, and that Arlington would reach it's capacity for new interments by 2050 if the eligibilities listed below are maintained.

In order to plan for the future, defined as 150 years by the Acting Army Secretary, several changes would need to be put into place. Instead of the eligibilities listed above, in-ground interment would only be available to:

- Servicemembers killed in action, to include repatriated remains
- Award recipients of the Silver Star and above who also served in combat
- Recipients of the Purple Heart
- Combat-related service deaths while conducting uniquely military activities
- Former POWs
- Presidents and Vice-Presidents of the United States
- Veterans with combat service who also served out of uniform as government officials and made significant contributions to national security

Above-ground display of cremains would be available to:

- World War II-era veterans, to include legislated active-duty designees
- Retirees from the armed forces eligible to receive retirement but not otherwise eligible for interment
- Veterans who served a minimum of 2 years of active-duty and who served in combat
- Veterans without combat service who also served out of uniform and made significant contributions to national security.

There are two current expansion plans for the cemetery. The first, known as the Millennium Project involves the acquisition of 27 acres in the northwest part of the cemetery. This was begun in the 1990s and dedicated on September 7, 2018 with the burial of two unknown soldiers from the Civil War. The dedication of the cemetery with individuals from the founding conflict is instructive here, and the use of unknowns from this conflict could be seen as tying the new section to the rest of the cemetery, formalizing the acquisition beyond the original scope. This expansion of the cemetery will allow for 27,282 new interment spaces and 16,400 new above-ground niches for cremains. The second expansion is known as the Southern Expansion and was begun in 2020. This plan will expand Arlington southward and provide 37 acres of additional space.

Here, embedded within the history of Arlington National Cemetery is the embodiment of the relationship between fallen servicemembers, the government, and the American people. Beginning with the Spanish-American War, there has been an emphasis on burial on American soil with honors. This desire to honor those who die in service to this country is indicative of a commitment on the part of the individual servicemembers and their families to honor this sacrifice to a higher goal. The US government has gone to incredible lengths to repatriate, and continues to do so through the work of the Defense POW/MIA Accounting Agency (DPAA). Identification efforts are on-going.

Burial within National Cemeteries, particularly Arlington, carries a greater meaning for those whose families are interred there and to the nation in general. These national cemeteries present a uniform appearance, are carefully planned, and through the use of these multiple graves become political monuments.

8.2 St. Louis Cemetery No. 1 (New Orleans, LA, USA)

St. Louis Cemetery No. 1 is the oldest surviving urban cemetery in New Orleans. It was established by royal decree from Spain in 1789 (Peters, 2002). Though small in size at one square block, it contains the graves of thousands of individuals in its 700 tombs. It has been owned by the Archdiocese of New Orleans since its creation and has been listed on the national register along with St. Louis Cemetery No. 2 since 1975. The cemetery consists of a variety of tomb types and styles, but most are made of locally made bricks and mortar covered in lime or stucco. New tombs were added through the middle of the nineteenth century along with the expansion of smaller tombs. Tombs were maintained by the families of the dead who white-washed the tombs with gatherings on All Saints' Day providing social pressure to keep up maintenance on the tombs (Peters, 2002). By the late nineteenth century, the cemetery began to fall into disrepair and neglect. Newer cemeteries were preferred places of interment, and many older families died out or migrated from New Orleans. Preservation efforts by concerned community groups and the Archdiocese included new stucco applications and modern alterations which have not necessarily held up well. The cemetery has become a focus for historic preservation efforts, as evidenced by of Peters' (2002) work modeling tomb decay as well as Curtis' (2002) work on transitional metalwork. Preservation of the cemetery is hampered by high humidity and the relatively soft nature of the brick.

The nature of the St. Louis No. 1 Cemetery is tied to the nature and founding of New Orleans as a multi-cultural and constantly changing city. Large influxes of immigrants, the development of Creole society, multiple epidemics (including yellow fever) are all reflected in the cemetery foundation and appearance through time. Initial burials associated with the city in the early eighteenth century were below-ground burials, but more socially prominent and wealthy individuals followed a European pattern of burial within the parish churches, including the parish church of St. Louis. This burial space was soon full and in 1784, additional interments within the church were banned. A new cemetery was ordered to be established outside the city limits, and came to be called St. Louis Cemetery No. 1. This was located on the northern edge of the city, consisting of a 300-square foot space that was officially approved on August 14, 1789 by Spanish Royal decree.

Population increases, coupled with the low elevation of the city, led to regulations necessitating above-ground burials. Above ground tombs were required to be built of brick and enclosed within walled cemeteries. The St. Louis Cemetery No. 1 was the primary burial location until the consecration of St. Louis Cemetery No. 2 in 1823, made necessary by high death rates resulting from Yellow Fever outbreaks (Peters, 2002). Individuals were still buried in St. Louis Cemetery No 1 after 1823, but this was no longer the most fashionable cemetery in the city (Fig. 8.2).

Fig. 8.2 Postcard of St. Louis No. 1, New Orleans' Oldest Cemetery. from 1930s—1950s (undated). Image Uncredited for "Post Card Specialties, New Orleans, La." "Curtich" / C. T. Art-Colortone"., Public domain, via Wikimedia Commons. https://upload.wikimedia.org/wikipedia/commons/0/07/Oldest_Cemetery%2C_St_Louis_No_1_New_Orleans_Curteich_Postcard.jpg

The cemetery consists three main tomb types: wall vaults, with multiple tiers of individual burial vaults forming an isolated block, usually as part of the perimeter enclosure wall; pediment tombs, consisting of multiple vault tombs with height greater than width; and simple tombs, or small mortuary structures containing one or more burial vaults within solid walls with length greater than width or height. Simple tombs can be either platform tombs (with a solid base or one on piers or columns), parapet tombs (with raised fronts creating a parapet that may have embellishment), sarcophagus tombs (resembling a sarcophagus with canted sides on a raised base), and step tombs (with steppe or corbelled top). Brick construction was the norm through the nineteenth century, but by the 1830s and 1840s, marble tombs were built intermixed with more modest brick monuments (*National Register of Historic Places Inventory—Nomination Form for St. Louis Cemetery No. 1*, 1975).

By the 1830s, St. Louis Cemetery No. 1 was a place for visitation by more than the friends and family of the dead. It was becoming a tourist location for those wanting to see above-ground tombs. Amongst these was Mark Twain (1957, p. 223), who described the cemetery as a city of the dead, noting that the dead are buried in vaults with "a resemblance to houses—sometimes to temples" that are "graceful and shapely; they face the walks and driveways of the cemetery." Twain also commented that the cemeteries were neatly kept, more neatly kept than the cities of the living.

By the 1870s, the socially prominent tombs had already been completed, and many of the single vault platforms had been altered to accommodate multiple burials within family vaults. The large society tombs were sites of the western side of the cemetery were commonly photographed and became the cultural touchstones for how the cemetery was viewed by those outside of New Orleans. But by 1900, with the loss of the older families through death and migration out of New Orleans, the cemetery began to fall into disrepair. By 1900, the tombs were beginning to fall apart and by the Great Depression, the Works Progress Administration was tasked with documentation of the tombs.

Tourism had been a feature of the cemetery since the end of the nineteenth century, but self-guided tours became more common in the mid-twentieth century. St. Louis No. 1 was featured on postcards such as that depicted in Fig. 8.2, further extending its reach as a touchstone for the culture of New Orleans. By this time, the cemetery had been rehabilitated, with ongoing restoration and maintenance projects that continue today.

As noted in the nomination form for the National Register of Historic Places, burials within St. Louis Cemetery No 1 span the history of New Orleans, from the times of Spanish control through the French and finally when New Orleans became part of the US. Also present is Marie Laveau, a notorious voodoo queen whose burial place is discussed more in Chap. 9.

It is important to place the St. Louis Cemetery No. 1 with the social context of New Orleans. New Orleans is a popular tourist destination, partially due the multi-ethnic interactions within its history. As one website says, "New Orleans is like no other place in the world, a city formed by the superstitions, traditions and history of Creoles, Spaniards, French, Irish, Italians, enslaved Africans and free people of color. The particular experiences to be had here have always been driven by independent thinkers, creative spirits and non-conformists. And that is never going to change"("New Orleans People & Cultures," 2021). This is identity that brings people in, one filled with Carnival, a never-ending party, and a place where on the surface, cultures can co-exist and be better for that experience. But identity is never as easy as a surface view, and the development of some elements of that New Orleans identity are founded in a history of unequal access and what Regis (1999, p. 472) identifies as "spatial apartheid."

According to Elliott and Pais (2006, p. 297) this can be drawn back to unequal economic development between the Gulf south and the North prior to the Civil War. Port cities in the Northeast grew more industrial in nature and attracted middle-class immigrants. By contrast, settlement of European aristocrats in the south "contributed to the development of a caste-like society… with an economy based almost entirely on agriculture (specifically cotton, sugar, and indigo), slave labor, and mercantilist exchange with Europe." The hierarchy built into the formulation of society in the South became more entrenched in the late nineteenth century, with railroads connecting the parts of the country and the decreased use of marine transport making southern cities more reliant on northern markets. This pushed the South into an economically peripheral status. Southern port cities such as New Orleans, Biloxi, and Mobile did not experience economic booms with de-industrialization in the North beginning in the 1960s, and been largely left "to unfold largely of their own inertia, undisturbed by mass in-migration from other parts of the country and the world" (Elliott & Pais, 2006, p. 297). White flight and uneven development in New Orleans has produced an "ever widening gap between rich and poor, which has its spatial correlates in the growth of depressed or blighted ghettos beset with widespread poverty and unemployment as well as environmental crises and crime" (Regis, 1999, p. 46).

Uneven development within New Orleans can be viewed through an environmental justice lens, with differential responses to environmental disasters such as Hurricane Katrina (Elliott & Pais, 2006) based on social race or socioeconomic status. African Americans, particularly African American women experienced an increase in disability, which Sastry and Gregory (2013) tie to post-traumatic stress disorder and generalized psychological stress. They identified a few potential factors in understanding differential experience. The first is that African Americans were more likely to live in areas that suffered the most damage from the hurricane, leading to higher rates of displacement and lowered likelihood of return, more adverse economic outcomes such as property loss and unemployment, and a general loss of community.

The hierarchical nature of society in the South meant that life was experienced differently based on both social race and socioeconomic status. This included the performance of funerals. Above-ground burials are the norm in New Orleans, necessitated by the low elevation of the region. This is true for numerous other locations around the world. But some funeral rituals, specifically the development of the Jazz Funeral or Second Line are unique to New Orleans. Coclanis and Coclanis (2005) tie this tradition back to funeral traditions within the African American community incorporating music in the decades following the Civil War. Also tied to this is the rise of benevolent and burial societies within the African American communities.

Fig. 8.3 "New Orleans Funeral" photo essay in 1969 Tulane University "Jambalaya", photos by Michael Smith and John Messina. From 1969 "Jambalaya" Tulane University yearbook, via [https://archive.org/details/jambalayayearboo74edit/page/400/mode/2up]. https://upload.wikimedia.org/wikipedia/commons/7/73/Jambalaya_1969_Jazz_Funeral_8.jpg

Second Line funerals consist of parades sponsored by benevolent societies and burial clubs. The formation of these parades is performative, with participants ordered according to their familiarity with the sponsoring club as well as the individual dead. The *First Line* consists of the club members and band, while the *Second Line* is composed of followers, members of other clubs, friends and family of the deceased, and neighborhood residents who walk next to or behind the club and band (Regis, 1999). Thus, the individual relationships as well as a community identity are celebrated with the parade (e.g., Fig. 8.3). Both groups are required for the performance to have its intended impact. Regis (1999, p. 472) argues that Second Lines "transform urban space, creating an alternative social order that private clubs actualize by 'taking it to the streets' in those very neighborhoods ordinarily dominated by the quotidian order of inner-city poverty and spatial apartheid." Regis (1999, p. 475) employs a specific understanding of landscape, examining landscape as "the relationship between two poles of existence, the quotidian and ritual life, foreground and background. Landscape is the 'cultural process' of mediation between these two dimensions of space." In effect, Regis is arguing for a phenomenological understanding of space. The ritual of the Second Line is positioned to change the interaction of individuals with each other and with their city within that Place of the Second Line. In 'taking it to the streets', participants in the Second Line "*become owners of the streets.* They experience a transcendent power in this collective celebration, which renews their strength and hope and allows them to face another Monday"(Regis, 1999, p. 478).

The landscape produced through these performances creates relationships that crosscut geographical and social differences, empowering individuals to walk though streets that might otherwise be off limits. It is also a way to incorporate newcomers. In discussing one parade in which she was a participant, Regis (1999, pp. 480–481) stated: "Their ritual work created a collective, embodied, and experiential knowledge of the area which educated me, and others who were new to the neighborhood, about its social structure... After hours of second lining, this place was remembered in the blisters on my feet, the fatigue in my calves, and the echoes in my ears of a tuba's driving rhythm and of conversations with fellow paraders, rather than in an external mapping whose abstract topography is experience distant."

The scale of the Second Line varies, but no matter the scale, the inclusion of both First and Second Lines is important for the creation of identity, and the performance of group identity. Whether it is a Second Line consisting of 3000–5000 individuals (Regis, 1999) or smaller, the parade routes take an individual to their final resting place in style, accompanied by their community. So the scale is not as important as the performance. When Lloyd Washington died in June of 2004, his family could not afford a cemetery plot or him. The remains of the former Tulane University groundskeeper and singer in the Ink Spots were cremated and kept in a Jazz club until October, when a space in St. Louis Cemetery No. 1 was donated (Coclanis & Coclanis, 2005). His Second Line consisted of a few dozen paraders that took him from St. Augustine's Catholic Church to the Cemetery, where he was interred in what became known as the "musician's tomb" ("Lloyd Washington was the first musician to be laid to rest in the new "musciians tomb" in St. Louis Cemetery No. 1," 2004).

8.3 Okunoin Cemetery (Mount Koya, Japan)

Okunoin Cemetery is the largest place of burial in Japan. Located in the forest of Mount Koya, the cemetery consists of tombs as well as a central temple complex that serves as the central location of the Shingon sect of Japanese Buddhism. The cemetery was founded by at least A.D. 816, and has been in use ever since. The area is centered on the temple dedicated to the veneration of Kūkai, the founder of Shingon Buddhism. Kūkai was entombed in the cemetery after his death in A.D. 835, and the journey to his temple is a journey organized through the burial place of numerous notables. The entrance to the cemetery is in front of a bus stop. Visitors enter the cemetery and follow a path along tens of thousands of graves. Famous samurai and Warlords are present in family plots. The monuments show family affiliation as well as occupation and change according to the cultural periods in which they were constructed. Modern monuments are more individualistic, including the grave marker for famous *fugu* chef Shimoneseki topped with a stone puffer fish. The area is incredibly scenic and often listed as a place for tourists to visit, in addition to more serious followers of Shingon Buddhism on pilgrimage to the temple. The overall path winds for approximately 2 km, and along this path more than 200,000 monks are said to be buried in the cemetery.

Fig. 8.4 Okunoin Cemetery, bodhisattva Jizō. Photo by Andrea Schaffer from Sydney, Australia, CC BY 2.0 <https://creativecommons.org/licenses/by/2.0>, via Wikimedia Commons. https://upload.wikimedia.org/wikipedia/commons/e/e5/Okuno-in_cemetery%2C_Koyasan_%283810261133%29.jpg

Common throughout the Space are numerous statues of the bodhisattva Jizō (Fig. 8.4). The Jizō cult is one of the most prominent folk cults in Japan, and statues associated with it are not rare in Japan. Jizō is associated with the earth or earthly realm, and is perceived to be a caretaker of living creatures (Donnere, 2019). Jizō is also considered to be a protector of children, both living and dead. Statues may be standing or sitting and usually depict a monk-like figure holding a ringed staff in the left hand. Seated figures are usually in a lotus position. These statues are commonly placed at the entrances to cemeteries. Often, these statues are dressed in red clothing, including hats and bibs. Red is a color of protection, and the bibs are placed on the statues by grieving parents as an offering for the Jizō to look after their lost children. One visitor described the decoration of the Jizō statues as "achingly beautiful" adding that "it was hard not to chock up over the quiet and heartbreaking dignity of it all. This was a landscape that had absorbed the pain and suffering of others" (Graham, 2016).

Continuing on the main path, the visitor reaches a temple called Gokusho. The farther from the entrance the visitor progresses, the more sacred the space becomes. Finally, the visitor crosses a bridge over the Tamagawa River to the main temple. Photography is allowed throughout the cemetery space with the exception of the sacred space beyond the bridge. Within the mausoleum, Kūkai is offered meals twice a day and he's said to remain in eternal meditation. The Torondo (or Lantern Hall) acts as a gateway to the mausoleum. This Space contains 10,000 donated lanterns, two of which have been burning continuously since A.D. 1088.

The social spaces and places created at Okunoin show a progression of access and an expectation that behavior will differ based on the location a visitor finds themselves. Visitors are encouraged to document and interact with the monuments throughout the majority of the cemetery, but access and activities are different around the Mausoleum at the center of the cemetery. That space, while democratic in the sense that it is open to the public, still follows Fisher's (2009) concepts of restriction of space in that social interactions and behavior are different in that space. The space is one of tradition, showing a continuation of use and interaction with the dead dating back to the ninth century. The tending of the lanterns for over 900 years shows the importance of this tradition, through world wars, natural disasters, changes in governments, those lanterns in that space have been maintained. The practice of pilgrimage to the cemetery are tied to the year 1088 as well, with the visit of Emperor Shirakawa. The earliest surviving map of the area dates to 1645, but the map does not show the entirety of tombs in the cemetery. Lukacs (2019) traces the history of creation of maps for pilgrims to the cemetery and Temple, noting the importance of iconography and the development of map making techniques. That maps were created from 1645 onward suggests that visitation to the site is an important element of Place making. Central to this is the transformation of the cemetery into a space for pilgrimage as well as a space for commemoration and interaction with the dead. Families visit for different reasons, asking the Jizō to look after their children, families visit their ancestors and important cultural personages. It is a way to reinforce community identity, both in a religious sense (through the temples and shrines throughout the cemetery) as well as a cultural and historical sense. In this cemetery, a visitor can come into contact with samurai, warlords, Jizō, and sushi chefs. It is a changing, evolving way of interacting the dead, the past and the present in a way that is identity-affirming yet different for each individual who visits.

8.4 Cimetière du Père Lachaise (Paris, France)

The Père Lachaise cemetery is the most visited cemetery in Paris, home to notables such as Jim Morrison, Edith Piaf, Oscar Wilde, Frédéric Chopin, Gertrude Stein and Amedeo Modigliani. It was established in 1804 and was envisioned as a place for the living. In fact, when it opened, it was considered to be too far from the city of Paris to be really useful, so administrators devised the transfer of the remains of Jean de La Fontaine and Molière in order to increase the numbers of people who wished to be buried in the cemetery. In 1805, 44 individuals were interred, with increasing numbers in 1806 and 1807. In 1812, 833 individuals were buried in the cemetery. Marketing campaigns increased in 1817, and the remains of Pierre Abélard and Héloïse d'Argenteuil were transferred to the cemetery. Portions of the abbey of Nogent-sur-Seine were used to make their monument. This monument has become a place for lovers or single individuals seeking love to visit and leave notes or letters at the crypt. These marketing efforts were successful, with increasing occupancy of the cemetery. By 1830, 33,000 individuals had been buried within the cemetery and it was expanded in 1824, 1829, 1832, 1842, and 1850 ("Père Lachaise: The Most Hauntingly Romantic Walk In Paris," 2020).

The changes in architecture of the monuments within the cemetery reflect changing tastes as well, including Eclectique, Haussmannien, Neobaroque, Second Empire, Académique, Empire, Monumental, Neobyzantine, Baroque, Gothique, Napoléon III, and Neoclassical ("Cimetière du Père Lachaise," 2020). Toussaint and Decrop (2013) conducted a series of interviews with visitors in an effort to understand the motivations for visiting the cemeteries. In their interviews, they found that the cemetery is viewed as an open-air museum of architecture and history. They also found it is a seen as a pilgrimage by visitors, both cultural or spiritual. They found the cemetery to be contested Space, a heterotopia along the lines of Foucault (1984 [1967]), "prized for its insulation and for its blurred sacred-profane dialectic"(Toussaint & Decrop, 2013, p. 419).

The cemetery contains a multitude of types of interments, including primary and secondary interments as well as at least one mass grave, known as the Communards' Wall (Mur des Fédérés). Along a wall of the cemetery, 147 Paris Communards were shot on May 28, 1871. The Paris Commune was a radical socialist, anti-religious, revolutionary government that ruled Paris from March 18—May 28, 1871. The capture of Emperor Napoleon III in September 1870 led to the collapse of the Second French Empire and the beginning of the Third Republic. Paris was under siege for 4 months and was known as a center of working-class radicalism. After surrendering to the Prussians on the 28th of January 1871, the chief executive of the French national government signed an armistice that disarmed the Army but not the National Guard. On the 18th of March, soldiers from the Commune's National Guard killed two army generals, and the Commune refused to accept the authority of the new French government. They ruled Paris for 2 months, until it was suppressed by the Army during "The Bloody Week", culminating in the killing of the 147 individuals killed within the Père Lachaise cemetery (Bassett, 2008). The army used cannon to demolish the gates of the cemetery and the naval infantry stormed into the cemetery. Fighting followed through the tombs until the defenders were surrounded and surrendered. They were taken to the wall of the cemetery in the area now known and commemorated as the Communards' Wall and shot.

These 147 individuals were not the total number of Communards killed in the cemetery, this number represents those who had surrendered and were subsequently massacred at the wall. Merriman (2014) presents a thorough account of the fighting and notes that hundreds bodies were laid in trenches (Fig. 8.5). Those killed at the wall, however, were placed in a natural hollow. They were not carefully placed, but instead thrown into pits and covered. Merriman, in tracing the battles that waged throughout the cemetery, notes that tombs and monuments had been damaged or destroyed by shells.

The cemetery was fundamentally seen as a place of interaction and visitation by the living. This allows for the examination of the social changes occurring during the time period as an *event*, which Bassett (2008, p. 898) defines as a rupture with the established order of things. These can happen in unpredictable ways. The cemetery can then be viewed as an "evental-site", or "the place where something new

Fig. 8.5 Memorial placed at the Communards' Wall. Photo courtesy of Père Lachaise Cemetery, CC BY-SA 4.0 <https://creativecommons.org/licenses/by-sa/4.0>, via Wikimedia Commons. https://upload.wikimedia.org/wikipedia/commons/3/32/P%C3%A8re-Lachaise_-_Division_76_-_Mur_des_F%C3%A9d%C3%A9r%C3%A9s_02.jpg

and decisive can happen" that concentrates the historical importance of the event. Essentially, this argues that context, both historical and geographical is important to how this is viewed within society. The cemetery was an important place celebrating large and small people, and the massacre and destruction of at least some of the structure in an aborted revolution provide synergistic meaning for the act.

Visitation of specific graves is conducted by individuals with different motivations. The grave of Jim Morrison (Fig. 8.6), for example, has become a place of visitation for music fans as well as a place where "his fans brought him to life again… and where his musicality, his lifestyle, and his poetic and philosophical legacy were evoked and propagated" (Margry, 2008, p. 143). The burial site was the focus on informal commemoration in the form of drinking, smoking, graffiti, and other activities that seem to go counter to the contemplative nature of a cemetery space. It was an opposition to the atmosphere that was intended for the cemetery. When Morrison died in 1971, a simple ceremony was conducted in less than 10 min and utilizing the cheapest available coffin. A simple marker was placed in 1972, though this was replaced later with a larger inscribed tombstone. Morrison's music and poetry began to increase in popularity with a new generation of listeners and readers, and with the inclusion of the Doors music in movies such as *Apocalypse Now*, more and more visitors began to alter the space around the grave. The 1991 film *The Doors* led to new mythologizing. The image of Morrison presented in this film presented

Fig. 8.6 Grave of Jim Morrison. Photo by ManoSolo13241324, CC BY-SA 4.0 <https://creativecommons.org/licenses/by-sa/4.0>, via Wikimedia Commons. https://upload.wikimedia.org/wikipedia/commons/9/9b/Jim_Morrison_tombe_1.jpg

Morrison as a figure who symbolized "the search for new forms of heightened consciousness and freedom"(Margry, 2008, p. 165).

In the 1990s, Margry (2008, p. 147) describes an "annexation" of the graves surrounding Morrisons as a Space in which visitors "drank excessively, smoked, took drugs, removed their clothes, had public sex, slept, and put into practice the nonconformist ideas and lifestyle championed by Morrison." The headstone functioned as a table, graffiti was common on not only Morrison's but also surrounding graves. In 2004, the cemetery's historian was quoted in a Guardian article as wishing to evict Morrison, but this was not possible. Morrison's grave was given in perpetuity to the family, and so he couldn't be relocated. In 2004, cemetery authorities placed a closed iron right and anchored barriers to keep visitors at a distance from the grave. While things have calmed down some at the grave, it is still the most visited grave and commonly makes lists of top tourist attractions (along with the Eiffel Tower and Louvre) to see in Paris (Margry, 2008).

Another notable grave is that of Oscar Wilde. Wilde died in November, 1900. According to Crowell (2012, n.p.), his tomb is "a living place; controversy still swirls around its conception, installation, suppression, desecration, veneration and, most recently, sanitation." The tomb's monument consists of Jacob Epstein's sculpture is that of an art-deco variation on a male Sphinx. According to Crowell, this monument "stands out like a sore thumb" an aesthetic that has prompted visitors to regard it as "future- rather than past-oriented, more modernist than Victorian, a monument to enlightened pride rather than retrograde shame." It should be remembered that Wilde was sentenced to 2 years' hard labor for "acts of gross indecency with young men". When he was released from prison in 1897, he moved to France. He died deeply in debt, with his funeral costs paid by friends. Wilde was initially buried in a leased gravesite at the Cimetiere de Bagneux, a modest cemetery outside of Paris. His friend and executor Robert Ross planned to move the remains to a different location later.

This site became a place of queer pilgrimage, as did the room where he died (Crowell, 2012). Ross began saving funds to secure the copyrights for Wilde's published works, and ultimately he was able to publish the *Complete works of Oscar Wilde* in 1908. The remains were moved to Père-Lachaise cemetery in 1909, with Epstein's monument unveiled and dedicated 5 years later. The choice of the Epstein as sculptor of the monument was contentious, but according to Crowell (2012), it made a certain kind of sense. Wilde had the reputation of being an "Irish sexual libertine", and Epstein had the reputation of being an American sexual libertine. The sphynx that Epstein created for Wilde's tomb borrowed heavily from Egyptian and Assyrian winged sphinxes that the sculptor had studied at the British Museum, while adding "an almost brutal simplicity of line and a sensual, otherworldly androgyny." The facial features resembled those of Wilde's, an inclusion that was regarded to be a memorial to his ill treatment in life.

One area of controversy was the genitalia of the statue, which upon installation in the Père-Lachaise cemetery had been covered in plaster by the cemetery staff. Ultimately, a "fig leaf solution" was installed over the genitalia. Ross pressed Epstein to allow for a bronze butterfly to be affixed to the monument. Epstein was enraged to the point of refusing to attend the unveiling in the cemetery. Legend says that the butterfly was removed after the unveiling ceremony by occultist Aleister Crowley and worn as a cod-piece. Whether this is true or not, the monument's genitalia was uncovered until 1961, when vandals chiseled the genitals off altogether. The desecration of the monument is identified as suppression by Crowell (2012, n.p.). She notes that "the suppression of Epstein's monument—and the way that suppression recreated the public shaming and silencing of Wilde through his trial and early death—influenced the particular rituals of mourning and admiration that grew up around the tomb at Père-Lachaise…. It is with Epstein's monument and its history of passionate admiration from pilgrims assumed to be either queer or queer-friendly that associate Wilde's memory; his status as queer martyr is inexorably bound up with the history of his tomb and its visitors." It was common practice to leave lipstick kisses on the monument, so much so that the practice was damaging the stone. In November 2011, a 100 years after the monument's installation, the Irish government partnered with Wilde's grandson to clear the monument and install a glass barrier that still stands. Admirers still leave kisses; they just now leave them on the glass barrier (Fig. 8.7).

Fig. 8.7 Oscar Wilde's tomb with modern glass barrier as it appeared in 2012. By Agateller—I took the photo myself with a digital camera. Previously published: 2012-05-22 <http://www.facebook.com/atkielski>, CC0, https://en.wikipedia.org/w/index.php?curid=35902755, via Wikimedia Commons

The Père-Lachaise cemetery is an example in the social construction of Space. It can be seen as sacred space for those with a religious motivation for visitation, but also as an important cultural Space for those that visit for different reasons. These need not be separate motivations, and a sense of community can be formed at the nexus between those with religious and cultural motivations. It is a place of contrasts, a living space, a turbulent space in which intersectional identities can clash and find expression. It is a contested space, and the synergistic elements of identity that can be gleaned through examination of the contests can lead to social understanding. As Gabriele (2006, p. 165) argues, this is a "liminal space… a place where life and death come together, past and present are fused, and permanence and transience may be said to coexist." Because the space is a cemetery, it is a place of memorialization, but this memorialization means different things to different visitors. Gabriele (2006, p. 175) sees the cemetery as reflective of our postmodern age, that being in this age is "by nature unfixed, and more importantly unfixable, ad that postmodern subjects are continuously crossing borders, repeatedly transgressing limits and constantly negotiating boundaries." This is a sentiment echoed by Toussaint and Decrop (2013, p. 419), who conclude their study by saying "it is the discovery of a very real place where the centuries mingle and merge together as one. Shut off from the outside world by its imposing walls, the symbolic and poetic cemetery offers an ever-changing, different kind of space and tourist experience."

8.5 Summary

Cemetery spaces are active, social spaces for the living. The four cemeteries presented here are sites that are important for nation building and group identity. Their histories include social strife and the development of new elements of national identity. The burial of the infamous and the famous within these cemeteries (such as at Cimetière du Père Lachaise) brings visitors in, and these spaces can be invaluable for cross-cutting social, religious, and class groups. The history of Arlington, for example, is one that starts out as a place of enslavement tied to the very founding father of the US (George Washington), with burials placed within the garden in order to completely change the landscape so that the mansion would never be livable again. Meiggs' intention in placing these burials close to the house was to effectively desecrate the house as a living space, but throughout the history of Arlington we see the development of this space as the most hallowed and most honored place of burial for US soldiers. By placing war dead in the garden, it had the effect of sanctifying the ground in a secular fashion. Continued burials in the cemetery only amplified this when the policy of concurrent return preferenced the quick recovery and repatriation of war dead. Arlington and other national cemeteries became an embodiment of the promise made to US soldiers not to be left behind, that their sacrifice was honored. Nation building can also be seen in Cimetière du Père Lachaise, with monument and wreaths left to those who were killed along the wall. These spaces become far more than simple places for the dead to rest, they become emblematic of how nations and communities have defined themselves. Okunoin cemetery has stood for hundreds of years, through massive amounts of social change. It has acted as a stabilizing and consistent monument to Japanese identity and spirituality, a place where connection with the past occurs, acting as a link between a modern industrial society and an ancient past.

References

Arlington National Cemetery, 2025. "Vietnam Helicopter Pilot and Crewmember Monument and Memorial Tree". Retrieved from: https://www.arlingtoncemetery.mil/Explore/Monuments-and-Memorials/Vietnam-Helicopter-Pilot-and-Crewmember-Monument#:~:text=Over%204%2C800%20helicopter%20pilots%20and,end%20of%20the%20Vietnam%20War. Accessed March 3, 2025.

Bassett, K. (2008). Thinking the event: Badiou's philosophy of the event and the example of the Paris Commune. *Environment and Planning D: Society and Space, 26*, 895–910.

Berman, R. (2017, August 18). The Nation's Official Memorial to Robert E. Lee Gets a Rewrite. *The Atlantic*. Retrieved from https://www.theatlantic.com/politics/archive/2017/08/a-national-memorial-to-robert-e-lee-gets-a-rewrite/537237/

Christensen, A. M., Passalacqua, N. V., & Bartelink, E. J. (2014). *Forensic anthropology: Current methods and practice*. Elsevier.

Cimetière du Père Lachaise. (2020). https://en.parisinfo.com/paris-museum-monument/71470/Cimetiere-du-Pere-Lachaise. Accessed December 9, 2020.

Coclanis, A. P., & Coclanis, P. A. (2005). Jazz funeral: A living tradition. *Southern Cultures, 11*(2), 86–92.

Crowell, E. (2012). Oscar Wilde's Tomb: Silence and the aesthetics of queer memorial. *Branch: Britain, Representation and Nineteenth-Century History*. Retrieved from http://www.branch-collective.org/?ps_articles=ellen-crowell-oscar-wildes-tomb-silence-and-the-aesthetics-of-queer-memorial

Curtis, S. O. R. (2002). *St. Louis I Cemetery, New Orleans: Transitional composite metalwork (19th-C)*. MS, University of Pennsylvania.

Donnere, A. S. E. (2019). Finding a place for Jizō: A study of Jizō statuary in the Buddhist Temples of Sendai. *Japanese Journal of Religious Studies, 46*, 151–172.

Elliott, J. R., & Pais, J. (2006). Race, class, and Hurricane Katrina: Social differences in human responses to disaster. *Social Science Research, 35*(2), 295–321. https://doi.org/10.1016/j.ssresearch.2006.02.003

Fisher, K. D. (2009). Placing social interaction: An integrative approach to analyzing past built environments. *Journal of Anthropological Archaeology, 28*, 439–457.

Foucault, M. (1984 [1967]). Of other spaces: Utopias and Heterotopias. *Architecture/Mouvement/Continuité*, 1-9.

Gabriele, J. P. (2006). Plotting postmodern being in Itziar Père Lachaise. *Studia Neophilologica, 78*, 165–175.

Gaughan, A. J. (2012). The Arlington cemetery case: A court and a nation divided. *Journal of Supreme Court History, 37*(1), 1–21. https://doi.org/10.1111/j.1540-5818.2012.01280.x

Graham, A. H. (2016). The eerie grave of 200,000 monks. BBC. Retrieved from: http://www.bbc.com/travel/story/20160308-an-eerie-night-with-200000-dead-monks. Accessed December 9, 2020.

H.J. Res. (2020). 105—Redesignating the Robert E. Lee Memorial in Arlington National Cemetery as "Arlington House". H.J. REs. 105 Stat. Retrieved from https://www.congress.gov/116/bills/hjres105/BILLS-116hjres105ih.pdf

Joint Resolution: Dedicating the Lee Mansion in Arlington National Cemetery as a permanent Memorial to Robert E. Lee, Volume 69, S.J. Res. 62, Pub. L. No. Public Law 84-107 190-191 (US Government Printing Office June 29, 1955). Retrieved from https://www.govinfo.gov/app/details/STATUTE-69/STATUTE-69-Pg190/summary

Labbe, S. (2018). Trampling Mrs. Lee's Roses: Union Soldiers at Arlington. The Gettysburg Compiler: On the Front Lines of History, 2018. Retrieved from https://cupola.gettysburg.edu/cgi/viewcontent.cgi?article=1390&=&context=compiler&=&sei-redir=1&referer=https%253A%252F%252Fscholar.google.com%252Fscholar%253Fstart%253D10%2526q%253Dmeigs%252Band%252Barlington%2526hl%253Den%2526as_sdt%253D0%252C25#search=%22meigs%20arlington%22

Lloyd Washington was the first musician to be laid to rest in the new "musicians tomb" in St. Louis Cemetery No. 1. (2004, October 24). The Times Picayune.

Lukacs, G. (2019). Mapping Mount Koya, one of Japan's most Revered Pilgrimage Sites. *The Cartographic Journal, 56*, 4–17.

Margry, P. J. (2008). The pilgrimage to Jim Morrison's grave at Père Lachaise cemetery: The social construction of sacred space. In P. J. Margry (Ed.), *Shrines and pilgrimage in the modern world: New itineraries into the sacred* (pp. 143–172). Amsterdam University Press.

Merriman, J. (2014). *Massacre: The life and death of the Paris commune*. Basic Books.

Murphy, R., & Stephens, T. (2020). *Section 27 and Freedman's village in Arlington National Cemetery: The African American history of America's most hallowed ground*. McFarland and Company Inc.

National Register of Historic Places Inventory—Nomination Form for St. Louis Cemetery No. 1. (1975). Retrieved from https://npgallery.nps.gov/GetAsset/83e93350-9bb3-4e30-bded-2ff73eeb3032/

New Orleans People & Cultures. (2021). Retrieved from https://www.neworleans.com/things-to-do/history/people-and-culture-of-new-orleans/

Père Lachaise: The most hauntingly romantic walk in Paris. (2020). https://www.perelachaisecemetery.com/history/

Peters, J. A. (2002). *Modeling of Tomb Decay at St. Louis Cemetery No. 1: The role of material properties and the environment*. (MS). University of Pennsylvania. Retrieved from https://repository.upenn.edu/cgi/viewcontent.cgi?article=1306&context=hp_theses

Poole, R. M. (2009). *On hallowed ground: The story of Arlington National Cemetery*. The Story of Arlington National Cemetery.

Regis, H. (1999). Second lines, minstrelsy, and the contested landscapes of New Orleans Afro-Creole Festivals. *Current Anthropology, 14*(4), 472–504.

Sastry, N., & Gregory, J. (2013). The effect of Hurricane Katrina on the prevalence of health impairments and disability among adults in New Orleans: Differences by age, race, and sex. *Social Science & Medicine, 80*, 121–129.

Torreon, B. S. (2019). Defense primer: Arlington Cemetery (IF11362). Retrieved from https://crsreports.congress.gov/product/pdf/IF/IF11362/2

Toussaint, S., & Decrop, A. (2013). The Pere-Lachaise Cemetery: Between touristic experience ad heterotopic consumption. In *Paper presented at the Marketings places and spaces: Shifting tourist flows. 5th advances in tourism marketing conference, Vilamoura, Portugal*.

Twain, M. (1957). *Life on the Mississippi*. Sagamore Press Inc.

Chapter 9
The (In)famous Dead

Abstract This chapter explores how death, burial, and commemoration shape historical narratives, social movements, and cultural identities through the mortuary performances and legacies of specific individuals. Using case studies, it examines how death rituals surrounding figures like Marie Skłodowska Curie, Frank Little, Marie Laveau, Richard Leroy McKinley, John Byres, Richard Legg, Emmett Till, and Elvis Presley have left lasting imprints on collective memory. The case studies here highlight the tension between individual and nationalistic needs as well as the use of the body for political purposes and as reflections of pop culture. Through these cases, the chapter explores how memorialization practices embody tensions between public commemoration, private wishes, and historical contexts, shedding light on the broader social forces at play in constructing historical legacies.

Just as the body can be manipulable as material culture and embody relationships, the body of well-known or public figures can embody large-scale social change. Some of these relationships will be discussed elsewhere in this volume, particularly the use of the bodies of political leaders for nationalistic and political ends. Here, I want to delve into the mortuary program used for a few famous and infamous individuals, all but one within the US. These individuals lived in different times, but all embodied significant social processes, political and social changes experienced by society. These are presented as examples of burials that would present as atypical in an archaeological sense, for different reasons. There are differences in how the body is prepared, when and where it is buried, and in the mortuary rituals used. The differences in mortuary program is indicative of a different import placed on these rituals by the living after their deaths, not necessarily representative of their social stature during life. Individuals have names, have identities, we can see parts of ourselves in them, and so their individual death and burial can be instrumental to understanding and *feeling* the social import of larger movements.

In selecting individuals for these case studies, I wanted to highlight social movements, technological changes, and the historical context of these individuals. In making these selections, there is a bias towards men in the case studies. Marie Curie is presented, as is Marie Laveau, but the rest are men. This is partly due to the social

A. Osterholtz, *The Bodies Poetic*, Bioarchaeology and Social Theory,
https://doi.org/10.1007/978-3-031-85998-4_9

processes that I wanted to highlight. In looking at the development of labor in Montana (where I was born) and the more general role of labor unions within US history, an obvious case study is the death and burial of Frank Little. In examining the development of (and tensions surrounding) nuclear technology, the death and burial of Richard McKinley, John Byres, and Richard Legg are also obvious choices as they were the first deaths due to an atomic accident within the US. In these cases, the focus on these individuals is due to their central role within these movements. An examination of the central role of lynching in the US would be remiss without an examination of the death and burial of Emmett Till. Another important aspect of choosing these case studies was the availability of popular sources relating to the death, burial, and postmortem influence of these individuals. Understanding the development of the Memorial Garden at Graceland, and the overall importance of Elvis Presley would not be possible without utilizing popular sources such as the New York Times and Rolling Stone articles. In choosing these individuals, I wished to highlight both the popular and academic viewpoints on their impact on culture and technology, without doubt, the impact of thousands of anonymous individuals have had significant impacts as well (many of them women and people of color), and in future years I plan to expand this approach substantially. The bodies of political leaders will be discussed in a separate space, Chap. 14, in relation to nation building with bodies.

9.1 Marie Laveau

Marie Laveau "is no myth; she is no mere legend. While her reputation precedes her as the notorious Voudou Queen of New Orleans, in reality, she was a free woman of color who ruled the city during antebellum New Orleans... She was a devout Catholic, an independent businesswoman, a mother, and healer who lived her life in accordance with the corporal works of mercy. Stories abound about her magickal prowess, freeing men from the gallows and healing the sick from the brink of death. Her belief in Catholicism guided her life as well as her magic in such a distinct way that people from all over the world are inspired by her spirit and her story. This phenomenon is evidenced by the fact that her grave site is purportedly one of the most visited pilgrimage sites in the United States, second only to the King of Rock and Roll, Elvis Presley"(Alvarado, 2020, p. 1). An amazing woman in life, she has become the focal point of what Long (2006) calls the *Laveau Legend*, tied in with the history of New Orleans. Laveau has been recognized as the Mother of American Voudou, a religious tradition drawn from the religious traditions of enslaved African-Americans, exploitation, and reconstruction. Long argues that part of the mystery surrounding this woman and her place in history are due to the fact that she was illiterate and left no records in her own words. There are also no known photos or paintings of her during her life, the portraits and drawings of her are based on speculation. As Fandrich (2005, p. 2998) notes, even name spelling and use during the period of Marie Laveau's life varies, and her name appearing variously as Marie Laveaux, Marie Paris, Marie Laveau, the Widow Paris, Veuve Paris nee Laveau,

Marie Glapion nee Laveau(x), Marie Glapion, Dame Christophe Glapion, and Maria Labeau. She has therefore become something of a blank slate for history to interpret.

Long (2006) produced an a biography of Marie Laveau based on historical documents dating back to the 1830s, including both church and municipal records. Marie Laveau was born a free woman in New Orleans. Her exact date of birth is unknown as there is no known birth certificate but both Long (2006) and Fandrich (2005) reconstruct her birthyear to be 1801. She first appears in official records with her marriage to Jacques Paris on July 27, 1819, where she was said to be under the age of 21 years. Marie Laveau and Jacques Paris were married at the St. Louis Cathedral in New Orleans. They had two children, both of whom likely died in childhood (Long, 2006). Jacques disappears from official records sometime around 1826, but whether he died or deserted Marie is unknown due to missing records. Marie supported herself as a hairdresser to wealthy white women, she would go to their houses to perform her work, and is supposed to have heard confidential information during those sessions that she could use to her benefit. After Jacques disappeared, Marie entered into a relationship with Louis Christophe Dominic Duminey de Glapion that lasted until his death in 1855. Together they had seven children. Marie was also heavily involved with raising her grandchildren when they came along; she was also known to care for orphans in New Orleans. Glapion was white, and while their children were minors, the couple began to circumvent the laws that would present a white man from leaving his wealth to his non-white descendants. The couple also purchased and sold enslaved individuals through the 1850s.

By spring 1854, the family was falling into debt. Christoph Glapion died on June 26, 1855, further endangering Marie Laveau's financial and social position. The home they lived in was mortgaged, and legally, he was prohibited from bequeathing possessions and property to his non-white family. Marie Laveau, though, was nothing if not resourceful. Long (2006) notes that the Laveau Legend tells that Marie Laveau leveraged the knowledge that she learned as a hairdresser to rich white ladies in conjunction with a thriving network of informants among their servants. Ultimately, these wealthy white ladies became coming to her for consultation and magical assistance. From these consultations, she was able to become wealthy. But the Legend doesn't map onto court records. Long (2006) argues that she was not impoverished, but that her status in the time before the Civil War was in flux as was the position of Creole women of color in general.

The Laveau Legend says that Marie Laveau was raised in the Voudou tradition, where she learned from her mother and grandmother. Voudou in New Orleans held an interesting position. It was both feared and respected, as Voudou took on elements of Catholicism, and practitioners held atypical positions of power within society. In the 1850s, with fears of a Civil War looming, white New Orleanians grew more fearful of practitioners of Voudou, and raids on Voudou ceremonies were not unknown. Marie Laveau's name does not appear on roster of those arrested, but she did lodge an official complaint that one of her wooden statues had been taken by the Municipality Guards in 1850. In newspaper articles regarding the complaint she was referred to as a leader of the Voudou women. Harassment of Voudou practitioners became more common throughout the 1850s, but it is not until 1859 that there is another official mention of Marie Laveau, where she was referred to as a Voudou Queen by a neighbor who complained of Voudou rituals occurring in a local grocery store at night (Long, 2006).

Oral Histories taken in the 1930s can be used to understand the role of Voudou in the latter parts of the nineteenth century. The original Voudou Queen was a woman known as the Widow Paris, who was an elderly woman in the 1870s and died in 1881. Some of those interviewed spoke of the Widow Paris, but many described a younger woman known as "Marie Laveau". One visitor to the house was Marie Dédé, who was born in 1866 and grew up in the neighborhood where she played with Marie Laveau's grandchildren. She described altars and saints' pictures. Others described alters in the front room intended for good luck magic. Others interviewed described small private ceremonies with offerings of herbs, food, alcohol, flowers, candles, and money. These ceremonies were attended by a diverse congregation. In addition to the ceremonies, Marie Laveau was known to give private consultations and to make and sell gris gris (charms). She had a universal kind of appeal, with her clients consisting of "women and men, black and white, enslaved and free, the rich and the famous and the poor and unknown alike "(Fandrich, 2005, p. 294). Those escaping slavery credited her intervention and charms. Marie Laveau seemed to have insight into everything, as an obituary stated, "[t]here were very few secrets of any nature which she did not know. Wherever there was a skeleton in the family closet Marie held the key. But she was true to the confidence reposed in her and did not turn her knowledge to any unjust advantage." ("Dead Voudou Queen," 1881, p. 2).

Marie Laveau died on June 15, 1881 of old age. She was interred within the Widow Paris tomb in St. Louis Cemetery No. 1 in an above-ground tomb that has become a place of pilgrimage for both Voudou practitioners and tourists alike (Fig. 9.1). After her death, six obituaries were published, five in Louisiana but also one in the New York Times. It states in part, "Besides knowing the secret healing qualities of the various herbs which grew in abundance in the woods and fields, she was endowed with more than the usual share of common sense, and her advice was oft-times really valuable and her penetration remarkable"("Dead Voudou Queen," 1881, p. 2).

Marie Laveau has become a cultural figure, her tomb a place where the curious can interact with the unknown. Practitioners of Voudou leave offerings of flowers, candles, perfume, liquor, carnival beads, tobacco, and money at her tomb (Fandrich, 2005). Curious tourists not only go to see her grave but go to see how those who follow her interact with that grave. It is a safe place to interact with a religion that is poorly understood and still feared by many. Tours through the St. Louis Cemetery No. 1 feature a stop at her grave. Chris Grant of New Orleans Spirit Tours estimated that his company took approximately 12,000 people to her grave in 2003, and that that number has more than doubled between 1993 and 2003 (Kinzer, 2003). The increasing number of visitors may also be linked to an increase in the number of Voudou followers, who see the grave as an important place of pilgrimage (Kinzer, 2003). Visitors often leave offerings at her tomb, adding to the material culture associated with her prominence and continued role in the community (Fig. 9.2).

Fig. 9.1 Photo of Marie Laveau's tomb. Photo by Dan Soto, CC BY-SA 3.0 <https://creativecommons.org/licenses/by-sa/3.0>, via Wikimedia Commons. https://upload.wikimedia.org/wikipedia/commons/6/64/Tomb_of_Marie_Leveau.jpg

A growing acceptance of Voudou may also be bringing about a new view of Marie Laveau, who Kinzer (2003, p. 28) describes as a "free woman of African, American Indian and European blood who thrived at a time when New Orleans was one of the most racially integrated cities in the United States," and notes that believers in Voudou see her as "a healer and spiritualist who cultivated an unparalleled network of informants among slaves and servants, and seemed to know every secret in town." She is seen as an excellent ambassador of Voudou to a diverse audience, with the characterization of Voudou as a positive way to deal with conflict, a way to engage with struggle and overcome it, just as Marie Laveau overcame all of the conflict and struggle in her life. She is thought by many to be the mother of Voudou within the United States.

From a Poetics point of view, the interaction with Laveau's tomb is a reflection of both individual and collective identity. A growing acceptance of Voudou is seen in the accumulation of offerings at her tomb. Laveau is also celebrated as an individual, with recognition throughout New Orleans of her resiliency and her ability to

Fig. 9.2 "Offerings left at the tomb of Voodoo Queen Marie Laveau" (June 2011). Photo attribution: Derek Bridges, CC BY 2.0 <https://creativecommons.org/licenses/by/2.0>, via Wikimedia Commons. https://upload.wikimedia.org/wikipedia/commons/9/92/Marie_Laveau_Tomb_2011.jpg

succeed in a time and place where the odds were stacked against her. In celebrating Laveau, Alvarado (2020) sees a legitimization of their own beliefs and a celebration of the legitimacy of oral history (particularly in African American and Indigenous contexts). This legitimization is expressed materially through the leaving of tokens and the visitation of the tomb. It is a celebration of local identity.

9.2 Frank Little

The death and burial of Frank Little in 1917 is tied to the contentious history of labor unions and the history of Labor Unions is tied to the history of Butte, MT and the Anaconda Mining Company. Butte was founded formally in 1864, but mining camps had existed on its current location prior to this. Founded in 1881 by Marcus Daly, the Anaconda Mining Company was a powerhouse in the production of copper that fueled electrification across the US. Butte was initially founded as a silver and gold mining camp in the late 1800s, but grew remarkably quickly when Daly shifted focus from these ores to copper sulfate. By 1885, the population had swelled to over 47,000 people, with about half of them listed as foreign-born (McGrath, 1976a). Copper sulfate required smelting of the ore before it could be used in industry, and instead of shipping the ore elsewhere decided to build a gigantic smelter in

Anaconda, MT, near Butte. As copper mining exploded and the number of mines increased in the area, workers flooded into Montana from all around the world, creating an early multiethnic and cosmopolitan environment.

By the turn of the twentieth century, Butte was the largest city between Chicago and San Francisco (Lee, 2015). In fact, the oldest functioning Chinese-American restaurant in North America is the Pekin Noodle Parlor in Butte, Montana (founded in 1911). Butte was booming, relatively well-paid positions could be had in the mines as well as the ancillary professions that the miners needed, such as rooming houses, laundries and restaurants. By 1870, nearly 10% of Montana's population was Chinese-American. Hum Yow and his grandfather Tam Kwong Yee opened a Chinese mercantile o the eastern edge of the city's Chinatown. By 1911, the top floor had become the Pekin Noodle parlor with the first floor housing first a gambling club and later an herbal shop (Fawcett, 2016). Immigration of individuals from around the world was spurred by relatively high wages in the mines, which led the nation in 1900 (Work, 2006). These groups tended to form ethnic enclaves within the city but worked together in the mines. Intersectional identities interacted through the common activities and rituals associated with mining activity. One of these common activities was food and cooking. The *Butte Heritage Cookbook*, a collection of recipes from Butte published to celebrate the bicentennial presents recipes from local Native American traditions, as well as the American South, England, Finland, France, Germany, Greece, Hungary, Ireland, Italy, Lebanon, the Netherlands, China, Korea, Japan, Poland, Scandinavian countries, Scotland, Spain, Wales, and Yugoslavia (still a geographical entity at the time of publication) (McGrath, 1976b). According to McGrath (1976a, p. vii), individuals of different nationalities were united in food. "This is a town where the Cornish pasty and Serbian povitica share equal billing… where an Italian immigrant family made famous a Spanish tamale… and the Chinese New Year drew hundreds of late night visitors into the "alley" to join in the celebration." These sections are headed with a brief history of the ethnic group represented within Butte.

Increased need for copper meant prosperity for Butte's mine owners. It also gave laborers a degree of power in negotiation that was uncommon for the era. In 1878, the Butte Workingman's Union was formed, the first organized union in Montana. They organized strikes and later reorganized as the Butte Miners Union (BMU), an organization that would form a counterpoint to the Industrial Workers of the World (IWW) later on. In 1903, the BMU, working with the Western Federation of Miners successfully negotiated to reduce the miners' shifts from 10 to 8 h. They fought for higher wages in 1907, seeking an increase to $4.50/day (an increase of $1/day). Butte Miners went on strike, ultimately settling for $4.25/day. Butte was widely viewed as a solid union city (Dubovsky, 2000). Disagreement between miners and the miners' unions somewhat limited union success in the early twentieth century. One large mining company opened a non-union mine, leading to a strike in 1913. The following year, the Miners Union Hall was destroyed by dissenters, while the Mining Company's employment office was dynamited. Private security in the form of hundreds of Pinkerton and other guards were brought in and martial law was declared. In 1916, wage negotiations resulted in a rate of $3.75/day but also tied wages to the price of copper, falling with the price of copper.

Tensions were incredibly high, particularly after the Granite Mountain/Speculator Mine disaster in June 1917. Karney, writing in 1917 (p. 223), claimed that "the greed of the mine owners" was responsible, continuing that "[t]he dividends of the shareholders are considered of more importance than the lies of the miners." Mine owners had not installed appropriate safety devices within the tunnels of the Speculator Mine 2000 feet below the surface. A fire broke out, sending smoke flames and poisonous gasses through the tunnels. During an ensuing rescue attempt, carbon monoxide contaminated the air supply. They attempted to put the fire out with water, which created steam, further burning those in the tunnels. At least 168 men died in this disaster, many of whom were unidentifiable. The Granite Mountain Memorial in Butte commemorates those who died in the accident. This disaster set the stage for general strikes, with over 15,000 miners actively involved in union activities (Karney, 1917).

It was in this fractious and violent environment that the IWW arrived in Butte in the form of Frank Little (Fig. 9.3). ittle was a well-known union organizer for the IWW who arrived in Butte just weeks before his lynching. On August 1st, 1917, six men broke into the boardinghouse where Little was staying, beat Little in his room and forced him to leave with them. He was tied to the car's rear bumper and dragged in the street. Ultimately, he was taken to the Milwaukee Bridge and hanged from a railroad trestle. His body was found bearing a note claiming that this was the first and last warning and containing the initials of other union leaders. Little's murder was never solved.

After his body was discovered the IWW determined that Larry Duggan, an undertaker active in with unions would prepare the body. This may have been so that the IWW would continue to have access to the body. Duggan laid out the body for public view and photographed the body extensively. A death mask and two postcard photos were taken of the body. These photos showed the extent of the beating and violence that had been inflicted upon Little before and during his lynching. These photos were displayed in public places, and when removed by the authorities were promptly replaced with new ones "emblazoning one more victim of the class struggle to enrage public opinion" (Botkin, 2017, p. 302). Copies of these photos were sold later in fundraising drives for the IWW.

Little's family in Oklahoma began to make funeral decisions. Little's father Lon asked to have Frank's body shipped back to Oklahoma, but news of the lynching had reached that charged area as well, and IWW board members worried that it would lead to violence. They urged, instead, that the body be buried in Butte. This began to be arranged for August 5 to be held at Mountain View Cemetery. The mining companies who controlled the newspapers attempted to downplay the announcement, placing it between baseball scores and want ads, but formal advertising was unnecessary. IWW members began streaming into Butte, quicker travel made possible by railroads. Thanks to faster communications by telegraph, demonstrations were planned across the country.

Fig. 9.3 Frank Little photo in Industrial Workers of the World Photograph collection. UW Digital Collections, No restrictions, via Wikimedia Commons. https://www.flickr.com/photos/8337233@N06/6942578308/

Despite having no roots in the community, the funeral is still considered to be the largest in Butte history. An estimated 10,000 workers lined the route of the funeral procession, which was followed by an additional 3500 more people. United in this funeral march were union members from various ethnic backgrounds, e.g., Irish followed by the Finnish. Little's death and mortuary processing (in the form of documentation, viewing, and publicization), brought together the disparate unions in a way that perhaps nothing else could: it gave them a common enemy in the funeral procession, committee members of the Butte Miners Union marched in front of the Electrical Workers' Union, behind them was the Street Railway Union, Hod Carriers' Union, Workingmen's Union, and other craft unions. Bringing up the rear was a large contingent of the Butte Miners Union, 2500 individuals walking in a tight formation directly in front of the hearse for the four-mile funeral procession.

No minister or family member was present for the graveside ceremony, instead a eulogy was delivered by a Union member who expressed outrage at the murder. Other speeches underlined that Little had died "for the solidarity of the working class"(Botkin, 2017, p. 308). He remains buried in the Mountain View Cemetery in Butte, MT, his tombstone reads "Frank Little: 1879-1917, Slain by Capitalist Interests for Organizing and Inspiring His Fellow Man"(Botkin, 2017, p. 347).

After Little's death, a general strike began in Butte, shutting down the largest copper mine in the world, the Anaconda Mine for 6 months. Frank Little holds an important place in the history of labor struggles, and in the history of the IWW. As late as 1950, Little's portrait was found in a Seattle, WA IWW office with the caption "Tell the Boys I died for my class"(Botkin, 2017, p. 346). The use of violence as a social communication is foundational to the Poetics model (as discussed elsewhere in this volume).

The violent action here drew disparate communities together in opposition to a common enemy (the Mining companies). They formalized and ritualized the funeral in a way that has not been rivalled in Butte History. They created a mortuary ritual that is contextualized to this time and space. Butte has a history of being a rough place, founded by tough men and women who came for opportunity in the Mines and who created a unique third space. *Cosmopolitan* is not a word typically associated with Montana, but in the early twentieth century, encountering individuals of different cultures and faiths would have been a very regular occasion. These individuals were bound together, largely due to organized labor, a recognition that despite differences in background (what came before), they had a new shared identity (the third space). This can be viewed through a postcolonial lens. Like most postcolonial spaces, it was not a calm one, but one that is continually negotiated and renegotiated. Chinese laborers faced significant racism, both structural and cultural, for example. But people still came together for a moment, a Space was created for these sectional identities. It is unclear, however, if without the death and particularly the funeral procession of Frank Little, these disparate groups would have formed as cohesive a group as they did. Visitations to the grave continue, with people leaving mementos such as work boots, union hats, coins, and every few weeks, a bottle of whiskey. Ellen Crain, the director of the Butte-Silver Bow Public Archives is quoted as saying "He is, in many ways, Butte" (Garcia, 2019). The multi-ethnic identity of Butte is wound up in the labor movement and in the death and burial of a single man who was not a native to the area. The celebration that continues argues that identity in a third space is a negotiated one, tied to a specific contextualized event.

9.3 Marie Skłodowska Curie

Marie Skłodowska Curie died On July 4, 1934 from aplastic anemia, the result of her long-term exposure to radiation during her lifetime of research. After her death, she was buried in the same cemetery as her husband, mother-in-law, and father-in-law, in Sceaux, France. Her sister and brother came from Warsaw along with native soil from Poland, which they sprinkled into the open grave during the service. The funeral and burial were private affairs, in accordance with Curie's wishes ("MME. Curie is Dead; Marter to Science," 1934). This was not to be her final resting place though. In honor of their achievements and role as a national icon, her remains and those of her husband Pierre were moved to the Panthéon, France's National mausoleum. She was the first woman to be accorded the honor based on her own achievements ("Marie Curie Enshrined in Pantheon," 1995). Of particular interest here is the nature of the secondary burial within the Panthéon (Fig. 9.4). The Panthéon was built from 1758—1790. It was therefore unfinished when the French Revolution began in 1789. The structure was repurposed as a temple devoted to liberty and set aside as a place of burial for notables. The ashes of Voltaire were placed in the Panthéon in a ceremony on July 21, 1791, followed over the centuries by artists, politicians, and scientists. It was here that the remains of Marie and Pierre Curie were interred. Their remains were radioactive and so were laid to rest in lead-lined caskets.

A private woman in life, focused on her science and lab work, she had a very quiet funeral, and little could be gleaned about it from her biographies. Pasachoff (1996) devotes little more than a paragraph to the actual event. What shines in

Fig. 9.4 Photo of the tomb of Marie Skłodowska Curie (And Pierre Curie) in the Pantheon. Photo by ManoSolo13241324, CC BY-SA 4.0 <https://creativecommons.org/licenses/by-sa/4.0>, via Wikimedia Commons. https://upload.wikimedia.org/wikipedia/commons/4/4f/Pierre_et_Marie_Curie_tombes.jpg

obituaries and her biographies is the impact and importance of her work and contributions to science and medicine. The discovery of radium had significant impacts on cancer research. A Dr. Barker is quoted in the New York Times obituary as saying "she accomplished a work which can truly be termed great. She made one of the greatest contributions to science in her domain. She was unquestionably one of the outstanding scientists. She was one who brought relief to countless thousands and passed on her knowledge to others of the scientific world"("MME. Curie is Dead; Marter to Science," 1934, p. 17). Aside from being the first person to win Nobel Prizes in two different categories (physics and chemistry), Curie was the first female professor at the University of Paris, a post she began in 1906 after the death of her husband in a carriage accident. She was instrumental in the development of battlefield radiology medical units that were set up during World War I, an innovation that allowed for better treatment, by the end of the War, over one million injured soldiers had utilized them. Since her death, she has come to personify the importance of modern science, the confluence of chemistry and medicine. She has also become a role model for women in the sciences and a way to introduce young girls into STEM ("3 Ways Murie Curie Would Fix the Plight of women in STEM Fields," 2020; "Beyond Curie: About the Project," 2020).

Of interest here is the secondary burial of a very private woman in a very public place, contrary to her own expressed wishes. She had indicated a desire for a quiet and simple burial, instead she became an attraction visited as a top tourist attraction in Paris. This is an excellent example of the concept that the dead don't bury themselves (Parker Pearson, 1999). Her remains, and the manipulation of those remains, took on a new role. She ceased to belong to her friends and family, and became the intellectual property of France, displayed along with heroes of the French Revolution and other luminaries. She is one of only five women buried at the Panthéon, and the first to be placed based on her own merit.

Her interment is exceptional in one other sense, the container in which she is buried. As her remains were still radioactive 60 years after her death, she needed to be buried within a lead-lined container. This is a protective element, so that visitors to the Panthéon are not exposed to radiation similar to that which caused not only Marie Curie's death but also likely led to her daughter's ultimate death duc to leukemia (Pasachoff, 1996). This allows the remains to be in the public without exposing the public to the dangers created by those remains. The contamination of Curie's remains is a reminder that modern science can have unintended consequences. These consequences are hidden from visitors but would be evident if the tomb were to be opened. This fear of contamination underscores the impact of science on mortuary processes. By the time of the reinterment of her remains, the dangerous nature of radioactivity was known. While it was important to memorialize her in a place of power (and by doing so emphasize her role within the French Academy), it was equally important to acknowledge a new scientific reality of the dangers of radioactive materials. This recognition of contamination further explored with the next case study, the SL-1 dead.

9.4 The SL-1 Dead: Richard Leroy McKinley, John Byres, and Richard Legg

On January 3, 1961, Richard Leroy McKinley became one of the first three men in the US to die in a nuclear reactor explosion along with John Byres and Richard Legg. Byres and Legg were buried in their hometowns, but McKinley was buried in Arlington National Cemetery (31-472). The three were working in the National Reactor Testing Station (now the Idaho National Laboratory) in Idaho. On the night of January 3, the three men were working in the SL-1 reactor that housed five control rods weighing 45 kg each and measuring 7 feet long. They were responsible for reconnecting the rods after some tests had been conducted during the afternoon of the same day. Alarms began sounding around 9 pm. Rescue crews reached the reactor at 9:10 pm. Rescue efforts were hampered by radiation. Byrnes had died in the explosion within the reactor, and the body of the third man, Legg, was found pinned to the ceiling by one of the control rods. Around 10:30 pm, rescuers entered the reactor room and saw Richard Leroy Mickinley still alive but severely injured. He was recovered and transferred to an ambulance but died before getting to a hospital while still in the ambulance. McKinley's body was incredibly radioactive, and anyone who came into contact with it was likely to suffer irreparable harm with radiation levels over 500 roentgens per hour recorded in the ambulance. After he died in the ambulance, the driver (who had been wrapped in a lead sheet) drove the ambulance to the middle of a field and abandoned it and the body (LeVay, 2008).

Despite the danger to the living caused by the radioactive nature of the remains of all three men, their bodies needed to be collected, decontaminated, and autopsied to help reconstruct the accident. To collect McKinley's body, workers approached the ambulance in minute-long shifts and moved the body from the ambulance, removed his clothes and wrapped the body in lead blankets. Roads were closed to other traffic, and the lead-sheathed body was placed in an ambulance and driven by a volunteer in a lead-lined jacket to the Testing Station's uranium reprocessing plant. The body was then placed in a steel bin containing ice and alcohol, in a hope this would leach radioactive contamination of the body. While this was going on, the body of Byrnes that had been lying on the reactor floor was also retrieved. Legg's body was more difficult to retrieve as it had been pinned to the ceiling by a control rod. Ultimately, 5 days after the accident, a crane with a long boom was brought to the side of the building and used to maneuver a net under Legg's body. Working again in 1-min shifts, volunteers used poles to attempt to drag the body from the control rod. The process took several days to complete. Legg's body, however, had been completely irradiated by the close exposure to the control rod, and showed no signs of decomposition (LeVay, 2008). The high initial dose of radiation essentially sterilized the bodies, creating an anaerobic environment that hindered decay (McKeown, 2003).

The bodies were effectively highly contaminated nuclear waste. After repeated washing, radiation levels still ranged up to 1500 roentgens per hour near the corpses. Officials from the Atomic Energy Commission (AEC) suggested the bodies be disposed of in the high-radiation waste pits and a small memorial be put up as a cenotaph to honor the men. If radiation levels of the remains could not be lowered, the bodies could not be released to the families for burial (McKeown, 2003). Autopsies and processing of the remains was done in 1-min shifts. Physicians from Los Alamos National Laboratory (LANL) were brought in to do this work. They worked behind lead shields using instruments attached to long pipes. In this way, they documented injuries caused by the blast. McKinley, not immediately killed by the blast had severe wounds to his face, eyes, and scalp as well as his left hand and left leg. Byrnes, whose body was found next to the reactor had severe crushing injuries to his face, throat, rib cage, left arm and leg, and pelvis. Legg, who had been impaled to the ceiling of the reactor room, had the most significant damage. The top of his head had been sheared off, exposing his brain. His upper body was twisted by 180 degrees compared to the lower body and internal organs had been displaced or destroyed as the control rod drove through his pelvis and abdomen. His left leg was also almost bisected.

The physicians from LANL were tasked with making the bodies suitable for burial as well. This created what McKeown (2003, p. 133) refers to as a "nuclear age conundrum". Significant processing of the remains had to be done to make them safe for funerals. One of the physicians is quoted as saying "You couldn't even go to a crematorium and cremate it and cremate it because this would contaminate the crematorium. The stuff [radiation] doesn't go anywhere. It's just there, and you have to deal with it. It would be terrible to say to the families, 'Here's the casket, but we're going to have to put it out in the middle of this field and have a funeral.' That was the consideration. What did you have to do to get the bodies' [radioactivity] down to where [the families] could not have open caskets but certainly no restrictions on the funerals?" (McKeown, 2003, pp. 133–134, brackets in original). They attempted to cut away flesh that was too contaminated to allow for burial within a cemetery. MCkinley's left arm needed to be removed completely. Legg's arms and legs were removed as well as his head. Byrnes' legs were removed. All of the removed body parts were placed in a drum and disposed of in a trench designated to receive high-level waste from the accident.

The remnants of the bodies were then prepared for burial. They were coated in drying powder, wrapped in cotton and plastic sheeting. The final wrapping consisted of lead sheeting, 1/8 inch thick for McKinley and Byrnes, ¾ inch thick for Legg. More than a ton of lead was used for this. The bodies were then placed in hermetically-sealed caskets that were also lead lined and containing to signs reading "Caution: High Radiation Area" on cardboard and "Caution: Radioactive Materials" on plastic (McKeown, 2003, p. 141). The caskets were placed within purpose-made lead vaults. This reduced the radiation levels outside the vaults to below 1 roentgen per hour, and the bodies were deemed safe to transport. The bodies were transported using military aircraft.

Byrnes and Legg were buried in cemeteries within their hometowns, but their burials would not resemble a typical funeral. Byrnes was buried in Rome, NY. His body arrived via military aircraft accompanied by AEC officials who took radiation measurements and deemed the detectable radiation low enough to allow for burial.

Cemetery workers opened a grave and laid a concrete pad 14 inches deep into the deeper-than-average grave. The weight of the vault meant that that it had to be brought to the cemetery via crane and put on a vault hoist over the open grave. Family and friends gathered a minimum of 20 feet away from the grave, and after a short Catholic service, the vault was lowered onto the concrete base. Cement was then poured over the vault, to a dept. of 12 inches on all sides. After this had set, the grave was filled in. Once the grave was filled, an AEC official took another reading on the grave, and found only background radiation (McKeown, 2003).

McKinley's funeral was to take place in Arlington National Cemetery. He was eligible for burial there due to his service in the Korean conflict. His body was flown to Bolling Air Force Base near Washington DC at 3:30 AM on January 23, where it taxied to a remote section of the airfield and was placed under guard. On January 24, an AEC official removed Mickinley's vault with a forklift and transferred it to a flatbed truck used to transport the body to Arlington. AEC officials continued to take radiation readings in the truck. The truck was driven to an unused chapel, where it stayed for two nights under guard with AEC officials continuing to monitor radiation emissions. On January 25, the vault was driven to the grave site and positioned on a hoist above a deeper-than-average grave. Like Byrnes' grave, it had been dug deeper and wider than the average grave and a 12-inch-thick pad of concrete had been poured prior to the day of burial. McKinley's graveside service lasted only 8 min, and the family was restricted to a distance of more than 20 feet from the graveside. After the mourners departed, cement was poured over and around the vault to a depth of 12 inches. Six days later, the Quartermaster of the military district ordered the superintendent of Arlington National Cemetery to include a note on McKinley's record of interment to state the following "Victim of nuclear accident. Body is contaminated with long-life radio-active isotopes. Under no circumstances will the body be moved from this location without prior approval of the Atomic Energy Commission in consultation with this headquarters" (McKeown, 2003, p. 145; "Richard LeRoy McKinley," 2006).

Legg's body had been pinned to the ceiling by the control rod, and so was the most contaminated. It had experienced the most post-mortem body processing as well, with the removal of all four limbs and the head. Legg was from Michigan and his family requested burial in his hometown. By the time of his burial, the disaster was known, and there was some controversy surrounding whether the body could contaminate the groundwater (McKeown, 2003). Legg's gravesite, like that of McKinley and Byrnes was deeper and wider than the average. Due to the higher degree of contamination of Legg's body, the concrete pad poured in this grave was 2-feet thick. Mourners were required to stand at least 20 feet from the gravesite during the short service. Legg's family insisted on viewing the casket, so it was removed from its vault, leading to an increase in radiation exposure and the service was necessarily reduced to 5 min. The vault was then lowered to their concrete pads. After the families departed, additional concrete was poured so that the vaults were encased in concrete.

The burial of these individuals is indicative of a specific time in our history as well as an increasing understanding of contamination. The cause and manner of their deaths necessitated significant post-mortem treatment and differential burial. Their burials would be immediately identified as atypical, even without the inclusion of the signs with their bodies indicating radioactivity. Their bodies, themselves, were altered significantly due to how they died as well. Irradiated and sterile, they had not decomposed by the time they were processed for burial. Their burial ceremonies were both truncated and elongated: truncated to 5–8 min of ritual with mourners, but a 5–6-day period between when the grave was dug and when it was covered to allow for radiation readings to be taken and concrete to cure around the vaults. Members of the AEC were present at all burials to take radiation readings. The bodies were transported atypically. They were treated atypically. All three were military veterans, but only McKinley's family chose burial at a national cemetery. This necessitated changes to the very formulaic and ritualized processes normally used at Arlington. The bodies cannot be moved without the approval of the AEC. It has often been said that when a person joins the military, they become the property of the US government, in the case of these three men, this is quite literally the truth. These remains are seen as individuals, but also as radioactive waste that falls under the purview of the AEC.

The contaminated debris and body parts that could not be returned for burial were buried in a specially identified area near the SL-1 reactor. The monument placed there is one that is designed to keep people away from the site (Fig. 9.5). It does not memorialize the remains of the individuals who died there that were included with the rest of highly contaminated waste; it simply serves as a warning.

Fig. 9.5 Photo of SL-1 Burial Ground for contaminated debris and human remains too contaminated for repatriation to family for burial. John R. Giles, EPA, Public domain, via Wikimedia Commons. https://upload.wikimedia.org/wikipedia/commons/9/96/SL-1Burial.jpg

The men who died in the SL-1 disaster can be seen as an embodiment of the conflict between medicalization (the very real fear of contamination) and the desire for a recognizable ritual that would allow for commemoration and visitation. The treatment of the SL-1 dead highlight the modern medicalization of death in important ways. The autopsies needed to be conducted to determine the nature of the disaster as well as what lead to the deaths of the individuals (proximate and ultimate causes). The bodies were treated as contaminated material, not individuals, a treatment that persists to the inclusion of their highly irradiated remains into the discarded waste buried at the SL-1 burial ground. The burials of these individuals also highlights the tension between the need to control the contamination of the bodies and the need for proper memorialization of the individuals who died. They were entitled to military burials, and so new rituals were designed to accommodate both the fears of contamination and the need to ritualize the burial in a way that would still be identifiable to their mourners. Though discussion was made about the possibility of denying military burial within the cemeteries chosen by the families, the decision to perform modified rituals was ultimately more powerful than a general fear of contamination. These modified rituals lead to different material culture and burial practices that would be identifiable if these graves were ever exhumed.

In the preface to her book *The Future of Nuclear Waste: What Art and Archaeology Can Tell us About Securing the World's Most Hazardous Material*, Rosemary Joyce (2020, p. xiv) notes that archaeology has a voice when it comes to nuclear waste. This "involves critical examination of the material traces of the development of nuclear power and nuclear weapons, and of resistance to them, and their normalization." The post-mortem modification of the remains of the SL-1 dead are examples of this tension. The dismemberment of the bodies, their wrapping in lead sheeting, their deeper-than-typical burial, and coverage in feet of concrete are all material correlates to this tension. The necessity for mourners to be distant from the open grave and the shortened burial services necessitated by concerns of contamination are new rituals, attempts to maintain the relationship between the living and the dead. Our relationship to nuclear power is a complex one, and the new methods for dealing with the atomic dead are indicative of the tensions between contamination and memorialization.

9.5 Emmett Till

Emmett Till was born outside Chicago, Illinois on July 25, 1941. His mother, MamieTill (later Bradley and then Till-Mobley), had moved north during the Great Migration from Tallahatchie County Mississippi and married Louis Till (his death and burial in 1945 is briefly discussed in Chap. 7 of this book). The Great Migration was a general movement of people of color from the deep south northward towards greater opportunity in industrial centers like Chicago, New York and Cleveland (Gorn, 2018). From Gorn's (2018) description, Emmett seems like a normal teenage boy, who loved being the center of attention, a joker, and an average student. In

August of 1955, Emmett went to visit cousins and family in Mississippi. The trip was designed as away to connect with Mamie's family and give Emmett time with his cousins and the experience of fishing and swimming during the Mississippi summer. On August 24, 1955, Emmett Till was visiting a grocery store with his cousins when they encountered Carolyn Bryant. Accounts differ on their interaction, but Whitaker (2005) states that Emmett asked for a date and informed Bryant that he'd dated white women before. Pool (2015, p. 414) says that Emmett "'talked fresh' or whistled." He was quickly escorted from the grocery store by his cousins, but the interaction had been witnessed by Juanita Milam, another woman in the store. The woman to whom Emmett talked, Carolyn Bryant, informed her husband Roy Bryant, who along with his half brother J.M. Milgram, went to the house where Emmett was staying and kidnapped him. They proceeded to beat the boy, and ultimately shot him in the head, weighting his body and throwing him into the Tallahatchie river.

On August 31, Robert Hodges discovered the body in the Tallahatchie River, and Mose Wright was brought in to give an identification. The body itself was severely beaten and had evidence of a bullet hole above the right ear. A few days in the river had also led to relatively advanced decay. The body was taken to Chester Miller's funeral home in Greenwood for preparation. Initially, burial was supposed to have taken place in Money, MS, but Mamie Till demanded the body be sent back to Chicago for internment. The body was then sent to C.F. Nelson's funeral home in Tutwiler for embalming, as this was a requirement if the body was to cross state lines. According to Whitaker (2005), C.F. Nelson was informed that the funeral would be a closed-casket, so pre-funeral preparation (involving intravenous embalming) was not attempted. Rather, the body was weighted and submerged in a vat of formaldehyde, with incisions made in the body to release gasses and admit the formaldehyde. According to Mamie Till (Till-Mobley & Benson, 2003), when the body was received at Rayner's funeral parlor in Chicago, she was informed by Mr. Rayner that the body had been packed in lime, possibly in an effort to speed decomposition. It is possible that both of these accounts are accurate, with lime being added to the body prior to shipment, but it is unclear. On September 1, Emmett Till's body began the trip home to Chicago. Pool (2015) relates that the body was only released for shipment to Chicago on the condition that the coffin would not be opened.

On September 2, Mamie made a decision that would influence the story right up to the present day. She decided that the funeral would be an open casket, and that Emmett's face would be uncovered. The body lay in public view at the funeral home from September 3—6th, where between 10,000 and 50,000 people made a visitation. The coffin had a glass cover, where visitors could view his body. According to Pool (2015), national news outlets covered the funeral, and all major African American newspapers published pictures of the body, furthering the impact of its display. Over a thousand people met the train with Emmett's body when it arrived in Chicago, a moment captured in a photo in Life Magazine in the September 12th issue, along with the caption "A grief-torn mother, Mrs. Mamie Bradley, received in Chicago the body of her son, Emmett Till, 14. Visiting a farm near Greenwood, Miss., he had been dragged from his bed, beaten and slain. Mississippi's governor deplored the deed and local authorities arrested two white men on kidnap charges. Emmett's alleged offense: he whistled at a white woman" ("Homecoming of a Lynch Victim," 1955). Till was ultimately buried in Burr Oak Cemetery in Alsip, IL (Fig. 9.6).

Roy Bryant and J.M. Malim were tried for the murder and kidnapping, but ultimately found not guilty, primarily due to the purported insecurity of the identification of the body. Sheriff Strider testified for the defense at the trial, saying that he didn't believe that the body pulled from the river was that of Emmett Till. He told the jury that he believed the body to have been that of an adult who had been in the river for 10—15 days. The medical testimony is somewhat confusing, as the defense did not employ a pathologist to bolster their claims, and the state did not refute the defense testimony offered by Sheriff Strider. Mamie Till had offered to allow the body to be exhumed for further analysis. This became the crux of the defense, that the identity of the body could not be established to be that of Emmett Till (who

Fig. 9.6 Grave marker for Emmett Till. Photo by Nick Number, CC BY 4.0 <https://creativecommons.org/licenses/by/4.0>, via Wikimedia Commons. https://upload.wikimedia.org/wikipedia/commons/3/35/Grave_of_Emmett_Louis_Till_%281941%E2%80%931955%29_at_Burr_Oak_Cemetery.jpg

the defense claimed was still alive). Whitaker attributes the failure of the prosecution to counter this two ways: first, the prosecution believed that the body was that of Till and that they had proven it (Whitaker notes that he interviewed the living jurors and only one of those interviewed believed the body not to be that of Till); and second, the district attorney knew the case was lost and was giving the jurors an out: "he did not wish to place all the blame on the jurors, most of whom served out of a sense of obligation and with deep regret. If the 'smoke screen' had been completely removed, 13 men would have become the victims of harassment that Sheriff Strider had faced for weeks"(Whitaker, 2005, p. 210). Whitaker (2005, p. 211) notes that in his interviews with the jury members, all actually agreed that Milam and Bryant had been involved in the murder of Emmett Till, but they found the killing to be justified: "The simple fact was that Negro had insulted a white woman. Her husband and his relative would not be prosecuted for killing him."

The phenomenological impact of viewing Emmett's body was noted by Mamie Till (Till-Mobley & Benson, 2003). She recalls that the odor of the body could be smelled from two blocks away from the funeral home. She wanted to identify the body of her son, and so took time to examine the body in detail at the funeral home. In explaining her decision to have an open casket funeral, she said, "It would be important for people to look at what happened on a late Mississippi night when nobody was looking, to consider what might happen again if we don't look out. This would not be like so many other lynching cases, the hundreds, the thousands of cases where families would be forced to walk away and quietly bury their dead and their grief and their humiliation" (Till-Mobley & Benson, 2003, p. 139). In Pool's (2015, p. 431) words, the glass-topped coffin allowed "her and the nation… the chance to *see* the effects of Southern racism."

Seeing the body creates witnesses to the aftermath of his beating, shooting, and the treatment of his body after death. That photos of the body were reproduced around the country meant that every person who viewed the photos had the experience of becoming a witness to it. A photo of Emmett from Christmas 1954, smiling and happy, was affixed to the lid of the coffin, allowing the viewer to see the juxtaposition between the living and the dead as well. This element of witnessing was transformed by Mamie Tiill from earlier forms of witnessing lynchings. As Edwin (2014, p. 714) notes, "Mamie not only broke through the historical silencing of survivors but turned the genre of the lynching photograph against the perpetrators. Moreover, challenging official attempts to dissociate the crime from the broader national history of racial terror, Mamie Till-Mobley insisted that the murder be understood as an act of lynching." The power of *seeing*, witnessing, has been noted in modern civil right struggles as well. Tracy (2023) draws a direct line between the cellphone video showing the death of George Floyd, the camcorder video of the beating of Rodney King in 1991, and the images of Emmett Till in his coffin, noting that what started with still photos continues with video in the current age where sharing on social media ensure wide-dissemination of images and video. Brandon Erby, a professor at the University of Kentucky, is quoted in the piece, saying "The circulation is sacred in a way… If we're not careful, we just continue to perpetuate the images of Black bodies under distress. We focus on the images and not necessarily on the actions, the justice and accountability." In effect, these images retraumatize the larger African American community but also act as the needed catalyst for change.

The name of Emmett Till and the images provoked by his murder and the display of his body are inextricably linked to the lived experience of African Americans. The name and image of Emmett Till have in recent years been linked to Travon Martin and George Floyd. The trial of J.M. Malim and Roy Bryant and the publicity surrounding it have been used as evidence to argue that racist structures are still very much in place in modern America (e.g., Onwuachi-Willig, 2016a). As Harold and DeLuca (2005, p. 265) put it, "the visceral imagery of the Till murder refuses to be filed away in a dusty archive of American civil rights history." Fernández Fernández (2020) traces the arc of the fight for social justice on a large scale from Emmett Till's murder through to the mass incarceration system that preferentially incarcerates African Americans over other populations. They trace the development of Jim Crow that developed in the south that allowed the killing of Emmett Till to happen to this larger carceral state. Building upon this, Fernández Fernández (2020) argues that this in turn lead to the development of the Black Lives Matter movement, looking at the deaths of Trayvon Martin and George Floyd as modern day lynchings. As potentially the most famous lynching victim, Emmett Till continues to be the face of violence perpetrated on African Americans in America. Part of this still centers on the physical body of Emmett Till.

As Rubin (1995, p. 57) notes, "it was a graphic reminder that all the Supreme Court decisions in the world were powerless in the face of concerted resistance." Again drawing on the language of lynching, Edwin (2014, p. 711) notes that if "lynching was less frequent and less visible than in earlier decades, and if young people raised in the North had been protected from some of the most egregious forms of racist assault, Emmett Till's murder made clear that African Americans, wherever they might live, were still vulnerable to racist violence." When the video of George Floyd's death went viral, allusions were made to the importance of witnessing and disseminating the video for others to see. McGuire (2021) draws a direct line between Floyd and Emmett Till, drawing on artwork painted by Kadir Nelson which portrays Floyd and Emmett Till together. This association was a common one, with Congressman John Lewis (2020) stating in a *New York Times* editorial "Emmett Till was my George Floyd. He was my Rashard Brooks, Sandra Bland and Breonna Taylor. He was 14 when he was killed, and I was only 15 at the time. I will never forget the moment when it became so clear that he could easily have been me."

In 2004, the Justice Department reopened the case in an effort towards reconciliation. This required the exhumation of Emmett in order to establish, once and for all, the identity of the body recovered from the river in 1955. The FBI announced plans for the exhumation on May 4, 2004, without notifying at least some of the Till relatives. Bertha Thomas, a distant cousin and the president of the Emmett Till Foundation said that some family members opposed the exhumation, quoted in a New York Times article as saying "They don't need his body or his remains in order to pursue them if they have solid proof that other people were involved" (Ruethling,

2005b). By the time of the exhumation, the family was united in approving the exhumation (Davey & Ruethling, 2005). Though few believed that any actual prosecutions would result from the reinvestigation, some Till family members supported the reopening of the case and the exhumation of the body in order to establish identity. The accounts of the murder suggested that numerous additional individuals may have been involved that were still living, though both Malim and Bryant were by this time dead (Russell, 2005). This was prompted in part by two different documentaries that brought up the possibility of further crimes that could be prosecuted (Ruethling, 2005a).

After exhumation, an autopsy was conducted, and the remains were returned to the family for reburial (Reuters, 2005). The original coffin could not be reused for the reburial, and so it was set aside at the funeral home, and should have been immediately returned to the family. In an interview with Smithsonian magazine (Callard, 2009), Simeon Wright (Emmett's cousin who was with him the night he was kidnapped) stated that the coffin was set aside in a storage shed after the exhumation until he inquired about it. Ultimately, the casket was returned to the family, and it was donated by the family to the National Museum of African American History and Culture in Washington, D.C (Pogrebin, 2009). When asked what he feels when he sees the casket, Simeon Wright responded "I see something that held the object of a mother's unconditional love. And then I see a love that was interrupted and shattered by racial hatred without cause. It brings back memories that some would like to forget, but to forget is to deny life itself"(Callard, 2009). He acknowledged the embodied materiality of the object that held the body of Emmett Till, saying that the casket was "going to speak louder than pictures, books, or films." The casket, used in any other context, would just have been a casket. The contact with Emmett transforms this object into a relic of sorts, one that must be protected to continue to tell its story for future generations.

In 2007, Tallahatchi County issued a formal apology to the Till Family. Congress passed the Emmet Till Unsolved Civil Rights Crime Act in 2008, cosponsored by Congressman Lewis. In 2022, President Joe Biden signed legislation into law that makes lynching a federal crime, the first time this type of crime was explicitly criminalized as a federal crime ("Biden Signs Bill Making Lynching a Federal Crime," 2022). In 2023, three new sites were announced as a National Monument to Emmett Till: Roberts Temple Church of God in Christ, the church where Emmett's funeral was held (Chicago), Graball Landing, Tallahatchie county, where Emmett's body was pulled from the river, and the Tallhatchie County Second District courthouse in Sumner, MS. The monument was established on what would have been Emmett Till's 82 birthday (Betts, 2023).

Performance runs like a thread through the life and death of Emmett Till. His life in Chicago was that of a typical, popular joker raised in a loving family. His lived experience typifies the difference in social expectations for young African American boys and men between those living in the South and the North. Emmett's willingness to stand up for himself was seen as an affront to the Southern white men who killed him, and they decided to teach him a lesson that would be visible for the young African Americans in their own Mississippi community. For Milam and

Bryant, in deciding to murder to make a point to the rest of his community, they were performing the cultural superiority that they felt and that was accepted by the community at large. Life and death were at stake for African American men in the South—to stand up for oneself placed one's life in danger, as evidenced by Emmett's cousins attempting to remove Emmett from the grocery store before he could get himself (and potentially them) into serious trouble.

Performance can also be seen in the editorials and newspaper accounts, both in Mississippi from a predominantly white audience and from black newspapers and the National Association for the Advancement of Colored People (NAACP). The NAACP utilized the Till killing as a catalyst for fund raising and for raising awareness of its programs. According to Whitaker (2005), the polarization arising from the case and the ultimate acquittal forced moderates in Mississippi to effectively pick a side. Following the verdict, Citizens' Councils throughout Mississippi proliferated. The first Citizens' Council was formed in Indianola, Mississippi shortly after the Brown v. Board of Education decision that desegregated schools. In the first decade of its existence, Council leaders organized opposition to the Civil Rights Movement, seeking local, statewide, regional, and national alliances to further their goals. (Rolph, 2018). In November of 1955, Citizens' Council membership throughout Mississippi was around 65,000. By January 1956, membership was around 75,000 in Mississippi with expansion into surrounding states as well (Whitaker, 2005). Onwuachi-Willig (2016b, p. 342) examined the African American community's response to the not guilty verdict as a case study in cultural trauma, noting that "African American residents of Leflore County and Tallahatchie County... continued on with their lives, damaged and demeaned even further in what had already been an explicitly oppressive system. African Americans also became forever altered by the verdicts official devaluing of African American life."

Showing a tremendous amount of resiliency, African American communities (particularly within Mississippi) found this verdict as a gathering point, strengthening a sense of community and identity formed in opposition to the white power structures. "Specifically, the shameful acquittal of Milam and Bryant caused African Americans to feel even more alienated and distant from the white community in the same area at the same time it brought them closer together into a tight-knit community that could lead a successful boycott against the two defendants' families. The verdict also brought whites closer together as a community and further apart from African Americans as a group intent on preserving the racial hierarchy in Mississippi" (Onwuachi-Willig, 2016b, p. 343). The strengthening of this community was facilitated through the actions of Mamie, and her desire for her son's death to be public, to share that pain with the world. "It was the unusually visible death of someone defined by his youth, and Till's apolitical youthful identity could more easily be used by agents to generate visibility and support for political ends. Tills murder and the publicity that followed during the display of mourning changed the interpretive context and provided whites with a new way to understand the growing political demands by African Americans for full citizenship" (Pool, 2015, p. 415). In this way, Pool frames the public mourning around Emmett Till as political mourning, which consists of numerous elements: contested identity, visibility, a failure of law, attribution of collective responsibility and political change (Pool, 2015, p. 417). In this way, while not central to the birth of the Civil Rights Movement, the death and funeral of Emmett Till can be seen as important and as a focus and a catalyst within the Civil Rights Movement (Pool, 2015).

The memorialization of Emmett Till has continued, in popular culture as well as through official memorialization through the establishment of the National Monument in 2023 and the placing of official signs at the same locations. In Mississippi, these signs in particular, have become the focal point of continued racial tensions. When a 32-mile stretch of US Highway 49 was renamed the Emmett Till Memorial Highway, the highway signs were defaced within a year (Editorial Board, 2019). The River site (one of sites in the newly proclaimed National Monument) was unmarked until 2008, when the first sign was put up. The sign was soon stolen and thrown into the river. A second sign was put up soon afterward, but this was shot multiple times. A third sign replaced this, and it too, was shot. Photos of fraternity brothers from the University of Mississippi (Ole Miss) surfaced in 2019 showing smiling Kappa Alpha members posing with the sign and their weapons (posed very much like a hunting photo). A spokesperson for Ole Miss stated that the photo was referred to both campus police and the FBI, but no disciplinary action was taken as what the students did was not a violation of the school's code of conduct (Vigdor, 2019). The Kappa Alpha fraternity, though did take some action, suspending the members identified in the photo for an unspecified length of time. The Kappa Alpha fraternity list Robert E. Lee as a "Spiritual Founder" ("Robert E. Lee," 2024), highlighting his definition of a gentleman. In contrast to the actions of the fraternity members, the Kappa Alpha website quotes Lee as someone who "strives for that nobleness of self and mildness of character which impart sufficient strength to let the past be but the past." This is, in effect, selective memory and selective historical recreation. This highlights the fact that history and historical figures are not objective. Stories about the past are told by individuals with intention and a point of view. As Manne (2024), a current Ole Miss student, opined, "as another professor of mine claims, nostalgia is about forgetting, not remembering. Selective memory and a painful lack of racial consciousness, however, are for the enjoyment of the privileged only. Minority students have no opportunity to forget, and it is irresponsible to tell them that they have no voice to criticize aspects of a place that cheerfully romanticizes a society that would have enslaved them."

In 2019, a fourth sign was erected at the River site, this time made of steel. Weighing over 500 pounds, the manufacturer says it is bulletproof (Ortiz, 2019). For Dave Tell, the bullet holes are important, and they're part of the story too. In an opinion piece, Tell (2019) states "While the cycle of violence has produced a succession of ever-more-pristine signs, we must not forget the three damaged originals, one of which has been lost. They convey the truth of Till's murder more completely than the pristine signs ever could because they refuse us the comfort of believing that racism is a thing of the past. That's why those signs belong in a museum."

If the burial had taken place in Greenwood, MS, or even been conducted as a closed-casket funeral, would Emmett Till's death have been the spark that it has become? The cultural trauma of viewing the body of Emmett till for the

10,000—50,000 that saw it in person or the millions who have since viewed photos is an integral part to why this death continues to resonate and continues to embody the individual struggle that often gets lost within the larger social struggles of the Civil Rights Movement. As the donation and curation of the original casket in the Smithsonian shows, the materiality and *thereness* of the objects associated with Emmett Till, whether they be photos or the casket itself, is an important touchstone for continuing to tell this story.

From a Poetics perspective, the material remains resulting from the exhumation and reburial of the body would point to an atypical burial process relative to cultural norms of the day. The potential use of both formaldehyde and lIme in the body preparation would also point to an atypical burial. The importance of the material objects that touched his life, death, and body become important when placed within their cultural context. The signs used to memorialize his place of death, body disposal, and funeral continue to be contested spaces used by different groups within society to perform their identities. The coffin that touched his body is preserved as an important cultural object and will continue to be a touchstone for the civil rights movement as long as the museum stands.

9.6 The Presley Family

Elvis Presley, the King of Rock and Roll, died on August 16, 1977 in his home known as Graceland in Memphis, TN. The time directly after his death is detailed in a Rolling Stone article (Flippo, 1977). According to Flippo, when his body was moved to the morgue, the morgue was sealed off and guarded during the autopsy. Very soon after the death was announced, mourners flooded into Memphis and began gathering outside Graceland. Elvis Presley was preceded in death by his mother Gladys Smith Presley, who was buried at the Forest Hill Cemetery Midtown. When Elvis' father Vernon allowed hundreds of mourners to file past the open casket in Graceland, the line of mourners stretched for miles.

Flippo (1977) describes the mourners as "as comprehensive a cross section of America as one could ever wish to see: bikers, businessmen, children, Shriners in clown shirts and phalanxes of middle-aged women, many of them sobbing." Elvis' music was playing from cars and surrounding businesses, but inside the grounds of Graceland, there a quiet atmosphere. Mourners left dozens of floral guitars, hound dogs, and hearts. Over 3000 floral arrangements were sent by everyone from the Soviet Union to Elton John to the Memphis Police Department (Flippo, 1977). Elvis' body was laid out in a 900-lb copper-lined casket under the crystal chandelier in the foyer of the house. Elvis was dressed in a pure white suit, light blue shirt and white tie. He was buried wearing a Taking Care of Business (TCB) ring. On the day of the viewing, the family had decided that the gates would close at 6:30 pm, leaving tens of thousands waiting and mourning in the streets. Mourners milled in streets and parking lots, leading to two fatalities.

The private funeral was rather plain and simple. Longtime friends acted as pallbearers. Two hundred individuals crowded into Elvis' music room at Graceland to hear eulogies from a TV evangelist, a comedian, and the reverend of a local Memphis church. The caravan, led by a silver Cadillac, followed by the white Cadillac hearse with Elvis' body and 17 white Cadillac limousines wound past bystanders to the nearby Forest Hill Cemetery Midtown. Elvis was entombed in a family crypt where his mother was buried. Friends and family returned to Graceland for a meal. Vernon Presley decided to give the accumulated flowers to fans who were still gathered at the gates. Mourners, fans, and curiosity seekers wanted a memento from the experience, and when the flowers were gone, any memento from the funeral or cemetery would do. One woman, Wenda Magyor, is quoted in the Rolling Stone article as saying "We'll stay out here all night just to get into the cemetery. We drove all night to get here. I will get a flower from the cemetery"(Flippo, 1977).

According to the application to place Graceland on the National Register of Historic Places prepared by Jody Cook (2004), the Meditation Garden (Fig. 9.7) at Graceland was constructed between 1964 and 1965 and intended as a private meditation area. After attempted break-ins to Elvis' tomb, Vernon Presley decided to use the space for reburial of remains from the family tomb. In 1980, Elvis' grandmother Minnie Mae Presley died at the age of 90. She was buried within the Meditation

Fig. 9.7 Photo of the Memorial Garden at Graceland. Photo by Thomas R Machnitzki, CC BY-SA 3.0 <https://creativecommons.org/licenses/by-sa/3.0>, via Wikimedia Commons. https://upload.wikimedia.org/wikipedia/commons/4/45/Graceland_Cemetery_Memphis_TN_3.jpg

Garden. Later a tombstone that Elvis had designed, complete with a Star of David to acknowledge his mother's Jewish heritage, was placed permanently at the Meditation Garden. The dedication of this tombstone that had been in storage occurred on the 60th anniversary of Gladys' death (Beifuss, 2018). An eternal flame was installed at the head of Elvis' grave. The large marble monument from the family plot at Forest Hill was relocated to the Meditation garden as well. According to this application, "Graceland drew large crowds while Elvis was still in residence, but since his death, it has literally become a pilgrimage site. Presley, his parents, and paternal grandmother are all buried here in the Meditation Garden on the south side of the house, and anniversaries of his birth (January 8) and death (August 16) are commemorated every year. Graceland is an extraordinary place where Presley's global community of fans can come together, connecting with Elvis and each other in their own way, and the site itself has played a critical role in fostering and maintain the living legend of Elvis Presley"(Cook, 2004, p. 44).

The Meditation Garden at Graceland contains both primary and secondary burials and is an excellent example of a family-based cemetery. It also contains a cenotaph commemorating Elvis' twin brother Jesse who was stillborn (DiRienzo, 2019).[1] Archaeologically, it has an atypical depositional history, with the primary occupant, Elvis Presley, as a secondary interment. An additional secondary interment is the body of mother Gladys. These are considered secondary interments because they have been moved from their initial burial location. In the case of Elvis, this occurred soon after death, and so the body would likely have remained intact but may have shifted during transit and may not have been in the same positioning as when initially interred. For Gladys, more time had passed between the primary and secondary interments, and the body may have been subject to decomposition, with the loss of articulation of skeletal elements.

Primary interments consist of family members who died after Elvis, and include his grandmother Minnie Mae Presley, father Vernon and his grandson Benjamin Storm Presley Keough (Beifuss, 2020) and daughter Lisa Marie Presley. The iconography and imagery is both secular and religious, and represents multiple religious disciplines. The Meditation Garden is a site of continued importance, not only as a burial location for Presley family members but also to celebrate the life and cultural significance of Elvis Presley. Every year, a week of activities are planned in August, culminating with a candlelight vigil on the night commemorating the night before Elvis' death at the Meditation Garden. It has become a place of pilgrimage and ritual that extends far beyond his initial family. Elvis' music served as a touchstone for an entire generation and was used as a bond between generations.

The impact of the Memorial Garden as a cultural space has outlived those who knew or saw Elvis in person. On my visits to Graceland, the demographics of visitors suggest that generational visits are important as well. For many, visiting the graves of the Presley family may become a way to connect with their own families.

[1] Jesse Garon Pressley is buried in the Priceville Memorial Gardens in Tupelo, MS in a grave marked only with a blank tombstone.

The tour of the Graceland mansion concludes with the Memorial Garden, and so the graves become a part of the overall experience for people of all ages. Over time, the importance and role of Elvis Presley as a musician, actor, and general pop culture figure changes as those who knew him or saw him perform live pass away. The focus of the displays at Graceland become more historical in approach, and we (those who were too young to have a more personal connection) begin to view Graceland as a time capsule of the 1970s, an example of excess and performative gaudiness (a visit the Jungle Room or the TV Room is highly encouraged). This lasting legacy is celebrated in the Space created by the Meditation Garden. Graceland is still privately owned, and the family allows visitation of the Meditation Garden free of charge between the hours of 7:30 and 8:30 am each morning (Shetler, 2020). This is an implicit acceptance of the greater role that the Memorial Garden holds for fans.

The free admission to the Memorial Garden is one of very few free areas within Graceland. The general approach to the legacy of Elvis at Graceland is one of commodification. While the celebrations of his life are available for viewers, including an amazing display of costumes and material culture related to Elvis' career and life, visitors must pay an admission fee to enter any of these displays. The Memorial Garden is both a tourist attraction and place of memorial used as a memorial by the Presley family. The development of the Memorial Garden adds to the tourist attraction, and so even though visitation to the Memorial Garden can be done without paying an admission fee, there is an element of commodification of the space given its location and association with the larger tourist attraction that is Graceland.

9.7 Summary

This chapter highlights how death, burial and commemoration are influenced by historical and cultural context and also become reflective of that cultural context in a reciprocal loop. Marie Laveau and the visitation of her grave can be seen as a celebration of resilience and the acceptance of new religious movements. The death, body processing, and funeral of Frank Little, a labor activist, became a powerful symbol for immigrant labor in Montana and as fodder for the larger labor movement. Marie Curie's body was transformed into a symbol of scientific achievement, co-opted by the state; the precautions surrounding her remains also speak to an understanding of contamination and the dangers of modern science. The SL-1 dead speak to the tension between an understanding of scientific contamination and a need for traditional methods of burial and memorialization. Emmett Till's open-casket funeral can be viewed as a catalyst for the Civil Rights Movement, by putting a face to the larger social issues of racism and lynching, this demonstrates how public displays of death can inspire social change. Finally, Elvis Presley's death and the creation of the Memorial Garden as a pilgrimage and tourist site reflect the importance of celebrity and how a single individual becomes a touchstone for social change in popular culture. Together these case studies reveal how funerary and memorialization practices reflect tensions between public commemoration, private wishes, and historical contexts. Together they offer insights into how death and remembrances shape both individual legacies and broader social movements.

References

3 Ways Marie Curie Would Fix the Plight of women in STEM Fields. (2020). https://www.melissallarena.com/3-ways-marie-curie-would-fix-the-plight-of-women-in-stem-fields/. Accessed December 11, 2020.

Alvarado, D. (2020). *The magic of Marie Laveau: Embracing the spiritual legacy of the Voodoo Queen of New Orleans*. Weiser Books.

Beifuss, J. (2018). *Gladys Presley's Elvis-designed tombstone returns to Graceland*. Memphis Commercial Appeal. Accessed at https://www.commercialappeal.com/story/news/local/2018/08/15/gladys-presley-tombstone-returns-graceland/991529002/

Beifuss, J. (2020). *Bejamin Keough, Elvis' grandson, interred at Graceland*. Memphis Commercial Appeal. Accessed at https://www.commercialappeal.com/story/news/2020/10/02/elvis-presley-grandson-benjamin-keough-buried-graceland-memphis-lisa-marie-presley/5894141002/

Betts, A. (2023, July 23). Biden to name national monument for Emmett till and his mother. *New York Times*. Retrieved from https://www.nytimes.com/2023/07/23/us/emmett-till-monument-biden.html?searchResultPosition=4

Beyond Curie: About the Project. (2020). https://www.beyondcurie.com/about. Accessed December 11, 2020.

Biden Signs Bill Making Lynching a Federal Crime. (2022, March 20). *New York Times*. Retrieved from https://www.nytimes.com/video/us/politics/100000008279262/biden-emmett-till-anti-lynching-law.html?searchResultPosition=22

Botkin, J. L. (2017). *Frank Little and the IWW: The blood that stained an American family*. University of Oklahoma Press.

Callard, A. (2009). Emmett Till's casket goes to the Smithsonian. *Smithsonian Magazine. Accessed at https://www.smithsonianmag.com/arts-culture/emmett-tills-casket-goes-to-the-smithsonian-144696940/*

Cook, J. (2004). National historic landmark nomination: Graceland (NPS Form 10-900). Retrieved from https://npgallery.nps.gov/GetAsset?assetID=5db64e4b-7360-4b57-9d04-fae564edd7fc

Davey, M., & Ruethling, G. (2005, June 2). After 50 years, Emmett Till's body is exhumed. *New York Times*, p. A12. Retrieved from https://www.nytimes.com/2005/06/02/us/after-50-years-emmett-tills-body-is-exhumed.html?searchResultPosition=33

Dead Voudou Queen. (1881, June 23). *New York Times*, p. 2.

DiRienzo, D. (2019). The forgotten Mississippi Gravesite that no one ever visits. Accessed at https://www.onlyinyourstate.com/mississippi/forgotten-gravesite-in-ms-no-one-visits/

Dubovsky, M. (2000). *We shall be all: A history of the industrial workers of the world*. University of Illinois Press.

Editorial Board. (2019, October 23). Opinion: Why the new Emmett Till memorial needed to be bulletproof. *New York Times*. Retrieved from https://www.nytimes.com/2019/10/23/opinion/emmett-till-memorial.html?searchResultPosition=15

Edwin, S. (2014). Remembering the Ancestor's image: Emmett Till and predicaments of witnessing. *Callaloo, 37*(3), 710–722.

Fandrich, I. J. (2005). The birth of New Orleans' Voodoo Queen: A long-held mystery resolved. *Louisiana History: The Journal of the Louisiana Historical Association, 46*(3), 293–309. Retrieved from http://www.jstor.org/stable/4234122

Fawcett, K. (2016). *The delicious history of America's oldest Chinese-American restaurant*. MentalFloss. Retrieved from https://www.mentalfloss.com/article/88648/delicious-history-nations-oldest-chinese-american-restaurant

Fernández Fernández, M. (2020). Tracing Emmett Till's legacy from black lives matter back to the civil rights movement. *Alicante Journal of English Studies, 30*(2020), 43–62.

Flippo, C. (1977). Elvis Presley: Funeral in Memphis. Rolling Stone. https://www.rollingstone.com/music/music-news/elvis-presley-funeral-in-memphis-61938/

Garcia, N. (2019). Legacy of slain labor organizer lives on in Butte. NBC Montana. https://nbcmontana.com/news/local/legacy-of-slain-labor-organizer-lives-on-in-butte

Gorn, E. J. (2018). *Let the people see: The story of Emmett Till*. Oxford University Press.

Harold, C., & DeLuca, K. M. (2005). Behold the corpse: Violent images and the case of Emmitt Till. *Rhetoric and Public Affairs, 8*, 263–286.

"Homecoming of a Lynch Victim." (1955). Life, 39(11), 47. Retrieved from https://books.google.com/books?id=41YEAAAAMBAJ&q=Emmett+till#v=onepage&q=Emmett%20till&f=false

Joyce, R. (2020). *The future of nuclear waste: What art and archaeology can tell us about securing the world's most hazardous material*. Oxford University Press.

Karney, J. (1917). Kaiserism in the copper industry. *Mother Earth, XII*, 222–223.

Kinzer, S. (2003, November 30). Interest surges in Voodoo, and Its Queen. *New York Times*, p. 28. Retrieved from https://www.nytimes.com/2003/11/30/us/interest-surges-in-voodoo-and-its-queen.html?searchResultPosition=6

Lee, D. (2015). 10 Biggest cities in Montana: How well do you know the Treasure State? *Newsmax*. https://web.archive.org/web/20170822094013/http://www.newsmax.com/FastFeatures/biggest-cities-montana-population/2015/04/13/id/638229/

Robert E. Lee. (2024). Retrieved from https://www.kappaalphaorder.org/about/history/robert-e-lee/

LeVay, S. (2008). *When sciences goes wrong: Twelve tales from the dark side of discovery*. Plume.

Lewis, J. (2020, July 30). Opinion: Together, you can redeem the Soul of Our Nation. *New York Times*. Retrieved from https://www.nytimes.com/2020/07/30/opinion/john-lewis-civil-rights-america.html

Long, C. M. (2006). *A New Orleans Voudou priestess: The legend and reality of Marie Laveau*. University Press of Florida.

Manne, S. (2024, August 8). Opinoin: Dear Ole Miss: Minorities are done being haunted by Confederate Ghosts. *Time Magazine*. Retrieved from https://time.com/3093672/university-of-mississippi-civil-rights/

Marie Curie Enshrined in Pantheon. (1995). *New York Times*, p. 3.

McGrath, J. (1976a). Foreword. In J. McGrath (Ed.), *Butte's heritage cook book* (pp. vi–vii). Butte-Silver Bow Bicentennial Commission.

McGrath, J. (Ed.). (1976b). *Butte's heritage cookbook*. Butte-Silver Bow Bicetennial Commission.

McGuire, D. L. (2021). Between seeing and believing, memory and forgetting, innocence and guilt. *Reviews in American History, 49*(1), 1–9.

McKeown, W. T. (2003). *Idaho falls: The untold story of America's first nuclear accident*. ECW Press.

MME. Curie is Dead; Marter to Science. (1934). *New York Times*, pp. 1, 17. Retrieved from https://timesmachine.nytimes.com/timesmachine/1934/07/05/issue.html

Onwuachi-Willig, A. (2016a). Policing the boundaries of whiteness: The tragedy of being out of place from Emmett Till to Trayvon Martin. *Iowa Law Review, 102*, 1113–1185.

Onwuachi-Willig, A. (2016b). The trauma of the routine: Lessons on cultural trauma from the Emmett Till verdict. *Sociological Theory, 34*(4), 335–357.

Ortiz, A. (2019, October 20). Emmett Till memorial has a new sign. This Time, It's Bulletproof. *New York Times*.

Parker Pearson, M. (1999). *The archaeology of death and burial*. Texas A&M University Press.

Pasachoff, N. (1996). *Marie Curie: And the science of radioactivity*. Oxford University Press.

Pogrebin, R. (2009, August 27). Emmett Till's coffin goes to Museum. *New York Times*, p. C2.

Pool, H. (2015). Mourning Emmett Till. *Law, Culture, and the Humanities, 11*(3), 414–444.

Reuters. (2005, June 5). Autopsy done, Emmett Till is reburied. *New York Times*. Retrieved from https://www.nytimes.com/2005/06/05/us/autopsy-done-emmett-till-is-reburied.html?searchResultPosition=78

Richard LeRoy McKinley. (2006). Retrieved from http://www.arlingtoncemetery.net/rlmckinl.htm

Rolph, S. R. (2018). *Resisting equality: The citizens' council, 1954-1989*. LSU Press.

Rubin, A. S. (1995). Reflections on the death of Emmett Till. *Southern Cultures, 2*(1), 45–66.

Ruethling, G. (2005a, May 5). F.B.I. Will exhume the body of Emmett Till for an autopsy. *New York Times*. Retrieved from https://www.nytimes.com/2005/05/05/us/fbi-will-exhume-the-body-of-emmett-till-for-an-autopsy.html?searchResultPosition=113

Ruethling, G. (2005b, May 6). Kin disagree on exhumation of Emmett Till. *New York Times*. Retrieved from https://www.nytimes.com/2005/05/06/us/kin-disagree-on-exhumation-of-emmett-till.html

Russell, M. M. (2005). Reopening the Emmett till case: Lessons and challenges for critical race practice. *Fordham Law Review, 73*(2005), 2101–2132.

Shetler, S. (2020). Visiting Graceland for free: See the meditation garden and graves of Elvis and his family. Retrieved from https://quirkytravelguy.com/visiting-graceland-for-free/

Tell, D. (2019, November 14). Opinion: Put the Vandalized Emmett Till Signs in Museums. New York Times. Retrieved from https://www.nytimes.com/2019/11/14/opinion/emmett-till-sign.html?searchResultPosition=17

Till-Mobley, M., & Benson, C. (2003). *Death of innocence*. One World Books.

Tracy, M. (2023, January 27). Video of violent encounters has often galvanized public opinion in ways other accounts have not. *New York Times*. Retrieved from https://www.nytimes.com/live/2023/01/27/us/tyre-nichols-memphis?searchResultPosition=105#video-of-violent-encounters-has-often-galvanized-public-opinion-in-ways-other-accounts-have-not

Vigdor, N. (2019, July 26). Emmett Till Sign Photo Leads Ole Miss Fraternity to Suspend Members. *New York Times*, p. A19. Retrieved from https://www.nytimes.com/2019/07/25/us/emmett-till-ole-miss-students-fraternity.html?searchResultPosition=51

Whitaker, H. S. (2005). A case study in Southern Justice: The murder and trial of Emmett Till. *Rhetoric and Public Affairs, 8*(2), 189–224.

Work, C. P. (2006). *Darkest before dawn: Sedition and free speech in the American West*. University of New Mexico Press.

Part III
Modern Interactions with the Dead

Part III (Chaps. 10–12) takes the Poetics model into the present and provides an interpretive framework for examining modern mortuary processes. These chapters examine the way in which modern mortuary performances are influenced by and reflective of a highly individualized understanding of identity. These chapters explore the increasing desire for personal agency in death practices, as seen in new rituals like living funerals and body composting, which emphasize the personalization and humanization of death rituals. Finally, these shifts underscore the ongoing negotiation between personal identity and collective memory, revealing how death rituals continue to evolve in a rapidly changing world.

Chapter 10 focuses on the modern methods of cremation and methods of dispersal or use of cremains. New rituals arise from the incorporation of cremains in memorial reefs, fireworks, tattoo ink, and the launching of cremains into space. Chapter 11 explores modern methods of mortuary performance as well as the introduction of pre-death rituals. Ultimately, increased agency of the dying has lead to more individualized treatment and memorialization. By exploring alternative body disposal techniques, living funerals, and body donation, I highlight the ways in which contemporary death rituals are shaped by modern values of environmental sustainability, individualism, and societal transformation. These shifts represent a desire for more personal, creative, and socially engaged ways of interacting with the dead. Chapter 12 provides an exploration of modern methods of memorialization, again with a focus on the individualized nature of these vernacular memorials. Vernacular memorials can be seen as personalized, sometimes informal memorials. Also discussed is the increasing popularity of digital memorialization. Some practices such as the evolving nature of the Día de los Muertos (Day of the Dead) tradition, become a larger celebration of resiliency and ethnic identity.

Chapter 10
Cremation and the Disposal of Ashes

Abstract This chapter investigates contemporary cremation methods, comparing traditional dry cremation with environmentally sustainable wet cremation (alkaline hydrolysis). Modern cremation reflects modern values surrounding death, remembrance, and environmental concerns. Finally, the use of cremated remains (cremains) in contemporary society, through practices such as using custom urns, memorial reefs, and even space burials. It highlights the growing trend of transforming cremains into personal, portable memorial objects, including glass sculptures and tattoos, illustrating the continued evolution of cremation rituals as a form of individual and collective identity. Through these diverse case studies, the chapter underscores cremation's role in shaping social memory, cultural continuity, and personal expression in both historical and modern contexts.

Cremation is the act of reducing the body to 'ash,' either through *wet* or *dry* cremation. In premodern contexts, this would almost entirely have meant burning the body on a culturally prepared pyre. Wet cremation is a relatively new process discussed below and primarily developed as an alternative to dry cremation that is argued to be more environmentally friendly. The product from cremation, whether it is modern or ancient, results in stable materials that can be stored or incorporated into reef structures, portable objects, even ceramics used for food service. Here, I will focus on the role of cremation as a social process in in our modern world in which cremains (the "ash" produced through the process of cremation) can be manipulated and disposed of in modern contexts.

The Poetics model prompts us to think about how a ritual performance is both *de*structive and *con*structive. Cremation is at its heart, the reduction of the body, and can be seen as destructive. In reality, this is not destructive, but transformative. In the transformation of the corpse into cremains, it allows for the creation of new rituals and new ways of interacting with the dead (and is therefore constructive). These new rituals tend to be reflective of intersectional elements of identity and are more personal than 'traditional' methods.

A. Osterholtz, *The Bodies Poetic*, Bioarchaeology and Social Theory,
https://doi.org/10.1007/978-3-031-85998-4_10

10.1 Modern Methods: Wet and Dry Cremation

Dry cremation involves the use of gas-powered crematory retorts that reach temperatures of 820—980 °C (Eckert et al. 1988). Temperatures are consistent and the exposure is timed. Bodies are placed within the retort in a consumable container, either a cardboard box or wooden casket and subjected to intense heat until all soft tissue has burned away. The body is then allowed to cool and removed from the retort. Bones are then placed within a cremulator, a rotating drum pierced with 4 mm holes that reduces the bones to a fine ash-like substance that is gathered and packaged for pickup by the family (see Fig. 10.1 for an example).

Wet cremation (Alkaline hydrolysis) uses water, alkaline chemicals, low heat, and sometimes pressure to reduce the body to bone that can then be crushed in a cremulator. Potassium hydroxide or sodium hydroxide (or a combination of the two) is introduced into the cylinder with the body, the hydroxides break chemical bonds in the body through hydrolysis, and this process is speeded by the addition of low heat. This is a newer process than flame-based cremation, and is limited in its availability across North America. The liquid, described as effluent, is sterile and can be disposed of with other waste water (Cremation Association of North America, 2021). Just like dry cremation, this destroys all DNA and soft tissue. Sometimes this is chosen as it is viewed as a gentler process and/or more environmentally friendly, using fewer fossil fuels due to the lower heat component.

Cremains are typically at least temporarily housed within an urn or container of some kind. These containers can be remarkably indicative of cultural identity, both individual and group, highlighting various elements of an individual's life. One of the founding members of the band Devo, for instance, was cremated upon his death and had his ashes split into two halves and placed in custom urns reminiscent of the

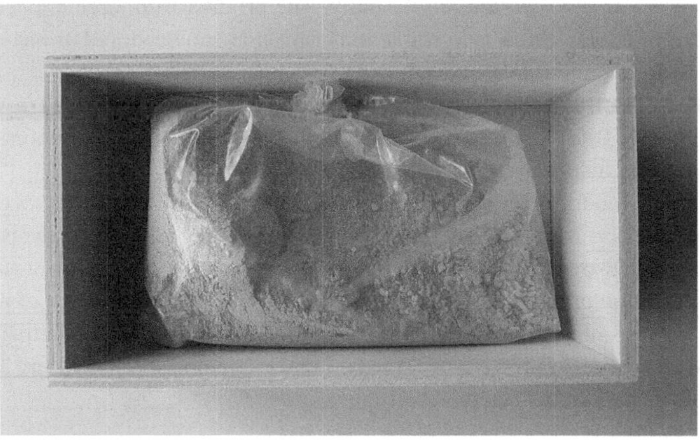

Fig. 10.1 "Labrador Amy's ashes after cremation." Image attribution: Asurnipal, CC BY-SA 4.0 <https://creativecommons.org/licenses/by-sa/4.0>, via Wikimedia Commons. https://upload.wikimedia.org/wikipedia/commons/b/b2/Labrador_Amy%27s_ashes_after_cremation-01ASD.jpg

famous Devo Red “Energy Dome” hat. The two urns were retained by his wife in Los Angeles and the other send to extended family in Ohio. Written on the urn were the words “Bob Casale. Hit recording artist. Devoted husband and loving father. Grill master 7/14/1952—2/17/2014” (https://foreverence.com/pages/bob-casale). Urns can be 3-D printed in the form of whatever the bereaved desire, including the heads of the deceased or famous individuals. A search on Etsy (an online store known for selling artisan objects) for “cremation urn” brings up thousands of designs available, from bowling balls to slot machines, to a replica of the Venus of Willendorf. Teaching at an Southeastern Conference School (SEC), I was particularly interested to find that branded university-themed urns (and coffins) could be purchased with official logos from multiple universities, including Mississippi State University (MSU). The SEC is known for particularly committed fans and devotion to one’s chosen University that can become important for family identity. I reached out to the legal department at MSU to determine which companies were officially licensed to produce MSU branded mortuary equipment, attaching a listing on a website offering both MSU coffins and urns. In an email, a representative from the university responded with “The University does not have any licensed venders that produce coffins or urns. This company seems to be infringing but caskets and urns are not infringements we pursue. We often approve these types of items on an individual basis and the marks are approved by us to ensure accuracy only and we do not collect royalties.”

Urns can be reflective of individual identity in important ways or highlight family and personal histories. Urns featuring universities can be found at multiple venders online, for example, and feature a variety of iconography including football helmets, likenesses of coaches, and baseball/football shapes with university logos. Frederic Baur, the inventor of the Pringles can was cremated upon his death in 2008, and his ashes were placed within a Pringles can. According to a QCD News article (2023), the family stopped at a Walgreens on the way to the funeral home to pick up the ashes. His son Larry is quoted as saying “My siblings and I briefly debated what flavor to use… But I said, ‘Look, we need to use the original.’” The Pringles can was then inurned (the term used for the burial of an urn within a cemetery) in his plot at a cemetery in Cincinnati, OH.

10.2 Modern Meaning

Cremations, or the remnants of the body left after the cremation process, are a stable substance that can be manipulated, altered, stored, and/or disposed of in any number of ways. Storage of the ashes can take place in urns for substantial lengths of time. These remains can be buried in burial plots (inurned), placed in cremation niches in above ground spaces in Columbaria (where they may be in sealed or unsealed niches), turned into memorial reefs, added to concrete structures to form artificial reefs, made into jewelry, or shot into space.

10.2.1 'Scattering' Ashes

The choice of where and when to scatter the ashes of friends or family members is a highly individualized decision. Cremated remains are sometimes scattered in places of special importance to the dead and/or their family members. In most states, ashes can be scattered on private land with permission of the landowner or on public lands with permission of the governing agency. For ashes scattered over water, the Federal Clean Water Act requires that cremated remains be scattered at least three nautical miles from land ("What are the laws on scattering ashes?," 2020). The important tie that threads through locations is that it typically is someplace that is important or meaningful to the living and the dead. One such place is Disney World. According to a Wall Street Journal article (Schwartzel, 2018), cremated remains are recovered from the park about once a month. Cremains have been recovered from flower beds, bushes, lawns, outside park gates and during firework displays, in the moat under the flying elephants of the Dumbo ride, and (most frequently) throughout the Haunted Mansion. This is not allowed by the management of Walt Disney World, and guests who are caught spreading ashes are escorted off the property. But this threat does nothing to stop the spreading of ashes. Ashes are brought into the park in Ziploc bags, prescription-pill bottles or makeup compacts. "When ash residue is discovered on a ride, Disney workers tell guests they must shut down due to 'technical difficulties.' Then a manager rides alone through the attraction looking for any ash piles while colleagues may hand out 'Fast Passes' to assuage guest who must leave before the custodians turn up with their high-powered vacuums" (Schwartzel, 2018). Interestingly, those who scatter cremains within the park do not seem to be concerned with what happens after they are spread. That they are vacuumed up with an industrial vacuum and disposed of is of less importance than the way in which they are deposited in the park. It is the mortuary ritual of placing them in the important Space, anything after that is not as important. The deposition of the ashes within Disney World, for some at least, transforms the park into a memorial Space. One woman quoted in the article, Kym Pessolano DeBarth, deposited her mother's ashes in the water underneath It's A Small World. She is quoted as saying "Instead of going to a grave… I go to Disney World." Sørensen (2009, p. 112) describes the expansion of what a place for disposal can be as the creation of "non-places—places of transience and transfer," acknowledging that the ultimate fate of the remains is less important than the act of scattering them in important places." Just as the body is transformed through cremation, the spatiality of what a memorial location (previously a cemetery) also changes.

10.2.2 Memorial Reefs, New Rituals, and Memorial Vacations

Cremation.com notes that memorial reefs have a growing popularity, being an affordable and sustainable option, easy to arrange, a way of giving back to the earth in a positive way, and leaving a unique and memorable legacy ("Cremation Memorialization," 2020). The Neptune Society offers some basic guidelines for how to safely and legally dispose of cremated remains. The Neptune society also maintains an area off the coast of Key Biscayne, FL known as the Neptune Memorial Reef. This area, when completed will encompass 16 acres of ocean floor. The Memorial reef is an artificial reef created by mixing cremains into concrete structures that will ultimately become a habitat for fish and other sea life. The concrete structures are shaped into forms such as sea stars or shells, and marked with identifications of whose ashes are contained within them (Fig. 10.2). The reef is designed as an artistic representation of the Lost City of Atlantis and approximately 40 feet below the surface (Fig. 10.3). After the structures are completed and placed, family members can scuba dive around them, "to visit their loved ones"("Neptune Memorial Reef," 2020). Memorial Reefs is another memorial reef company that operates memorial reefs around the world, from Victoria Island, British Columbia to Venice, Italy. Their website states that they "help families, the community and the environment by building artificial reefs using cremated remains. When you dedicate your loved one's cremated remains as an undersea memorial, you enhance coral generation and marine biomass. It's an eco-friendly alternative to traditional burial"("Memorial Reefs International," 2020). This company also mixes cremains into a concrete structure and affixes a brass plaque to the structure.

As of 2020, 'burial' by Memorial Reefs International begins at around $3000 for the creation of a 500-lb structure standing 2 × 3 feet. They also offer 'Remembrance Vacation' packages for an additional $3000, packaging a new type of ritual to go along with the formation of the reefs. Eternal Reefs offers a variety of packages, with similar sizing and pricing. The Aquarius Eternal Reef is approximately

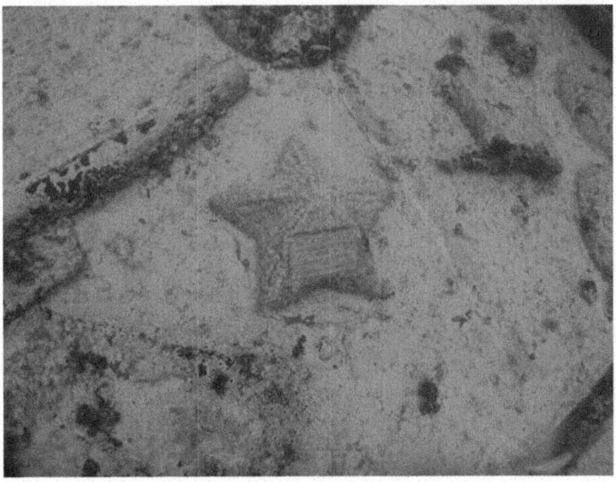

Fig. 10.2 Starfish memorial with attached plaque within Neptune Memorial Reef. Photo by Todd Murray from Bloomington, MN, USA, CC BY-SA 2.0 <https://creativecommons.org/licenses/by-sa/2.0>, via Wikimedia Commons. https://upload.wikimedia.org/wikipedia/commons/b/b6/Neptune_memorial_reef_07.jpg

Fig. 10.3 Abhishek at Neptune Memorial Reef diving near a memorial reef in the shape of a lion with arches visible in background. Photo by AbhiSuryawanshi (May 2020). Photo attriution: AbhiSuryawanshi, CC BY-SA 4.0 <https://creativecommons.org/licenses/by-sa/4.0>, via Wikimedia Commons. https://upload.wikimedia.org/wikipedia/commons/f/fc/Abhishek_with_Dolphins_at_Neptune_Memorial_Miami.jpg

550–700 lbs., and costs approximately $4000 to place. They offer placements in one of two locations off the coast of Florida. Interestingly, they also offer an "unattended" option, where the cremains are collected, converted into an artificial reef, and placed without any attendance by the friends and family of the deceased. This comes at a substantial discount, for the Aquarius Unattended option, the cost is roughly $3000, a savings of 25% percent. Though offered, it is an option that is rarely used. Eternal Reefs will also include the cremains of family pets into the structures of their owners at no additional charge, reflecting the importance of family pets in modern life. They note "It is rare that we have a placement without at least one family member and their beloved companion(s)"("Eternal Reefs," 2020).

The creation of memorial reefs, with their new rituals is a major shift in mortuary ritual. These involve extensive processing of the body, making it into something completely different. In some cases, combining the remains of a human body with that of pets as important elements of identity (e.g., 'dog person' or 'cat person'). This is combined with the placement of a monument at sea, necessitating a very different set of rituals, dress, and actions than a 'typical' interment in a cemetery space. Designated as a Remembrance Vacation as opposed to a memorial service, this reframes the mortuary process as a celebration. While seen as environmentally responsible and friendly, the rise popularity of this type of mortuary program signifies a different and highly individualized view of death and remembrance. The space used and Space created by the memorial reef is a celebration of life, creating new habitats and providing a view into the mindset as one of regeneration.

10.2.3 Fireworks

Humans can be remarkably creative in how we dispose of cremains. Author and Gonzo-journalist Hunter S. Thompson had his remains packed with fireworks and shot from a 153-foot tower into the sky above his Colorado home ("Ashes of Hunter S. Thompson blown into sky," 2005). The 15-story tower was constructed and modeled after Thompson's personal logo (a clenched fist with two thumbs rising from the hilt of a dagger). The invite-only celebration included Bill Murray and Johnny Depp and 350 guests and had music provided by the Nitty Gritty Dirt Band and Lyle Lovett. Depp, who organized the celebration claimed that he was following Thompson's last wishes, to be shot out of a cannon ("Hunter S. Thompson's ashes were blasted out of a cannon on this date 15 years ago," 2020). The cremains were mixed into some of the fireworks, the entire display took approximately 10 min and consisted of red, white, blue, and green fireworks, something very fitting according to his wife Anita Thompson, who is quoted as saying "He loved explosions"("Ashes of Hunter S. Thompson blown into sky," 2005). The tower height violated local building ordinances, and so was only a temporary construction. It was removed, and Anita Thompson said that she wanted to construct a pond nearby as a permanent sanctuary space. She also planned to erect a government-issued tombstone which Thompson was entitled to as an Air Force veteran with the inscription "It never got weird enough for me"(Seelye, 2005).

10.2.4 To Boldly Go… Reaching for the Stars, Becoming Comets

Space burials involve the launching of cremated remains into space. Capova (2016, p. 309) refers to these as "posthumous space tourism." Whether this act is seen as secular or a religious experience for those that choose it remains an open one. For some whose family chose space burial, it does take on a religious overtone. For example, for Carolyn Shoemaker, the deposition of her husband Eugene's ashes on the moon took on religious overtones, "He always said that every crater was a sacred site for him"(quoted in Volante, 1998). This is in line with Cutting's (2009, p. 367) conclusion that space burials "reveal fundamental tensions in the cultural imaginary of technoscience."

These can take several forms but began with a simple idea of sending cremains into space. In 1992, an astronaut brought a sample of the ashes of Gene Roddenberry aboard the Space Shuttle Columbia as a "personal effect" ("Shuttle bore

Roddenberry's ashes," 1994), fitting for the creator of Star Trek. This is the first known Space burial to have occurred. In 1997, Celestis began conducting private launches, opening the market significantly for this type of disposal. They offer several different types of space-based services: Earth Rise, Earth Orbit, Luna, and Voyager (www.celestis.com). The Earth Rise service involves a trip via rocket from New Mexico to an altitude of over 100 km in a steep trajectory. After breaking into space, the rocket and capsule payload is weightless and then begins to descend back towards the earth. The capsule, upon reentering the atmosphere produces "audible sonic booms… heralding the return to Earth"("Touching the Sky," 2021). The capsule is recovered, and returned to friends and family as a keepsake "suitable for display." The Earth Orbit service offers a different experience. It involves a gathering at friends and family of the launch site. Small amounts of cremains (approximately 1 gram (Moskowitz, 2008)) are loaded into the capsule payload of a rocket and the launching of the rocket becomes a part of the memorial service. The rocket launches and releases its payload at a higher orbit, where it "continues on its mission for months or years (mission dependent), eventually re-entering and burning up in entirety, blazing like a shooting star in final tribute"("The High Frontier-An Earthview Like no Other," 2021). This service includes interaction for the friends and family with "space experts and astronauts", combining the elements of a memorial service and a con-type atmosphere. For individuals fascinated by space and exploration, and those who love and mourn them, this would create powerful memories and group identity.

In 1998, Celestis was chosen to assist in the memorial service of Dr. Eugene Shoemaker aboard the Lunar Prospector spacecraft. Shoemaker was a geologist and is known as one of the founders of planetary science (Brown, 1997). The Lunar Prospector was part of the Discovery Program, a 19-month mission designed to investigate the moon. Shoemaker's ashes were placed in a capsule inscribed with the image of the Hale-Bopp comet and a quote from Romeo and Juliet on the spacecraft and at the end of the mission; it was steered into a collision in Shoemaker crater near the lunar south pole (Porco, 2000). When cremains are sent to the lunar surface, they are in capsules and do not disperse. They land and will remain as permanent monuments to the dead. As the Celestis website states, "What could be more compelling than looking up in the night sky at our glowing neighbor, knowing your loved one has completed a journey accomplished by so few? The Celestis spacecraft is a permanent lunar memorial for adventurous souls" ("Touching Another World," 2021).

The plan to place Shoemaker's ashes on the moon was publicized in 1998, and when informed about the deposition, the President of the Navajo Nation (Albert Hale) is quoted in the *Albuquerque Journal* (1998) as saying "The Moon is a sacred place in the religious beliefs of many Native Americans. It is one thing to probe, to study, to examine and even for men to walk upon the Moon. But it is sacrilegious, a gross insensitivity to the beliefs of many Native Americans, to place human remains on the Moon." Hale reached out to NASA to express concern, and NASA's director of public affairs apologized for the lack of prior consultation. Carolyn Shoemaker, widow of Eugene is quoted in the Spokesman Review (Volante, 1998) as being astonished to hear that there was controversy or that the deposition of the cremains on the moon was seen as a desecration, adding that "It [the moon] was just an important part in his life, and he would never have thought about desecrating it." The commitment to consult with tribal entities was codified in Executive Order 13175 in November of 2000.

A private mission to the Moon has revived this controversy. Eighteen teams from 12 countries around the world competed in the Google Lunar X Prize (GLXP), which tasks contestants with developing a private moon launch. Alver (2014, p. 30) refers to this as "Uber for space," noting that the goal of the prize was to develop a reliable way to get to the moon and return back. Another, smaller scale, project also looked at developing a sustainable Lunar Colony. For the NASA RASC-AL competition, Wertz (2015) developed a plan with students from the University of Southern California about how to build a sustainable lunar colony. Key to the start up economy of a sustainable lunar colony would be the development of a cemetery within Tycho crater, "visible and identifiable at full Moon from anywhere on Earth"(Wertz, 2015, p. 3). While this is simply a thought experiment, calls such as this have brought about renewed criticism from the Navajo Nation. A statement issued by current Navajo Nation President Buu Nygren is quoted in a CNN article: "The suggestion of transforming it [the moon] into a resting place for human remains is deeply disturbing and unacceptable to our people and many other tribal nations"(Fisher, 2024). The CEO of Celestis, Charles Chafer, responded "Just as permanent memorials for deceased are present all over planet Earth and not considered desecration, our memorial on the moon is handled with care and reverence, is a permanent monument that does not intentionally eject flight capsules on the moon. It is a touching and fitting celebration for our participants—the exact opposite of desecration, it is a celebration."

In their dissertation, Harvey (2024, p. 39) details the process that NASA went through in 1999, developing in-person gatherings with Navajo Nation Citizens and Leaders with an eye towards "establish long-term culturally sensitive modes of communication with the Navajo Nation, fostering stronger outreach activities, forging new connections, and fostering healing into the twenty-first century." According to Harvey (2024, p. 44), NASA developed a policy of no longer sending cremains to the moon, but concludes "Whether this is a soft policy or if it applies to commercial activities that NASA supports is unknown." Private companies such as Celestis may not be subject to any agreements that NASA makes (Fisher, 2024), and in January of 2024, Celestis launched cremains from 68 humans and one dog to the moon at a cost of $12,500 for each participant (Celestis.com). The Navajo Nation office of the President's press release of February 8, 2024 (Office of the Navajo Nation President, 2024) gives appreciation for NASA science performed in space, but highlights the need for tribal consultation and communication with regards to sacred sites (including the Moon). The press release particularly highlights the private launch of the Vulcan Centaur carrying payloads from two companies that specialize in space burials.

Launching cremains into deep space has also occurred, with more planned for the future. NASA's New Horizons spacecraft carries a small amount of the cremains of Clyde Tombaugh, the discoverer of Pluto. It launched in 2006 with the primary mission of a flyby of the (then) planet Pluto in 2015. The canister with his cremains bears the inscription "Interred herein are the remains of American Clyde W Tombaugh, discoverer of Pluto and the Solar System's 'third zone'. Adelle and Muron's boy, Patricia's husband, Annette and Alden's father, astronomer, teacher, punster, and friend: Clyde W. Tombaugh (1906–1997)"(Buckley, 2006). Additional deep space missions are planned by Celestis and will include samples of the ashes of Gene Roddenbury and his wife Majel. The Enterprise Flight (January 8, 2024) included not only the remains of Gene and Majel Roddenberry but also those of Jimmy Doohan (Mr. Scott from the Original Series), Jackson DeForrest Kelly (Dr. Leonard "Bones" McCoy from the Original Series) and Nichelle Nichols (Lt. Nyota Uhura from the Original Series). Doohan and Gene Roddenberry also have the distinction of having their ashes sent into space multiple times. In addition to the Enterprise flight, Doohan's ashes were sent on Celestis missions in 2007, 2008, and 2012.

There is a common thread amongst those whose ashes have been shot into space. Amongst the list of these individuals are physicists, scientists, and lovers of science on a grand scale. Cultural icons such as Star Trek writers and actors, and cultural icons such as Timothy Leary (whose ashes were on the same Celestis mission with Gene Roddenberry's in 1997). Essentially, the thread is an adventurous nature, a desire to expand the known into the unknown, a desire to boldly go. The disposal of their ashes in this particular manner is a reflection of their lives and their loves. And it is a reflection of how they were viewed by those who honored them in this manner. The relationship between the living and the dead and the desire for their memory to go on, to be inscribed on a capsule and shot into deep space, shows a profound connection that continued after physical death. Take, for example, the incorporation of Clyde Tombaugh's ashes on NASA's New Horizons spacecraft. Those who knew Tombaugh thought the best way to honor his memory was to have even a small piece of him come closer than any other human has ever been to Pluto. The amount of ashes sent on these journeys is purely symbolic, mostly consisting of 1 gram. The body, transformed by cremation into ash, is further transformed by exposure to space. In some cases, these remains were returned to the family, in other cases, they become shooting stars as they burn up on reentry into the atmosphere.

As the Moon and space become commercialized spaces, particularly by private entities outside of government control (and therefore not subject to consultation guidelines, ineffective as they may sometimes be), discussions of cultural sensitivity will become more common. The deposition of Shoemaker's ashes on the moon is viewed very differently depending on the cultural perspective of the viewer. Is this a celebration? A desecration? Poetics tells us that how we interpret this particular form of memorial is culturally and historically contingent. As a twenty-first century expression of identity highlighting particular elements of an individual's intersectional identity, it is interpreted as a celebration. This emphasis on the individual, though, comes into conflict with others' individual identities and collective cultural understanding of the proper interaction with the dead. For Harvey (2024) this boils down to understanding the differences and commonalities in Indigenous and

Western astrophilosophies, the development of research partnerships that incorporates Indigenous research methodologies and methods and highlights the value of Traditional knowledge and continuing conversation, no matter how difficult that process can be.

10.2.5 Cremains Can Become Many Things

Cremated remains can be included in numerous objects. Here, I want to focus on the incorporation of cremains into portable objects such as glass objects, tattoo ink, stones, and ceramics. For McCormick (2015, p. 212), "their materiality gives ashes creations an established place to be located in the home: ashes-paintings are located on the walls; ashes-pottery on display on shelves; and ashes-paperweights are on desks.... This a painting or paperweight that incorporates cremation ashes does not resemble traditional material culture associated with death, such as traditional urns. This enables ashes creations to be on display without being a constant reminder of the death of a loved one, either to commissioners or to visitors to the home." We can interpret this in different ways. This viewpoint could be in line with Walter (1993) and Ariès (1974) assertion that death and mourning have become hidden in modern society. But we can also see these as evidence of a continuing relationship between the living and the dead that is expressed through new material forms. Mathissen's (2017) study showed that for some individuals, objects made from cremains allowed for continued attachment with the dead and for others, it was the very quality of portability and freedom that were meaningful in the created objects. "Human ashes are embedded in material practices and entangled in social relationships, evoking a discourse of care among the bereaved. Since both the temporal as well as the final destination of human ashes are important to people, the cremated remains become an object of solicitude that evokes moral obligations"(Mathijssen, 2017, p. 41). In effect, the objects have significant agency, in line with Heessels et al.' (2012) assertion that objects made with cremains blur the boundaries between the living and the dead.

Cremains, after all, are stable objects. They do not decompose, they do not disappear until they are intentionally dispersed or transformed into other objects. Heessels et al. (2012) argue that this mutability is important, as the remains of the dead become incorporated into the daily life of the living. They also highlight the phenomenological aspects of interacting with objects made with cremains, noting that "people's beliefs about the dead can only be investigated by considering not only what people say, but also what they bodily and materially express through their practices with ash objects"(Heessels et al., 2012, p. 483). They conclude, "ash objects are endowed with a power to influence the bereaved as much as the bereaved influence the objects" (Heessels et al., 2012, p. 484). This is very much in line with the concepts of postmortem agency put forth by Crandall and Martin (2014). This concept of postmortem agency and how it articulates with Poetics is discussed further in Chaps. 13 and 14 in this volume.

The website "Art from Ashes" ("Art from Ashes: Handcrafted rememberences," 2020) offers a service where one can mail cremains to glass artists who will then include them in glass sculpture or glass beads. Offering glass pieces ranging from beads on keychains to suncatchers, they note with each product offered where the cremated ash is visible. The ashes are transformed into an art object but never hidden within these sculptures. A testimonial from Stephen, Laurie, Vincent, and Thomas from Springfield, MA says "The hearts are truly works of art. We will cherish the hearts knowing that a part of our mom will always be with us. The hearts remind us of the wonderful life we shared with our mom." Remembrances memorialized in glass aren't limited to human remains, but can include the remains of beloved non-humans as well. Says D. Fouts of the Chimpanzee and Human Communication Institute "When Washoe, our beloved friend and family member, passed away we began looking for a way to honor her memory. Washoe was the first chimpanzee to learn a human language (American Sign and pass it on to her adopted so. Her accomplishments over the last four decades have helped humans better understand their responsibilities to our fellow animals. Because of her outstanding and unique qualities, we were delighted to learn about the one-of-a-kind remembrances that Art From Ashes creates. The pieces that Art From Ashes created of Washoe are all beautiful. Our Board members were touched and surprised when each was presented with an exquisite piece of art memorializing Washoe." Bead prices begin at around $100 (in 2020). A search on Etsy performed in 2020 (etsy.com) for the term "Cremains" returns 2880 results, including numerous glass workers who will incorporate cremains into sculpture and glass beads, pens with glass cylinders for cremains storge, necklaces with storage areas for cremains. Many of these pieces of jewelry or sculpture provide discrete ways of keeping cremains with the holder, making the relic a very personal expression of the relationship between the living and the dead.

Cremains can be converted into everyday objects such as ceramics or commemorative "stones." Based in Santa Fe, NM, Chronicle Cremation Designs produced dinnerware and other ceramics containing cremains within their ceramic glass. Finished pieces are dishwasher and microwave safe and range in price from around $200 for a mug to over $600 for bottles. Customers are sent a kit for ash collection and orders take between 30 and 60 days to complete. Justin Crowe, the ceramicist behind Chronicle Cremation Designs, began creating pieces with cremains in the glaze after the death of his grandfather. He told Hirsh (2016) "Being with my grandfather when he died in his home had a profound impact on my outlook. Home is a place that is routine, average and familiar, and experiencing such a huge emotional event in that environment helped normalize his passing for me… Based on this experience, I began to search for a way to give others the ability to remember passed loved ones in everyday life." He went on to say that the response has been mixed with some people "questioning religious values, ethics and respect." They focus on the gratitude they receive from their clients, noting that "the people who buy our memorial products are looking for a way to keep their memories close to them." In creating every-day products such as coffee mugs, Crowe created a way to not only normalize death but to continue to include the dead in every-day activities. Looking at the pieces created online, they in no way resemble the dead. They have a soft blue tone to the glaze and are beautiful art pieces with simple lines. As he pieces are designed to be used in daily life, Crowe understood that chips might happen. He kept a supply of the glaze created from the cremains for 2 years in case a piece needs to be replaced or fixed (Stern, 2016).

Crowe also used the mixed ashes of more than 200 bones (purchased online from a medical supply company) to create a dinnerware collection that he called "Nourish" and then used it at a dinner party as a modern day memento mori. The goal of the performance piece which was videoed and later uploaded to YouTube (https://www.youtube.com/watch?v=IW2YpfgK_d0) was to spark a conversation about death. Crowe said, "It was really beautiful… It's a conversation we don't typically have. To create objects that foster that conversation was one of my goals. (My guests' attitudes) ranged from sarcastic to really thoughtful and analytical. The mood was never dark. It was never grim. It was a very candid exploration of different viewpoints" (quoted in Rodgers, 2016).

Justin Crowe has expanded from cremains ceramics into creating Partingstone, a company that specialize in converting cremains into commemorative "stones". Partingstone.com, currently run by Justin Crowe, offers a service to convert cremains into objects they identify as "stones", their website states that the "appearance of the solidified remains varies naturally from person to person in shape, color, texture, making each collection uniquely beautiful. Solidified remains are an alternative to conventional ash." Customer reviews on the website highlight that the idea of a stone is both a way to commemorate an individual love of travel and a discrete way to keep an individual present. One customer, identified as Garth Clark, noted that "I've had my mother's ashes for over a decade. I kept them in my clothing closet. They didn't make me feel warm or sentimental. Getting these has been transformative" (Partingstone.com, 2024). Dutch artist Sofie Boonman creates what are known as *cuddle stones*, individually designed to complement the contours of the customers' hands. According to Heessels et al. (2012), Boonman's cuddle stones "make the intangible tangible." It creates a material object that can be handled that embodies grief and loss. The tactile nature of the stones is an especially important element here, the materiality of embodied grief.

Cremains can be mixed with tattoo ink and used for tattoos. Cremationink.com will combine cremains into tattooing ink for what they describe as "the ultimate memorial tattoo" (cremationink.com, 2024). McCormick (2015) interviewed several individuals with ashes-tattoos. In one case, cremains from an individual were mixed into the ink used for a memorial tattoo depicting the lyrics of a song sung to the wearer as a child that they then sung back to their grandmother as she was dying. Others chose to memorialize a landscape or object that was important emblem of the relationship between the living and the dead. Heessels et al. (2012, p. 479) note that for ashes tattoos, "the bodies of the deceased and the bereaved are literally interconnected. The border between living and dead matter is crossed as cremated remains become part of a living body. When the bereaved touch their tattooed skin they also touch the remains of their deceased relative." Both McCormick and Heessels et al. note the frequent touching of such tattoos when their interviewees discussed them. As with stones made from cremains, there is a tactile quality to ashes tattoos that seems to be an important element for memorialization. In Chap. 12, the concept of memorial tattoos is discussed, and in Chap. 16, the preservation of tattooed skin as a personal relic is discussed. It is conceivable for memorial tattoos made with cremains ink to become preserved and passed down for generations.

The incorporation of cremains into personal relics draws upon the transformational power of cremation (Oestigaard, 1999) (for more on relics, see Chap. 16). Cremation creates a stable substance that can be split and provide a concrete connection between the dead and the living, and the distribution of the cremains as personal relics can reinforce social bonds amongst family members every bit as much as has been hypothesized for the reinforcement of social bonds between Bronze Age peoples in Britain (e.g., Booth & Brück, 2020; Brück, 2004; Brück, 2014).

10.3 Modern Viking Funerals (the Legal Way in the US)

This section is included as a case study in one of the more interesting ways of dispersing ashes. During a class discussion in my "Death and American Culture" class, a student asked whether it was legal to perform a Viking funeral in the United States. This is my effort to answer that question, and my student wasn't the only one curious about the possibility of a Viking funeral. The cultural perception of a Viking funeral is one from movies and film, where the body of a warrior is set adrift in a long ship surrounded by fine goods. A fiery arrow is shot to light the boat on fire. It burns to the waterline and disappears. In reality, those described as Vikings performed a wide variety of mortuary customs. These customs changed with time and location and will not be discussed here. Outdoor cremation is only currently (as of 2024) legal in a couple of localities in Colorado, so modifications to the perceived ceremony need to be made. Cremation can be conducted in a licensed crematory, with the ashes returned to the family. The cremains can then be placed in a ship-shaped urn to which timer-based incendiaries or fireworks can be added. As of 2022, the Etsy seller SailshipUrns had sold 114 such urns. The counter was removed from the listing as of 2024 (https://www.etsy.com/listing/835307170/viking-funeral-ship-viking-ship-urn?click_key=45f96629e2bdf505553474263e7bdc1a87ffe751%3A835307170&click_sum=1008e3e9&ref=search2_top_narrowing_intent_modules_etsys_pick-1&frs=1). The product description notes that the urn is biodegradable and engineered to float about 5 min before sinking. The fact that by 2022, 114 such urns had been sold illustrates the creative ways that the dispersal or retention of cremains occurs in our modern era.

10.4 Summary

This chapter explored contemporary cremation methods as well as the many ways that cremains can be transformed into objects or disposed of in culturally or personally relevant ways. Here, modern cremation practices can be seen to reflect shifting cultural values around death, remembrance and the environment. Also highlighted was the growing trends of using cremains in creative ways, such as custom urns, memorial reefs, space burials, and even transformed into personal memorial objects like glass sculptures and tattoo ink. Overall, cremation results in the creation of a stable material that can be stored long term or incorporated into various objects. By examining these modern methods, the chapter underscores how creation continues to shape and be shaped by social memory, collective identity, and personal expression.

This chapter has set out a variety of different ways that the living and the dead interact using cremation and the cremains that result from this process. In looking at twenty-first century methods for dispersal, disposal, and transformation, there is no functional way to cover all of the different methods that are meaningful to those that utilize them. This chapter is not meant to be exhaustive on this topic, but to show some of the different approaches taken with cremation and how the Poetics model can be used to understand how these can be meaningful methods of remembrance and an indication of identity formation particularly through the creation of new rituals and new methods of ash dispersal.

References

Alver, J. (2014). Marketing the moon. *Harvard Political Review, 41*(3).

Ariès, P. (1974). The reversal of death: Changes in attitudes toward death in Western Societies. In D. E. Stannard (Ed.), *Death in America* (pp. 134–158).

Art from Ashes: Handcrafted remembrances. (2020). Retrieved from: https://www.artfromashes.com

Ashes of Hunter S. Thompson blown into sky. (2005). *New York Times*. Retrieved from https://www.nytimes.com/2005/08/21/world/americas/ashes-of-hunter-s-thompson-blown-into-sky.html?searchResultPosition=7

Booth, T. J., & Brück, J. (2020). Death is not the end: Radiocarbon and histo-taphonomic evidence for the curation and excarnation of human remains in Bronze Age Britain. *Antiquity, 94*(377), 1186–1203. https://doi.org/10.15184/aqy.2020.152

Brown, G. E., Jr. (1997). In remeberance of Dr. Eugene Shoemaker and Dr. Jurgen Rahe. *Congressional REcord, 143*(106), np. Retrieved from https://www.govinfo.gov/content/pkg/CREC-1997-07-24/html/CREC-1997-07-24-pt1-PgE1508-2.htm

Brück, J. (2004). Material metaphors: The relational consruction of identity in Early Bronze Age burials in Ireland and Britain. *Journal of Social Arhaeology, 4*, 307–333.

Brück, J. (2014). Crematio, gender, and concepts of the self in the British Early Bronze Age. In I. Kuijt, C. P. Quinn, & G. Cooney (Eds.), *Transformed by fire: The archaeology of cremation in cultural context* (pp. 119–139). The University of Arizona Press.

Buckley, M. (2006). Happy 100th birthday, Clyde Tombaugh [Press release]. Retrieved from https://www.jhuapl.edu/PressRelease/060203

Capova, K. A. (2016). The New Space Age in the making: Emergence of exo-mining, exo-burials and exo-marketing. *International Journal of Astrobiology, 15*(4), 307–310.

Crandall, J. J., & Martin, D. L. (2014). The bioarchaeology of postmortem agency: Integrating archaeological theory with human skeletal remains. *Cambridge Archaeological Journal, 24*(03), 429–435.

Cremation Association of America. (2021). Cremation process. Retrieved from https://www.cremationassociation.org/page/CremationProcess

Cremation Memorialization. (2020). Retrieved from: https://www.cremation.com/cremation-memorialization/eternal-reefs/

cremationink.com. (2024). Cremation Ashes Tattoo Ink. Retrieved from https://cremationink.com

Cutting, A. (2009). Ashes in orbit: Celestis spaceflights and the invention of post-cremationist afterlives. *Science as Culture, 18*(3), 355–369. https://doi.org/10.1080/09505430903122992

Eckert, W. G., James, S., & Katchis, S. (1988). Investigation of cremations and severely burned bodies. *The American Journal of Forensic Medicine and Pathology, 9*(3), 188–200.

Eternal Reefs. (2020). Retrieved from: https://www.eternalreefs.com/eternal-reefs-donation-levels/

Fisher, K. (2024). Navajo Nation's objection to landing human remains on the moon prompts last-minute White House meeting. CNN. Retrieved from https://www.cnn.com/2024/01/05/world/peregrine-moon-mission-navajo-nation-objection-human-remains-scn/index.html#

Harvey, A. D. (2024). *The cosmos as a uniting pillar of relationships: A Diné & Hózhó perspective*. Massachusetts Institute of Technology.

Heessels, M., Poots, F., & Venbrux, E. (2012). In touch with the deceased: Animate objects and human ashes. *Material Religion, 8*(4), 466–488.

Hirsh, S. (2016, December 1). Cremation ceramics allow you to immortalize and drink out of your loved ones. *Mashable*. Retrieved from https://mashable.com/article/cremation-ceramics#9EQ_RMtUEmqm

Hunter S. Thompson's ashes were blasted out of a cannon on this date 15 years ago. (2020). The conversation. Retrieved from https://www.nyswritersinstitute.org/post/hunter-s-thompson-s-ashes-were-blasted-out-of-a-cannon-on-this-date-15-years-ago

Mathissen, B. (2017). The ambiguity of human ashes: Exploring encounters with cremated remains in The Netherlands. *Death Studies, 41*(1), 34–41.

McCormick, S. (2015). *Ashes to art, dust to diamonds: The incorporation of human cremation ashes into objects and tattoos in contemporary British practices*. Manchester Metropolitan University.

Memorial Reefs International. (2020). https://www.memorialreefs.international/?gclid=Cj0KCQiA5bz-BRD-ARIsABjT4ngx3ncZQUdDaeHjDzfc8I5WUWeU9jeuzx726S3w34hTiyfS2ENCSLIaAruREALw_wcB

Moskowitz, C. (2008). Fly me to the moon… forever. Retrieved from https://www.space.com/5184-fly-moon.html.

Neptune Memorial Reef. (2020). Retrieved from: https://www.nmreef.com

Neptune Society. (2020). What are the laws on scattering ashes? Retrieved from: https://www.neptunesociety.com/cremation-information-articles/what-are-the-laws-on-scattering-ashes

Oestigaard, T. (1999). Cremations as Transformations: When the Dual Cultural Hypothesis was Cremated and Carried Away in Urns. *European Journal of Archaeology* 2(3): 345–364.

Office of the Navajo Nation President. (2024). Navajo President Buu Nygren, Navajo Council Speaker Crystalyne Curley [Press release]. Retrieved from https://opvp.navajo-nsn.gov/navajo-president-buu-nygren-navajo-council-speaker-crystalyne-curley-urge-protection-from-desecration-of-moon-celestial-sites-to-nasa/

Partingstone.com. (2024). Solidified remains. Retrieved from https://partingstone.com/products/human-solidified-remains

Porco, C. (2000). Destination moon. *Astronomy*, 53–55. Retrieved from https://astronomy.com/~/media/Files/PDF/Magazine%20articles/Destination%20Moon.ashx

Press, A. (1998, January 13). Moonbound ashes "sacrilege" to tribes. *Albuquerque Journal*.

Rodgers, B. (2016). Art + design | "Nourish" series creates dinnerware from ashes of the deceased. Retrieved from https://cfileonline.org/chronicle-cremation-contemporary-ceramic-art/

Schwartzel, E. (2018). Disney world's big secret: It's a favorite spot to scatter family ashes [Article]. *Wall Street Journal*. Retrieved from https://www.wsj.com/articles/disney-worlds-big-secret-its-a-favorite-spot-to-scatter-family-ashes-1540390229

Seelye, K. Q. (2005). Ashes-to-fireworks send-off for an 'outlaw' writer. *New York Times*. Retrieved from https://www.nytimes.com/2005/08/22/us/ashestofireworks-sendoff-for-an-outlaw-writer.html?searchResultPosition=4

Shuttle bore Roddenberry's ashes. (1994, April 29). *Rome News Tribune*, p. 8A. Retrieved from https://news.google.com/newspapers?nid=348&dat=19940429&id=bUMvAAAAIBAJ&sjid=CjMDAAAAIBAJ&pg=1725,8689414

Sørensen, T. F. (2009). The presence of the dead: Cemeteries, cremation and the staging of non-place. *Journal of Social Archaeology, 9*, 110–135.

Staff. (2023). Did you know… Inventor of Pringles Buried in a Pringles Can. *QCD News*. Retrieved from https://quadcitiesdaily.com/did-you-know-inventor-of-pringles-buried-in-a-pringles-can/#:~:text=Not%20Pringles%20inventor%20Fredric%20Baur,buried%20in%20a%20Pringles%20can

Stern, C. (2016, December 2). Drink a cup of coffee out of grandma's remains! Ceramics company creates a range of mugs, bowls, and vases out of your late loved one's ashes. *The Daily Mail*. Retrieved from https://www.dailymail.co.uk/femail/article-3995562/Drink-cup-coffee-grandma-s-remains-Ceramics-company-creates-range-mugs-bowls-vases-late-loved-one-s-ashes.html

The High Frontier-An Earthview Like no Other. (2021). Retrieved from https://www.celestis.com/experiences-pricing/earth-orbit/

Touching Another World. (2021). Retrieved from https://www.celestis.com/experiences-pricing/luna/

Touching the Sky. (2021). Retrieved from https://www.celestis.com/experiences-pricing/earth-rise/

Volante, E. (1998, January 15). Navajos upset after ashes sent to moon; Nasa apologizes. *The Spokesman-Review*. Retrieved from https://www.spokesman.com/stories/1998/jan/15/navajos-upset-after-ashes-sent-to-moon-nasa/

Walter, T. (1993). Death in the new age. *Religion, 23*(2), 127–145.

Wertz, J. R. (2015). *NASA 2015 RASC-AL Compentition*. University of Southern California: Low-Cost, Large-Scale, Near-Term Sustainable Lunar Colony.

Chapter 11
Modern Interactions with the Dead

Abstract This chapter explores modern methods of interacting with the dead that move beyond burial and cremation, reflecting broader shifts in societal values toward individuality and personalized memorialization. It examines alternative body disposal techniques such as body composting, as well as emerging practices like body donation, display, and pre-planned memorial services, including living funerals. These practices are framed as responses to the deritualized nature of contemporary life, offering new ways to express grief, honor identity, and shape the relationship between the living and the dead. The chapter highlights how these changes in death rituals are particularly shaped by the Baby Boomer generation's experiences with major social movements and mass death events such as the HIV/AIDS epidemic and the 9/11 attacks, which led to the creation of new mourning practices. Additionally, it contrasts these shifts with the perspectives of younger generations, such as Gen Z, who tend to find meaning in more traditional forms of memorialization. The chapter underscores the ongoing evolution of death rituals and memorial practices as they adapt to both personal desires and societal transformations. Specific practices explored include the role of funeral preplanning, living funerals, body donation as a mortuary process, and body composting.

This chapter centers on modern methods of interacting with the dead outside the avenues of 'traditional' funerals and cremation. There are a wide variety of ways to interact with the dead; in keeping with the concept that these new methods of memorialization and body disposal reflect heightened individuality in society. The emphasis on individual identity or the highlighting of some elements of identity can be seen as a form of deritualization (Palgi & Abramovitch, 1984); it is this very deritualized nature of modern life that can lead to the incorporation of new rituals and ways of interacting with the dead that can carry great meaning for the participants. Each of these different methods can be seen to highlight some elements of intersectional identity over others, to form community, and to reflect how the deceased wished to be remembered. I'll focus on a few general themes: body donation and display, body composting, and modern methods of memorialization and planning. Preplanning has become far more common in recent years, with the living planning

A. Osterholtz, *The Bodies Poetic*, Bioarchaeology and Social Theory,
https://doi.org/10.1007/978-3-031-85998-4_11

their own memorial service and, in some cases, throwing living funerals. All of these are important ways of reimaging the relationship between the living and the dead and leave different material traces. Some, particularly things such as body composting, can be seen as mechanisms for active forgetting.

Central to these new ways of interacting with the dead is a fundamental shift in our relationship between the living and the dead. The examples listed in this chapter focus primarily on body disposition methods that run counter to the concept of a 'traditional' funeral. The methods described here tend to focus on the individual over the group collective and highlight specific elements of intersectional identity. To put it in Putnam's (2020) terms, these new methods build bonding capital at the expense of bridging capital. What brings about this change?

Ellens (2014, p. 38) examines the role of the Baby Boomer generation in changing perceptions of death and dying. While hesitating to lump all Baby Boomers together, Ellens notes that this generation lived through some truly monumental social changes, including the Civil Rights Movement, Women's Rights Movement, the American Indian Movement, and an increasing recognition of environmental issues, just to name a few. The Baby Boomers include individuals born between the years of around 1946 through 1964, and would have been teenagers and young adults during these times of social change. As Ellens (2014, p. 38) puts it, "They started out in the 1960s with the intent of remaining forever young. Now they are wondering what it means to grow old. They are trying to figure out how such a generation grows old sensibly and whether there is any point in growing old." Schachter and Finneran (2014) argue that for Baby Boomers, two main events influenced their attitudes towards proper mourning and memorialization: the HIV/AIDS epidemic in the mid to late 1980s and the events of 9/11. Together these two instances of mass death were overwhelming, there was too much grief to keep hidden at home. Public displays of mourning spilled into streets and public spaces (such as public memorials at parks, the creation of the AIDS quilt, and so on). With a lack of prescribed rituals for mass mortality events, this generation had to create them.

In creating new rituals of public mourning, it created the flexibility to create new rituals for personal or private memorials as well. O'Rourke and colleagues (2011) conducted a study of funeral satisfaction and found that participating in the sharing of stories, expressing condolences, and otherwise interacting with others at a funeral had a robust relationship with whether the respondents found a funeral to be meaningful. Interestingly their sample consisted of college undergraduates, and so likely represents the viewpoint of a younger generation than the Baby Boomers (likely Gen Z given the publication date). It is interesting to note that current undergraduates find more meaning in "traditional" mortuary performance and suggests that the relationship with death that began to change with the Baby Boomers is likely to continue to change.

11.1 Mortuary Ritual Before Death

In recent years, a focus on interacting with death prior to an individual's death has grown. This section will look at the processes of funeral preplanning and living funerals as indicators of agency in the dying process, with the dying individual taking active control of their own dying process, mortuary performance, and how they will be remembered.

11.1.1 Funeral Preplanning

Funeral preplanning has gained more acceptance in recent years, driven in large part by an ageing Baby Boomer generation and their more open approach to discussing and interacting with death. Numerous mechanisms exist online to help with funeral preplanning and to plan conversations surrounding end of life care. One such website, The Conversation Project (https://theconversationproject.org), provides downloadable pdfs for interested individuals to begin their funeral plan, including things like a guide to choosing a health care proxy, being a health care proxy, and developing Do Not Resuscitate (DNR) and Do Not Intubate (DNI) documents. Another website envisions a dinner party where the central topic of conversation is end of life planning. For Death over Dinner (https://deathoverdinner.org), the website helps to determine who should be on the guest list as well as setting the intention and topics for the conversation. The website helps you to assign homework for guests, including something to read, something to watch, and something to listen to. After going through the "getting started" survey, the website says "You are now ready to activate your Death Dinner! After you enter your email, we will send you a personalized email with invitation language to send to your guests, the read/watch/listen homework, post-gathering next steps, your conversational prompts, and a quick overview on hosting and moderating this important discussion." In envisioning the discussion occurring over a communal dinner with friends and family, it creates a ritual in itself.

Preplanning guides are also offered by the Federal Trade Commission (https://consumer.ftc.gov/articles/planning-your-own-funeral) as well as various funeral homes, insurance companies, private companies, and the American Association of Retired Persons (AARP, https://www.aarp.org/home-family/friends-family/info-2020/funeral-planning-tips.html).

Who is likely to preplan their own funerals? Kemp and Kopp (2010) developed a model for a marketing journal that seeks to identify the motivations for end of life planning. They found that there was a positive relationship between perceived behavioral control (the amount of control an individual actually has over end of life planning) and the desire to develop those plans. They also found that death avoidance (the fear of discussing death) was negatively correlated, meaning that individuals who did not have a fear of discussing death were more likely to engage in end of life planning; this effect was influenced by the perceived level of control over the planning process. The conclude with suggests for marketers. Marketers should supply "messages that simply acknowledge the resistance that individual feel about engaging in the behavior, or that encourage planning through the initiation of small incremental steps (e.g., sharing wishes with family and friends)" (Kemp & Kopp, 2010, p. 88) to reduce negative feelings around end of life planning.

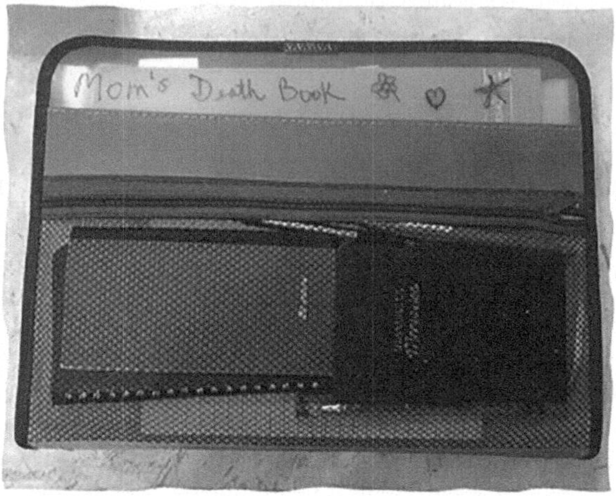

Fig. 11.1 Example of a binder created as part of end of life planning (created by Janet Osterholtz) containing information about her living will, DNR, DNI, account information, and her wishes for a memorial service. Photo by the author, 2023

My mother spent her career as a registered nurse, ending by providing end of life care via Hospice. End of life questions are central to hospice care, and as part of her own end of life planning, she developed her own plan for what will happen to her body and how she would like to be memorialized (including what to serve at the dinner and songs that should be played). Her planning included conversations with her family (my brother and I) and the gift of her "death book" a binder containing copies of her living will, advance directives, DNR and DNI orders, will, agreements for medical and financial proxies, and information about bank and utility accounts (Fig. 11.1). Her experience in Hospice, as well as the end of life decisions that she had to make for her parents and husband, predisposed her to desire more control over her end of life plan. She also expressed her desire that her children not have to make those decisions for her, particularly at a time of intense emotional hardship.

11.1.2 Rituals for the Dying

It is not uncommon to begin the memorialization and mourning prior to the death of an individual when that death is imminent, such as individuals with terminal cancer. These pre-death rituals can provide benefits for both the dying and their loved ones (Moneymaker & Traeger, 2007). Schachter (1999) performed a phenomenological research study of dying cancer patients and found two main themes in the creation of rituals: the importance of participating in and review of their lives and the need for patients to come to terms with and find meaning in their lives and their deaths. Many of these rituals revolve around established and newly created family traditions, such as enjoying a meaningful final thanksgiving dinner or creating new traditions of cookie making. The dying individual is actively engaged in designing these new rituals, and so has a significant impact in how they will be remembered, to "self-sooth and comfort themselves as they create the ritual that is most meaningful to them"(Schachter & Finneran, 2014, p. 174). This agency on the part of the dying, linked to both the rise in preplanning and the rise in palliative care is traced to patient empowerment stemming from consumer rights and Civil Rights Movements of the 1960s and 1970s (Beider, 2005).

New mortuary specialties are also becoming more common. One such specialty is that of a Death Midwife or Death Doula (deathmidwife.org). These practitioners perform both functional and social work for the dying and their loved ones. Functional aspects include understanding paperwork required to honor end-of-life wishes, researching state laws to determine what is required for the desired funeral to proceed (such as laws surrounding green burial or ash dispersal), helping to plan a living funeral/funeral/memorial service, and assisting with body processing after death (including washing and laying out the body). Social aspects of their work involve working with the family and the dying, creating an environment meeting the needs of everyone, providing support, and assisting with the mourning process. According to Rawlings et al. (2019), death doulas perform valuable services in areas with gaps in palliative care of community based care. Effectively, the role of a death doula is to demystify death, to take the conversation out into the public, to make it "everyone's business"(Rawlings et al., 2019, p. e87).

One service performed by the death doula may be the organization of a living funeral. These can take numerous forms; they are designed by the dying person and those close to them and so reflect the variety of identities that they wish to express. Living funerals began gaining popularity in the 1990s in Japan, where they're known as *seizensô* (Breen, 2004). Kawano (2004) ties this to the ageing nature of the Japanese population. They can also be seen as a negative response to the overt commercialization of mortuary performances in the 1980s. According to Kawano (2004, p. 156), those "choosing to have a pre-funeral challenge the common assumption that survivors take charge of mortuary arrangements. The pre-funeral does not replace a regular funeral, yet planners of pre-funerals indicate their desire to take charge of their mortuary ceremonies and say that they do not need a regular funeral." Kawano puts the pre-funeral or living funeral into performative terms, regularly referring to the performance of the pre-funeral as a way to maintain control over the memorializing process. Suzuki (2013, p. 103) states, "As a new ritual performance, the funeral-while-alive is significant for the deceased-to-be to acknowledge their life's worth, find identity amongst others, and to understand mortality and the life beyond in their own way to create permanence." In Japan, these also reduce stress on the survivors, both emotionally and financially.

While there is no culturally proscribed method for holding a living funeral, there are some trends. Living funerals offer an opportunity for the dying to say goodbye and thank you to those present. They offer an opportunity for the living to tell the dying important things as well, so that nothing is left unsaid during life. Georgia Martin, founder of a company that facilitates living funerals is quoted in a Guardian article, "I think when you say these things out loud, it takes away an element of grief, because you've said everything you wanted to say. It gives you clarity that you otherwise wouldn't have" (Aron, 2024). Living funerals can also be thrown for individuals in good heath but advancing age. For Millie, a woman quoted in the

Guardian, it was the experience of visiting other funerals that solidified her desire to have a living funeral when she was 87. "I decided to have one after we went to a funeral for a family friend at the local crematorium. On the way home, we talked about how the whole thing felt like a conveyor belt; one family would be ushered out while another was led in. It seemed so impersonal" (Aron, 2024). In general, we can see a desire amongst the participants and actors to have an individualized experience, something unique to them.

One widely publicized living funeral was that of Andrew Henderson (Glamdrew), a performance artist in Canada. Designed as a piece of performance art, Henderson's living funeral included a manicure bar, "lots of glitter and Henderson seating inside a champagne bottle" (Beaudette, 2016, n.p.). Even planning the living funeral was important, as Henderson is quoted for CBC News (Beaudette, 2016) saying "as soon as we tell people about the project, they're already having an immediate reaction. They're already experiencing a cathartic moment right then and there and they haven't even seen the piece yet." The event took place inside of an art space (aceart-inc) in Winnipeg, and was photographed by Michelle Panting and Adam Kelly. Eroca Nicols acted as emcee for the event. Panting (2016) wrote a blog post about this event, and notes that they felt "anxious and excited. This second emotion doesn't feel right for a funeral. But nothing about Andrew Henderson's living funeral performance piece, Taking it to the Grave, smacks of the ordinary."

Emotions were strong, "the manifestations of grief are palpable and intriguing. The young woman running the bar's hands are shaking… some sit quietly, others talk and laugh as if we were at any other social event. Old friends from out of town catch up" (Panting, 2016). Eroca, the choreographer and planner for the event, acted as a master of ceremonies as well:

> She leads us into Ace [the event space] where Glamdrew lounges shirtless, dressed in white, wearing a glittering gold cat-eye mask and makeup. He's spread out on a beige chaise crowned by a tule champagne bottle topper, encircled by an alter of objects: candles, cloth roses, glitter tchotchkes, a Roman bust, a framed photo of Kim Kardashian. To his right, a shroud of green fabric ghosts dangle above floor pillows, inviting guests to relax and take cover. Dance music throbs through the speakers. There's gold everywhere….
>
> Eroca sweeps gold glitter inside a diamond demarcated by tape between Glamdrew and the audience. She invites us to join her. Behind her, Glamdrew waits for people to approach.
>
> The come forward one-by-one to say goodbye, chat, joke, and sob. Each audience member is invited to confide a regret or secret to Glamdrew. Then Glamdrew, the individual, and tattoo artist Carly Boyce will agree upon a symbol to represent the regret. Glamdrew then has these symbols tattooed on his body: Carly on one side with her stick-n-poke, the mourner on the other….
>
> As the performance draws to a close, Eroca showers Glamdrew in champagne while he stands in a kiddie pool, then she invites audience members forward to pour water over him.

Henderson passed away days after their living funeral (CBC News, 2016). Nicols (2023, pp. 65–66) reflected on the performance years later. In planning the event with Henderson, Nicols wrote of the unknowns, particularly not knowing when

Henderson would actually die. "The project was a once-in-a-lifetime idea. We were very aware of the epic nature of our proposal, and it took a fair amount of effort to be casual about it. Everyone is going to die. Casual, epic, and inevitable. *Taking It to the Grave* called us and our audience to approach the undeniable fact *glamorously*." Nicols (2023, p. 69) revisits the relationship with Henderson in a physical sense: "I call back to my body the sensation of the material weight of how my love used to hit Glam's body and bounce back to me. and now it just goes and goes with no body to bounce it back. New old loss revisited." Henderson was ultimately buried in Clandeboye United Church Cemetery in Manitoba, Canada. The inscription on his tombstone reads "In Loving Memory of Andrew Carl Henderson 'Glamdrew' Sept. 26, 1998—Oct. 26, 2016. Embrace your true self with boldness and without hesitation. Share yourself, your burdens and your joys with others, celebrate love, life and all its wonderful, messy, amazing moments. Live a glamorous life, and drink champagne! Beloved Son, Brother, Friend to All. It bears an infinity symbol and a representation of the tattoo Henderson wore on their chest, a depiction of Venus on a half shell (Find a Grave, 2018).

Henderson's living funeral was extravagant, a true reflection of their personality and their approach to both life and death. The experience of Millie (a woman living in the UK) was significantly different. She described a sunny May day in a family orchard. Millie's daughter decorated the trees with ribbons, scattered lavender and rose petals on the ground and set out wooden mushrooms for the children to sit on. There were activities for the children to keep them busy, including cake decorating. Millie played her favorite records on a gramophone and had pastry and coffee served on her wedding china. They looked through photos and the children asked questions about them. She concluded saying "the whole occasion made me feel more connected to everyone around me. some people might not understand the concept of a living funeral, but it doesn't have to be called that—you can call it whatever you like. The main thing that I took from it was that life needs to be shared. It gave me the chance to share stories from my life that my friends didn't know about. I felt that I was very loved" (Aron, 2024).

11.2 Modern Inhumation: Green Burials as Emblematic of Social Relationships and Identity

Just as burial practices can serve as links to the past (such as the Yazidi faith-based use of cemetery space, discussed in Chap. 7), burial practices and locations can serve as ways to make new traditions and highlight different elements of identity. One such method of burial with increasing popularity is that of Green burial.

Interest in Green burial practices has grown in recent years, perhaps tied to the aging Baby Boomer generation, a generation that was and continues to be environmentally engaged. These burials and cemeteries are centered around the idea of minimal intervention in the body, meaning a lack of embalming and coffin-free or biodegradable coffin burials. Bodies tend to be shrouded or wrapped in biodegradable shrouds and buried within 24 hours of death unless refrigeration is an option. Burials tend to occur at 4 feet in depth. The Green Burial Council (https://www.greenburialcouncil.org, GBC) was founded in 2005 in the US to provide certification for providers and information for consumers of Green burial (Green Burial Council, 2020). The Natural Death Centre (http://www.naturaldeath.org.uk, NDC) provides similar information in the UK (Natural Death Centre, 2020a). Information can be found on both websites about approved cemeteries as well as provides for Green burial services. Both of these organizations see the Green burial movement as a way to both open a conversation about the nature of death and as a way to minimize environmental impacts due to burial with traditional embalming and burial services. But there are differences in how Green burial is approached; by its nature, Green burial is an individualistic process. In the US, burial within identified cemeteries using Green methods is allowable (e.g., Fig. 11.2) while the NDC in the UK argues that Burial Grounds should not be identifiable as such, nor should individual graves be identifiable or marked.

According to the GBC, within the US 4.3 million gallons of embalming fluid are used each year as well as 20 million board of hardwoods, 1.6 million tons of concrete, 17,000 tons of copper and bronze, and 64,500 tons of steel (Green Burial Council, 2020). The development of the funeral industry in the US shifted the responsibilities of preparation and burial to a set of specialists, separating the living from the dead. Lee Webster (2020) notes in a blog post that "so much of what goes on during standard burials is hidden—the opening of the grave, the removal of the soil, the setting of the vault."

Proponents of Green or Natural burial argue that in addition to a more environmentally responsible method of burial, this method challenges long-held perceptions of death and care. In anthropology, we tend to view mortuary ritual as very conservative, so changes in mortuary ritual reflect large scale social changes. Funerals are an industry, and one of the few industries that in which individuals rarely comparison shop. There is also a general understanding that the amount of attention and love an individual had in life is reflected in their mortuary kit (e.g., the coffin, type of service, etc.). Proponents of Green burial argue that this perception of worth and love needs to change. In a blog post, Ruth Faas argues that Green burial not only unlinks the idea of money spent on funerals with the feelings towards the dead by the living but gives "more options in how we grieve and honor someone"(Faas, 2020). According to the NDC, a green funeral: 1) rejects cremation; 2) burial in a site with a conservation purpose; 3) creates an environment not visually identifiable as a Burial Ground; 4) avoids embalming; 5) requires a coffin or shroud made from natural, sustainable materials; 6) forbids demarcation of the

Fig. 11.2 Green Acres Natural Burial Cemetery in Boone County, Missouri. Photo by Grey Wanderer, CC BY-SA 3.0 https://creativecommons.org/licenses/by-sa/3.0, via Wikimedia Commons. https://upload.wikimedia.org/wikipedia/commons/8/8d/Green_Acres_natural_burial_cemetery_in_Boone_County%2C_Missouri_.jpg

grave; 7) forbids marking or personalizing the grave with any sort of permanent memorial; and 8) forbids tending of the grave. The website states "It is not the grave that commemorates the life lived, it is the entire site"(Natural Death Centre, 2020b). The focus here is not on the individual within a Green Cemetery, but in the creation of a Space of commemoration for all the dead. Their website cautions the reader to consider this aspect carefully, as some find it difficult to not be able to locate specific friends or family within the cemetery space.

Green burial also puts the mourners back into the mortuary ritual as more than just observers of the ritual. During Green burial, family members and mourners are involved with dressing and preparing the dead as well as digging the grave, lowering the body into the grave, and closing the grave. It provides a more visceral relationship with the process of burial, a return to a time before the industrialization of death and development of death industry specialists. In a blog post, Buller (2020) describe walking individuals through this new process, reminding them to bring water and sturdy shoes for the walk, to lift with their legs and to be careful of their own bodies during the lowering of the deceased's body into the grave, essentially that there is a more physical experience with a Green burial than can be expected in a modern twentieth century burial where one is expected to drive to the cemetery in dress clothes and observe the lowering of the body into a mechanically dug grave. This physicality and participation in the mortuary ritual is seen as a tribute to the dead, a sacrifice of labor to send the dead off well, and a loving tribute. For Spade (2013), Green burials offer a secular approach to burial that is focused on connection with the dead and the earth and the environment. These approaches seek to find a meaningful way to inter the dead while expressing larger social concerns about environmentalism and connections to nature. According to Bechhoefer and Ochsner (2009, p. 2), "Society must evolve to respond to global environmental challenges. Natural burial is chemical-free and can have a negligible carbon footprint."

Green burial is mostly a rural or suburban phenomenon. For Bechhoefer and Ochsner's (2009) feasibility study in Cambridge, MA, one reason may be perception of ground water contamination related to decomposition. They note that water "bodies in cities are more susceptible to pollution by chemicals, oil, sediment, and litter than water bodies in rural areas, due to the fact that much of the ground surface in urban areas is not permeable. Water flows quickly over impermeable surfaces, and pollutants move too fast to settle out of the water"(Bechhoefer & Ochsner, 2009, p. 4). They go on to argue that open spaces within cities, including those used for Green burial could help to protect water quality be providing a filter for water. Soil in Green cemeteries is not compacted artificially, allowing for more permeation of water through it. They argue that topsoil can be added through mulching or soil importation for Green burial. Green cemeteries in urban spaces could provide a breathing space, perfect for visitors (Bechhoefer & Ochsner, 2009). They would provide different interaction with landscape, in effect bringing elements of nature into a built environment. For Bechhoefer and Ochsner (2009, p. 10), this incorporates the creation of the space with views is "important to visitors' aesthetic experience and sense of safety". In built environments such as urban landscape, this connection to nature created by Green cemeteries may highlight environmental stewardship as a responsibility owed by humans.

11.3 Body Donation (or How Grover Cleveland and His Wolfhounds Got to the Smithsonian)

In this section, a particular form of mortuary performance will be addressed: body donation programs and the creation of 'known' skeletal collections. Body donation programs can be viewed as intentionally giving one's body to medical or anthropological research; here I focus on body donation within anthropologically-oriented research facilities, in these cases the bodies are used for research and the bones are retained for future osteological research. For body donation programs with more of medical school focus, the retention of the bones is not necessarily an element; those remains may be cremated, buried, or returned to the donor's family. Within anthropologically-focused body donation, the long-term research opportunities of skeletal remains is integral to the decision to donate. Having demographic data as well as personal and health histories of the donors allows for the postulation of anthropological questions relating to lived experience, health and disease. The 'known' nature of the collections also allows for the establishment of new methods for the estimation of age at death and sex.

Body donation forms the basis for medical education, an opportunity for experiential learning for future physicians and nurses. The first voluntary dissection within the US took place in 1804 when a physician named Lewis Leprilete died near Boston. He had given a directive that his body should be given to a fellow Boston physician for anatomical dissection (G. F. Clark, 1859). Leprilete was one of the first surgeons to offer cataract surgeries in the US (Leffler et al., 2015), and his familiarity with surgical procedures is likely one reason that he wished his body to be used as a teaching tool after his death. Up until the 1870s, the majority of bodies willingly donated for medical education were those of physicians and/or surgeons or medical students (Lederer & Lawrence, 2022). Lederer and Lawrence (2022) trace the history of body donations for medical education, noting that by the mid 1980s, nearly all American medical schools relied on voluntary body donation programs. Such programs gained in popularity with an increasing cultural awareness of organ transplantation beginning in the 1930s (with corneal grafts) and blood donation in the 1940s.

Here we focus on voluntary body donation, not the acquisition of human remains acquired through processes leading to the creation of anatomical collections from indigent individuals who died in State care (e.g., poorhouses or almshouses) whose family were not able to claim the bodies of their loved ones at death. For more on this type of skeletal collection, Carlina de la Cova has published extensively on this (e.g., de la Cova, 2020a, 2020b). In particular, she focuses on the issues of structural violence inherent in these collections. The role of cadavers in medical education, including the history of structural violence is further discussed in Chapter 15.

Grover Kranz and his wolfhounds provides an interesting case study. As he said to David Hunt (then at the Smithsonian Institution), "I've been a teacher all my life, and I think I might as well be a teacher after I'd dead, so why don't I just give you my body" (Caputo, 2009). Kranz was an anthropologist who devoted his life to teaching. When he died in 2002 of pancreatic cancer, he reached out to David Hunt to facilitate the donation of his body for display at the Smithsonian. His body was used in research at the Forensic Anthropology Center (FAC) at the University of Tennessee, Knoxville. At the conclusion of the research study utilizing his remains, his skeleton was sent to Washington where it was articulated along with the skeleton of one of his Irish Wolfhounds (Clyde) and placed on display as part of an exhibition titled "Written in Bone: Forensic Files of the 17th-Century Chesapeake" at the Smithsonian's National Museum of Natural History (Fig. 11.3). Krantz's only stipulation for the donation was that he be stored along with his dogs at the Museum of Natural History. In a conversation with David Hunt in 2021, I was able to ask him about their interaction surrounding Krantz's donation. Hunt described the collection

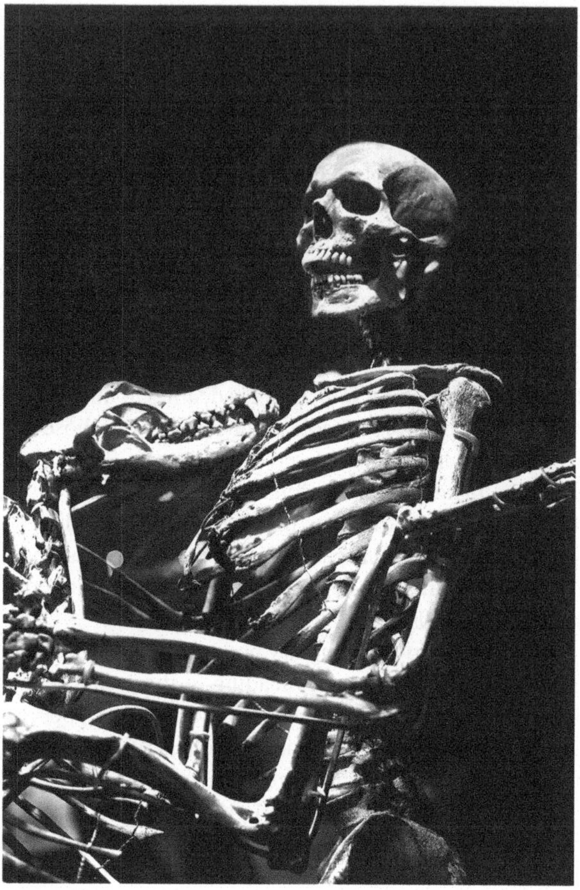

Fig. 11.3 Anthropologist Grover Krantz articulated hugging one of his Irish Wolf Hounds. Articulated for display at the Smithsonian's National Museum of Natural History. Photo attribution: Strickwerker, CC BY-SA 3.0 https://creativecommons.org/licenses/by-sa/3.0, via Wikimedia Commons. https://upload.wikimedia.org/wikipedia/commons/f/f0/Washington_-_Smithsonian_Museum_-_Grover_Krantz_%26_Clyde.jpg

of Krantz's wolfhounds from the family backyard (Claude, Dicky and Yahoo; Ralph is still buried in the family backyard). To honor the request to be stored with the wolfhounds, the dog skeletons needed to be cataloged as exemplary domesticated dogs for the Vertebrate Zoology section. While co-registered for both sections, the wolfhounds are housed in the Anthropology section. If Krantz's body is ever disarticulated, it will be stored with the skeletons of the wolfhounds, but for now, he remains articulated with Clyde and on display in an educational library.

Another anthropological example is that of Richard and Sheilagh Brooks. Sheilagh Brooks is well known for co-developing a primary method for age at death estimation in forensic anthropology and bioarchaeology (S. T. Brooks & Suchey, 1990). She was an important founding member of the Department of Anthropology at the University of Nevada, Las Vegas (UNLV) and ultimately became the first distinguished professor at UNLV in 1987 and received the Distinguished Faculty award. One of the Anthropology labs at UNLV is named for her. Sheilagh died in 2008. Richard Brooks was an accomplished archaeologist, founding the Archaeo-Nevada and Nevada Archaeological Association. As a researcher at UNLV and then the director of the Desert Research Institute, he led the way to create the Marjorie Barrick Museum of Natural History on the UNLV campus where he served as the director. His work at Tule Springs laid the groundwork for its designation as a National monument as well. He was an emeritus Professor of Anthropology when he passed away on June 23, 2011.

After their deaths, their bodies were sent to the FAC for use in research projects. At the conclusion of the projects involving their remains, their bones were sent to the Smithsonian's National Museum of Natural History to be housed in those collections. Their remains are stored side by side on a wide storage unit. The same section also contains another couple who donated their remains to the Physical Anthropology Division (stored together) and the remains of Grover Krantz's wolfhounds Dicky and Yahoo (David Hunt, pers. comm). Richard and Sheilagh had a true and lasting commitment both to anthropology and to each other. They were lifelong teachers and their bodies continue to teach. Their commitment to each other is also evident in their desire to be stored together. When I teach osteology, I teach the disposition of their remains as example of what love looks like in anthropology. While it may not may not be for everyone, body donation for Krantz and the Brooks completely and perfectly suits their lives and personalities.

Body donation is growing in popularity for several reasons. In a study of prospective doners to the Forensic Anthropology Center (FACTS) at Texas State University, Martinez and Brunson (2019) identified three main motivations: 1) an interest to participate in a scientific process; 2) a desire to have their bodies be helpful after death; and 3) a desire for an alternative to a traditional funeral. In the next section, I will discuss the body donation program at the Forensic Anthropology Center (FAC) at the University of Tennessee, Knoxville, as a case study. The FAC is culturally known as the "Body Farm" and has operated a body donation program for decades.

11.3.1 Body Donation at the FAC

Body donation to skeletal assemblages is gaining in popularity in the twenty-first century. Leading this, with the longest history and most prominent cultural placement is the Anthropology Research Facility (ARF), a part of the Forensic Anthropology Center (FAC) at the University of Tennessee, Knoxville (UTK), better known as the Body Farm. The ARF was initially developed by William Bass, then professor and head of the Department of Anthropology at UTK (Mission statement, 2024). Bass approached the UTK administration about developing an outdoor research facility to address issues of time since death estimation (post-mortem interval or PMI) and evidence security issues. The establishment of the facility was a matter of being the right person to ask the right administration (right time, right place, right people) (Meadows Jantz & Jantz, 2008). They broke ground on the facility in 1980 in a wooded site that needed to be cleared and access was gained through a road to a new facility built by Bass and his students consisting of a 16 ft^2 concrete pad and storage shed. They enclosed the facility (including overhead) with chain-link fencing. By the spring of 1981, the first donated body was received. Over the years, the ARF increased in size, ultimately to 1.3 acres of usable land. Better fencing was added, including an 8-ft privacy fence and razor wire.

As of March 2021, there are over 4000 individuals identified as 'pre-donors', these individuals have made the active decision to donate their physical remains after their deaths (Dawnie Steadman, pers. comm). They have filled out the forms and had them witnessed. One such pre-donor described the process in a New York Times article, "The process reminded me of applying to colleges—except that my alma mater didn't ask about my height and weight, medical and dental history, shoe size or scars and tattoos, nevermind my childhood socioeconomic status and every place I'd ever lived for more than a year" (Fitter, 2019). Additionally, families may make the decision upon the death of a family member to donate their remains without prior documentation. In order to be considered as a post-humous, family-directed donation, the request must come from the primary next of kin. Individuals who have infectious diseases such as Hepatitis C are encouraged to pursue cremation and donate the cremains to the FAC, who uses those in research projects as well.

11.3.1.1 Why Do People Donate?

The FAC receives over 100 bodies per year, consisting of both pre-donors, self-directed donations, and family-directed donations. The reasons given for donation tend to fall along a spectrum with a few notable patterns. First, teachers and educators commonly donate. Choosing a life-time of teaching can naturally be extended past death through body donation, and these individuals choose a path (or their families make the decision for them) that allows them to continue to contribute to teaching with their physical remains. Second, a common reason given is a desire to contribute to knowledge that helps the criminal justice system. Individuals giving this reason generally have some contact with the criminal justice system during life, either a victim of crime or the family member of a victim. This is a desire to give back to the criminal justice system and a desire to increase the ability of the criminal justice system to bring those responsible to justice in a larger sense than their own cases. Finally, a growing number see body donation as a form of green or natural burial as well. These individuals don't want to be buried or cremated, and see this as a way a method of green burial that has the additional benefit of contributing to science and knowledge.

Fitter (2019) sees the entire process as comforting, and gives an overview of the process in for the New York Times, contemplating an afterlife spent in the climate controlled storage of the Bass Collection: "There I'll stay forevermore, cared for like a rare book or stack of historical documents, in a library where students can check out my disarticulated bones, study me, then return me to the shelf for the next scholar." Fitter sees this mortuary process as permanent, but with continued engagement with the living, finishing the piece "I hope they return me to the shelf with the satisfying sense that even though I'm long gone, there are still plenty of stories I can tell."

11.3.1.2 Who Donates?

The FAC has been in operation for over 30 years (FAC Mission Statement, 2024). The Bass Collection is the result of those individuals who donate their bodies and skeletons, as very few individuals who go through the FAC are stored in other locations. The Bass collection is therefore going to reflect the demography of the individuals who choose to be donors (or whose family choose donation for them based on an understanding that it would have been their wish). Currently, the Bass Collection is skewed towards individuals of male sex of European descent. This pattern will likely change, though, since the pre-donor pool consists mostly of individuals with sex of female. Because of this, female skeletons are more likely to enter the collection in coming years and this will change the makeup of the Bass Collection significantly. Why is this change occurring? Dawnie Steadman (pers. comm) suggests that two factors may be at play: body donation is becoming more socially acceptable, and women are taking advantage of that; and the interest in forensic sciences amongst women. In recent years, women have increased in representation in forensics, a trend has been evident both in forensic sciences in general and forensic anthropology in particular. This increased interest is reflected in an interest in donation to contribute the further development of the science that they love.

11.3.1.3 At the FAC: The Process

Upon death of the donor (whether they be a pre-donor or family directed), the Department of Anthropology is notified by the family. The administrative assistants quickly transfer the calls to senior faculty who walk the family through the next step of the process. If the donor dies within 100 miles of Knoxville within Tennessee, representatives (normally consisting of graduate students from the Department of Anthropology) from the FAC collect the body. These individuals cannot enter nursing homes or residences. If death occurs within nursing homes or residences, the body must be transferred to a location that the FAC representatives may enter. Outside of this 100-mile radius or outside the state of Tennessee, families are responsible for transportation costs. The fame of the FAC is widespread, and so when families call to ask about donations for a family member who has died (family-directed donations), the faculty of the FAC may suggest that they contact a similar, more closely located facility. In some cases, the faculty may contact the research facility and help facilitate the donations. In some instances, though, the family may wish to donate specifically to the FAC for personal reasons (in one case, a family wished to donate to the FAC as the donor was a UTK graduate and specifically wanted to remain at UTK) (Dawnie Steadman, pers. comm).

When the body is received at the Center, intake procedures begin. These consist of taking hair, blood and fingernail samples for potential future research projects. Samples are also taken for current research projects and the body is evaluated for which projects that it can be enrolled in. Body weight and stature are measured, and a photographic record is created (Meadows Jantz & Jantz, 2008).

Once enrolled in a project, the active research may take from 1 to 2 years or longer depending on the nature of the research questions being asked and trainings offered to law enforcement and university students (Dawnie Steadman, pers. comm).

By the end of the research projects, the remains consist primarily of skeletal remains. The remains are then collected by undergraduate students as part of a field-based training exercise consisting of laying out a grid, documentation of the field conditions, and collection of the remains. The remains are then accessioned into the Bass Collection, a permanent skeletal collection that can be used to formulate new methodology to estimate age at death, sex, and stature (Meadows Jantz & Jantz, 2008; Shirley et al., 2011). These remains form one of very few skeletal collections that consists of modern individuals who provided informed consent for research and teaching with their remains. If we want to understand the effect of secular change (the change we've undergone over the last few hundred years in forehead height and stature, for example), this collection constitutes an important element of the baseline data used in those analyses. These remains can be used to understand more about antemortem trauma and disease patterns as well. Remains are not returned to the family members; in effect, the collection becomes a memorial to their contribution to science.

11.3.1.4 Donation and Memorialization

Every step of the process is presented as an educational one, from the interface with pre-donors and their families to the collection of the remains. Individuals who donate their bodies can expect their bones to become teachers, to enrich the lives of the students and researchers who come in contact with them. Students become more proficient excavators and analysts while gaining an understanding of the important nature of the gift they're given in working with these remains. As Dawnie Steadman notes, "there is no end to the gift of body donation" (pers comm).

Family members do on occasion ask to see and interact with the bones of the donors. Family members are not allowed to visit the facility during active research. But they may visit the remains once they are accessioned into the Bass Collection. When this request is made, the senior faculty (usually Lee Jantz) will facilitate these interactions, which may range from wanting to visit the building that the bones are housed in to interacting with the remains directly. They may want to simply see the box, or may want to hear from researchers what they've learned based on the remains. The faculty accommodate the wishes of the family members in these cases, they may leave the family with the remains, or lay them out in anatomical order, they answer questions, listen to stories about the life of the donor, and so on. Dawnie Steadman notes that this often becomes an impromptu memorial service, often involving the entire family. This may be a solemn affair or a joyous occasion, but it is always treated with reverence and respect by the faculty at the FAC (pers comm). As Fitter (2019) put it, “the part that I contemplate with a smile is that once I’ve entered the permanent holdings, the Forensic Anthropology Center will let my loved ones check me out too. I don’t really expect them to, but if they do, I picture them spreading me across a tabletop and reassembling me like a jigsaw puzzle. I image them asking how I’ve helped to solve murders, identify missing people or exonerate the wrongfully accused.”

11.3.1.5 Living Subjects Research

These individuals may be involved in Living Subjects Research as well. Currently, living research programs center on the creation of antemortem data that can then be compared with their postmortem remains. Iris patterns and fingerprints are collected from pre-donors that will ultimately be compared with changes to their physical remains after they pass.

11.4 Body Composting

If body donation creates a stable research-oriented collection of skeletal individuals, body composting can be thought of a process designed to quicken decomposition and return the body to earth as quickly as possible. Body composting modifies the body into a substance that can be partitioned to multiple mourners or can be used as rich topsoil. It can be seen as an alternative to green burial or traditional burial and an act of active forgetting.

Composting of the body is central to the experience offered by Recompose, a human-composting funeral home in Kent, WA. This is the adaptation that Spade (2013) proposes for Green burial within a truly urban environment. According to their website (www.recompose.life), they offer a service that is environmentally responsible while providing a Place of connection for the friends and family of the departed. The opening of this center in Washington was the result of almost ten years of research, fundraising, and campaigns to change state law (Kiley, 2021). The facility is located in a warehouse district, “inside it feels like an environmentalist’s version of a sleek, futuristic spaceship: spare, calm, utilitarian, with silvery ductwork above, a few soil-working tools… on racks, bags of tightly packaged straw neatly stacked on shelves, fern green walls, potted plants of various sizes”(Kiley, 2021). Small elements of nature are added to a very urban environment, the display of soil-working tools and straw brings the visitor to a space with the scents of a rural environment, but in a controlled and limited way.

Recompose is the brainchild of Katrina Spade, who began to explore the idea of an urban cemetery concept based on composting for her Master's degree in architecture focused on urban communities in the twenty-first century, who she describes as "decidedly urban, and made up of people for whom the city is the chosen site for living. This city dweller loves the bustling, complicated, concrete and steel metropolis reality. I posit they would find a deep comfort in becoming part of the city after dying… it is my position that a deep connection to the cycles of nature is critical in order for the dead to rest peacefully, and for the living to properly grieve. Therefore, I propose that the space I am designing—and its processes within—will be deeply rooted in the cycles of nature, for it is only by truly comprehending our part in these cycles that we can grieve and heal"(Spade, 2013, p. vii). Her work predisposes a need for connection to the earth as an element of human identity, and that disconnection from natural processes causes a kind of cognitive dissonance within individuals and groups. Connections to the physical world are easier in rural settings, in her view. So in creating the Recompose facility, she is bringing in a connection with nature and the physical world to an urban space. Her work is rooted in a view of death as part of the natural world, a process that shouldn't be ignored or feared. She argues that casket use and embalming are emblematic of a societal fear of death (Spade, 2013).

Recompose describes their approach "as practical as it is meaningful" and claim that it "connects the end of life to the natural world." In their process, the nude body is laid out and surrounded in a "cradle" by woodchips, alfalfa, and straw. This is then placed into a "recompose vessel" and covered with additional "plant material." The body and plant material are in the vessel for 30 days during which "microbes break everything down on the molecular level, resulting in the formation of nutrient-dense soil." In her thesis, Spade (2013, p. 31) calls this the "Process space," and states that "it is important that the process space is not simply a back-of-house area, rather, it will be designed as a sacred space." One cubic yard of soil is created during the composting process. This soil can be taken by the families after the process is complete, but families are also encouraged to donate this soil to Bells Mountain, a 700-acre land trust in southern Washington ("The Process," 2021). The handover of soil might take place in what Spade (2013, p. 31) calls the "Resolution Space, where the end product of the process of decomposition will be revealed and celebrated. This space will function as memorial to the deceased, a place where people will come to visit and remember… the Resolution Space might, at times, be boisterous and light, just like the city itself." According to their website, bones and teeth are also broken down during the composting process along with "blending" and "aeration" during the decomposition time. Non-organic materials such as dental fillings, medical devices and artificial limbs/joints are found during screening and recycled when possible.

Recompose is designed to be a "sacred space that is non-religious" (Spade, 2013, p. 34). It is, in effect, a space that celebrates the circular nature of existence, from life to death and decomposition. For Spade (2013, p. 34), the soil that is created is "deeply meaningful and should be celebrated." For secular individuals, Green burial may offer funeral services which place the death of friends and family within an ecological framework, giving meaning and purpose to the modification of the body as it nourishes the earth. "This project is both a space and a ritual, and its purpose is to connect us to the earth. In doing so, it helps to facilitate the grieving process, reminding us that we are a part of something much bigger than ourselves"(Spade, 2013, p. 45). In this way, it can be seen as regenerative, and community-building for those whose friends and family members' bodies have been composted.

As of early February, 2021, several obituaries are listed on the Recompose website, including that of Robert Michael (Amigo Bob) Cantisano. Cantisano died at his home in December 2020 of cancer. His obituary describes as a "beloved farmer/gardener/mentor and early pioneer of the organic farming movement." It goes on to say, "Amigo's wife and Adam Nuber of the Felix Gillet Institute plan to pick up the resulting cubic yard of compost… Being an organic farmer, Amigo knew that compost is the foundation of all the best organic farming. It is the living, microbiology that makes minerals and nutrients most bioavailable, contributing to robust, productive and happy plants. It is only fit that he should be composted and become the living biology that inoculate and nourish the composts, farms, and gardens of others"("Robert Michael (Amigo Bob) Cantisano," 2021). His family sees the composting of his body as a natural continuation of his passions in life. Their relationship with Cantisano will continue through the activities that he loved, partially through the use of his body in their own gardening. He is converted to a material substance that can be used, given away, and continue to nourish those around him. The act of taking the soil that held the body and controlling its dispersal and disposal is an interaction with transformed body that carries tremendous meaning for those involved, and helps to turn individual gardens into Places of interaction with the dead.

Another individual who chose this new method of body transformation was Paulie Bontrager, a committed environmentalist and vegan nature lover. Charlotte Bontrager discussed the option with her mother Paulie, and told a reporter that "she'd be thrilled to know she was among this first group of pioneers"(Kiley, 2021).

Also in the theme of composting is the use of the Infinity Burial Suit, or "mushroom suit" developed by Jae Rhim Lee. This is a suit that the body of the deceased is dressed in after death made from organic cotton fibers embedded with mushrooms and microorganisms that speed decomposition, "work to neutralize toxins found in the body and transfer nutrient to plant life" (Coeio, 2016). Brought about by Lee's question, "Could mushrooms be the symbol and tool for a cultural shift in how we think about death and our relationship to the planet?"(Coeio, 2016), she began imagining a shift in the conceptions of death and how we approach "our relationship with death." The idea here is to approach death as a natural process of returning to the earth in the least environmentally harmful manner possible. The suit and process is very much based in art and design, with the fashion designer Daniel Silverstein responsible for the overall design of the suit. The company also produces a shroud and coffin liner with the same mixture of biomix embedded to allow for viewings and funeral services, though they suggest that biodegradable coffins should be used. Burials using the Infinity suit are designed to be buried at approximately 4 feet in depth, shallower than a traditional burial and without a vault.

The first individual to be buried in the Infinity Suit was a carpenter named Dennis White from Woburn, Massachusetts. White had dementia, and contacted Coeio after seeing a Ted Talk by Lee on an early version of the Infinity Suit. Coeio produced a documentary that can be watched on vimeo (https://vimeo.com/145882693). This documentary is a fascinating view into one man's concept and planning of his own death as well as how his family approached death and plans for burial. This combines what began as an art project with the pre-death mourning process that is involved with having a terminal family member.

A more recent example of the use of the Infinity Suit was the burial of Luke Perry in 2019. Perry died in March of 2019 of a stroke at the age of 53. Sophie Perry, his daughter, wrote in a social media post that mushrooms held a whole new meaning for her (lemonperry, 2019), writing that "They are truly a beautiful thing for this beautiful planet, and I want to share it with all of you." The public forum of Instagram allowed for the sharing of this death experience and burial within a Infinity Suit and the creation of a new kind of community centered on this death and burial experience. In a broader sense, the innovation of a 'new' burial method such as the Infinity Suit creates a space to discuss death in a very secular and international sense. Perry's use of this suit made the news worldwide, with articles such as Sini's (2019) *Would you get buried in a mushroom suit like Luke Perry?* on BBC News, Kaur's (2019) *Luke Perry's daughter says he was buried in a mushroom suit*, and Pesce's (2019) *Luke Perry was buried in a mushroom suit—which could become a trend*. In response to Perry's use of the suit, Lee noted that "the power of the suit is that it creates the need for meaningful planning and discussion around death"(quoted in Pesce, 2019).

11.5 Summary

This chapter has examined modern alternatives to traditional burial and cremation, focusing on the practices of body composting and body donation. Pre-death rituals such as living funerals and funeral pre-planning are also discussed, particularly in relation to agency and as a reflection of broader societal shifts towards individuality and personalized memorialization. These new rituals offer new ways to express grief, honor intersectional identities, and reflect the changing relationship between he living and the dead. The development of these new rituals is influenced by the Baby Boomer generation's experiences with social movements and mass death events such as the HIV/AIDs epidemic and the 9/11 attacks. In contrast, younger generations (such as Gen Z) appear to find more meaning in traditional forms of memorialization. Overall, this chapter highlights how death rituals continue to evolve, adapting to personal preferences and societal transformations.

References

Aron, I. (2024). 'I didn't realise I was so loved': the people hosting their own 'living funerals". *The Guardian*. Retrieved from https://www.theguardian.com/society/2024/jan/28/i-didnt-realise-i-was-so-loved-five-people-stories-of-a-living-funeral

Beaudette, T. (2016). 'Taking it to the Grave': 28-year-old with terminal cancer plans performance art living funeral. *CBC News*. Retrieved from https://www.cbc.ca/news/canada/manitoba/taking-it-to-the-grave-28-year-old-with-terminal-cancer-plans-performance-art-living-funeral-1.3799098

Bechhoefer, R. & K. Ochsner (2009). Urban Ground: An Urban Natural Burial Guide + Site Sustainability Study, Conway School of Landscape Design. Spring 2009. Retrieved from https://issuu.com/sundewdesign/docs/urban_ground_natural_burial_guide.

Beider, S. (2005). An ethical argument for integrated palliative care. *Evidence-based Complementary and Alternative Medicine, 2*(2), 908416. https://doi.org/10.1093/ecam/neh089

Breen, J. (2004). Introduction: Death issues in 21st century Japan. *Mortality, 9*(1), 2–12.

Brooks, S. T., & Suchey, J. (1990). Skeletal age determination based on the Os pubis: A comparison of the Acsadi-Nemeskeri and Suchey-Brooks methods. *Human Evolution, 5*, 227–238.

Buller, J. (2020). "Green Burial During Covid-19." GBC Blog. Retrieved from https://www.greenburialcouncil.org/blog/green-burial-during-covid-19. Accessed December 2, 2020.

Caputo, J. (2009, February 11). Grover Krantz donated his body to science, on one condition…. *Smithsonian Magazine*. Retrieved from https://www.smithsonianmag.com/smithsonian-institution/grover-krantz-donated-his-body-to-science-on-one-condition-38726179/

Clark, G. F. (1859). *A history of the town of Norton, Bristol County, Massachusetts, from 1669–1859*. Crosby, Nichols, and Company.

Coeio. (2016). *The infinity burial suit*. Retrieved from https://coeio.com

de la Cova, C. (2020a). In C. M. Cheverko, J. R. Prince-Buitenhuys, & M. Hubbe (Eds.), *Making silenced voices speak: Restoring neglected and ignored identities in anatomical collections. Theretical approaches in bioarcaheology* (pp. 150–169). Routledge.

de la Cova, C. (2020b). Processing the destitute and deviant dead: Inequality, dissection, politics, and the structurally violent legalization of social marginalization in American anatomical collection. In A. J. Osterholtz (Ed.), *The poetics of processing: Memory formation, cosmology, and the handling of the dead* (pp. 212–234). University Press of Colorado.

Ellens, J. H. (2014). From death denial to death control: The new baby-boomer approach. In C. Staudt & J. H. Ellens (Eds.), *Our changing journey to the end: Reshaping death, dying, and grief in America* (New paths of engagement) (Vol. 1, pp. 29–44). Praeger.

Faas, R. (2020). "Money, Grieving & Burial Amid Coronavirus." GBC Blog. Retrieved from https://www.greenburialcouncil.org/blog/money-grieving-burial-amid-coronavirus. Accessed December 2, 2020.

Fitter, F. (2019). My afterlife on the body farm. *New York Times*. Retrieved from https://www.nytimes.com/2019/02/28/well/live/my-afterlife-on-the-body-farm.html?action=click&module=Discovery&pgtype=Homepage

Find A Grave (2018). Memorial page for Andrew Carl "Glamdrew" Henderson (26 Sep 1988—26 oct 2016). Retrieved from https://www.findagrave.com/memorial/191372419/andrew-carl-henderson

Green Burial Council (2020). Green burial Council. Retrieved from https://www.greenburialcouncil.org. Accessed December 2, 2020.

Kaur, H. (2019). Luke Perry's daughter says he was buried in a mushroom suit. *CNN*.

Kawano, S. (2004). Pre-funerals in contemporary Japan: The making of a new ceremony of later life among aging Japanese. *Ethnology*, 155–165.

Kemp, E., & Kopp, S. W. (2010). Have you made plans for that big day? Predicting intentions to engage in funeral planning. *Journal of Marketing Theory and Practice, 18*(1), 81–90. https://doi.org/10.2753/MTP1069-6679180106

Kiley, B. (2021, January 22). Recompose, the first human-composting funeral home in the U.S., is now open for business. *The Seattle Times*. Retrieved from https://www.seattletimes.com/life/recompose-the-first-human-compositing-funeral-home-in-the-u-s-is-now-open-for-business/

University of Tennessee, Knoxville-Forensic Anthropology Center (2024). Mission statement. Retrieved from https://fac.utk.edu/mission-statement/

Lederer, S. E., & Lawrence, S. C. (2022). Rest in pieces: Body donation in mid-twentieth century America. *Bulletin of the History of Medicine, 96*(2), 151–181.

Leffler, C. T., Schwartz, S. G., Grzybowski, A., & Braich, P. S. (2015). The first cataract surgeons in Anglo-America. *Survey of Ophthalmology, 60*(1), 86–92. https://doi.org/10.1016/j.survophthal.2014.08.002

lemonperry [S. Perry]. (2019, May 3). In December I went to San Francisco with two of my best friends. One of them, had never been to California, so we went to show him the Redwoods. I took this picture while we were there, because i thought, "damn, those mushrooms are beautiful." Now, mushrooms hold an entirely new meaning for me. Any explanation i give will not do justice to the genius that is the mushroom burial suit, but it is essentially an eco friendly burial option via mushrooms. All i can say is that you should all look into them at coeio.com or just by googling "mushroom burial suit" . My dad discovered it, and was more excited by this than I have ever seen him. He was buried in this suit, one of his final wishes. They are truly a beautiful thing for this beautiful planet, and I want to share it with all of you [Instagram]. Retrieved May 20, 2020.

Martinez, H., & Brunson, E. K. (2019). Motivations for whole body donation. *Practicing Anthropology, 41*(1), 7–11.

Meadows Jantz, L., & Jantz, R. (2008). The anthropology research facility: The outdoor Laboratory of the Forensic Anthropology Center, University of Tennessee. In M. W. Warren, H. Walsh-Haney, & L. E. Freas (Eds.), *The forensic anthropology laboratory* (pp. 7–21). CRC Press.

Moneymaker, K. A., & Traeger, J. (2007). Creating space and ritual for the time of death. *Journal of Palliative Care, 10*, 270–271.

Natural Death Centre (2020a). The Natural Death Centre. Retrieved from http://www.naturaldeath.org.uk/index.php?page=home. Accessed December 2, 2020.

Natural Death Centre (2020b). What are Natural Burials and Green Funerals? Retrieved from https://www.goodfuneralguide.co.uk/what-is-a-green-funeral/. Accessed December 2, 2020.

News, C. (2016). Anderw Henderson dies days after his living funeral. *CBC News*. Retrieved from https://www.cbc.ca/news/canada/manitoba/andrew-henderson-dies-days-after-his-living-funeral-1.3822182

Nicols, E. (2023). Funerally yours: A collective re-membering of taking it to the Grave. *Canadian Theatre Review, 195*, 65–75.

O'Rourke, T., Spitzberg, B., & Hannawa, A. (2011). The good funeral: Toward an understanding of funeral participation and satisfaction. *Death Studies, 35*(8), 729–750. https://doi.org/10.1080/07481187.2011.553309

Palgi, P., & Abramovitch, H. (1984). Death: A cross-cultural perspective. *Annual Review of Anthropology*, 385–417.

Panting, M. (2016, November 21). *Taking it to the Grave: Glamdrew's living funeral at aceartinc*. Retrieved from http://michellepanting.com/archive/2016/11/21/taking-it-to-the-grave-glamdrews-living-funeral-at-aceartinc

Pesce, N. L. (2019). Luke Perry was buried in a mushroom suit—which could become a trend. *Market Watch*. https://www.marketwatch.com/story/the-woman-who-made-luke-perrys-mushroom-burial-suit-is-shaking-up-the-death-care-market-2019-05-07

Putnam, R. D. (2020). Bowling alone: The collapse and revival of American community (Revised and Updated). New York, Simon & Schuster Paperbacks.

Rawlings, D., Tieman, J., Miller-Lewis, L., & Swetenham, K. (2019). What role do death Doulas play in end-of-life care? A systematic review. *Health & Social Care in the Community, 27*(3), e82–e94. https://doi.org/10.1111/hsc.12660

Robert Michael (Amigo Bob) Cantisano. (2021). *Recompose*. Retrieved from https://recompose.life/2021/robert-michael-amigo-bob/
Schachter, S. R. (1999). *The experience of living with a life-threatening illness: A phenomenological study of dying cancer patients and their family caregivers* (Ph.D.). The Union Institute, United States -- Ohio. Retrieved from https://go.openathens.net/redirector/msstate.edu?url=https://www.proquest.com/dissertations-theses/experience-living-with-life-threatening-illness/docview/304570890/se-2?accountid=34815
Schachter, S. R., & Finneran, K. M. (2014). Expansion of new rituals for the dying and bereaved. In P. N. Stearns (Ed.), *Our changing journey to the end: Reshaping death, dying, and grief in America* (New paths of engagement) (Vol. 1, pp. 165–192). Praeger.
Shirley, N. R., Wilson, R. J., & Jantz, L. M. (2011). Cadaver use at the University of Tennessee's anthropological research facility. *Clinical Anatomy, 24*(3), 372–380.
Sini, R. (2019). Would you get buried in a mushroom suit like Luke Perry? *BBC News*. https://www.bbc.com/news/48140812
Spade, K. M. (2013). *Of dirt and decomposition: Proposing a resting place for the urban dead. (master of architecture)*. University of Massachusetts, Amherst.
Suzuki, H. (2013). Funeral-while-alive as experiential transcendence. In H. Suzuki (Ed.), *Death and dying in contemporary Japan* (pp. 102–122). Routledge.
The Process. (2021). *Recompose*. Retrieved from https://recompose.life/our-model/#the-process
Webster, L. (2020). About the Vault.... GBC Blog. Retrieved from https://www.greenburialcouncil.org/blog/about-the-vault. Accessed December 2, 2020

Chapter 12
Modern Memorialization: Vernacular Memorials

Abstract This chapter explores the evolving nature of memorialization and grief in the twenty-first century, focusing on vernacular memorials—informal, personalized forms of remembrance that exist outside traditional, sanctioned practices like tombstones and monuments. It examines a variety of contemporary memorial practices, including roadside memorials, spontaneous shrines, memorial tattoos, and virtual memorials, all of which reflect a shift toward individualized expressions of mourning and memory. Roadside memorials, such as ghost bikes, offer public spaces for grief while advocating for social change, blending personal loss with activism. Memorial tattoos provide a permanent, embodied tribute, allowing mourners to carry the memory of the deceased on their bodies, marking grief in a deeply personal way. Virtual memorials, such as Facebook memorialized profiles and online cemeteries, blur the boundary between life and death, enabling ongoing interaction with the deceased through digital platforms. This chapter also explores how digital spaces have introduced performative aspects of mourning, raising questions about the authenticity of online grief through phenomena like "bandwagon mourning." The rise of secular and personalized forms of memorialization is situated within broader societal shifts, such as a decline in religiosity and a growing desire for individual agency in death rituals. Additionally, the chapter discusses traditional commemorative practices, such as Día de los Muertos, which continue to evolve as expressions of cultural identity and community building. Through these diverse forms of memorialization, this chapter highlights how the interaction between personal mourning and collective memory is changing, providing new opportunities for connection, social identity, and public grieving in an increasingly digital and individualistic society.

Twenty-first century mourners have a plethora of options for memorialization and interaction with the dead. Those covered in this chapter are not meant to be an exhaustive list, but to provide some examples of how we continue the relationship with the dead. Here, I focus on vernacular memorials (those memorials that are not

A. Osterholtz, *The Bodies Poetic*, Bioarchaeology and Social Theory,
https://doi.org/10.1007/978-3-031-85998-4_12

officially sanctioned) as well as the processes that go into the decorating of graves and the Día de los Muertos (Day of the Dead) celebrations.

The term "vernacular memorial" refers to those memorials that occur outside official forms of memorialization such as tombstones, monuments, and so on. These may take the place of roadside memorials, tattoos, and virtual memorial spaces amongst other things, "limited only by the limits of individual innovation"(R. Dobler, 2010, p. 4). Here, I want to focus on a limited number of these vernacular memorials: roadside memorials, spontaneous shrines, tattoos, virtual memorials, and the modern material culture that comes out of all of them. As Dobler (2010, p. 6) notes, the spontaneous or personal nature of such memorials can be viewed as intensely meaningful, on both personal and cultural levels, they are "richly constructed, informed by ritualized behavior, and complexly functional." In many ways, vernacular memorials are the expression of the Social Body, mediated by the individual experience.

12.1 Roadside Memorials

Roadside memorials and spontaneous shrines can be placed in the location of death, and are not officially placed or maintained by responsible governmental agencies. Dobler (2010) traces the use of roadside memorials to the eighteenth century within the American Southwest as markers of those who died along the route of travel, predating the use of automobiles in many cases. These memorials allow mourners to have a public expression of grief and a regaining of personal control. As death has become ever-distanced from the daily experience of most individuals in American society, the positive actions of creating a memorial on the side of a road or, as will be discussed below, in front of a Las Vegas casino, allows the mourner to perform an action—to be actively involved—to contribute through labor or material objects to communal grief. As Dobler (2010, p. 9) notes, the "physical act of constructing a spontaneous shrine often helps the mourner to symbolically reconstruct his or her own life around the absence of the departed. It also allows the mourner to create a meaningful narrative of the deceased's life, giving structure and meaning to a death that may seem chaotic and meaningless in the immediate aftermath." Roadside memorials are often placed at the site of an individual's death, which has for their mourners, become sacred ground (Collins & Rhine, 2003).

Collins and Rhine (2003) comment on the pains that those they interviewed took to identify the exact location of death, often times placing the memorials directly next to blood stains on the concrete. Clark and Franzmann (2006, p. 579) see these memorials as evidence of the "overwhelming empowerment of grief", the bereaved taking authority to erect their own shrines in the way they see fit. Placing these memorials at the site of the death is a recognition that the nature of that particular bit of space is forever changed for them, it has become sacred. The dead are believed to be more present in these locations, and capable of interaction with the living. This sacredness of space at roadside shrines, along with the personal design of the memorial allows for the memorialization of individual elements of identity in ways that might not be sanctioned by official church or administrative doctrine. The erecters of such shrines do not ask permission to place them and frequently revisit them, adding new mementos to update the dead on the event in their lives (Fig. 12.1). For the builders of those shrines, they indicate "a willingness to take spiritual matters out of the hands of the church and divert spiritual energy in new directions" (Clark & Franzmann, 2006, p. 595; see also Collins & Rhine, 2003). They may have some religious iconography present (such as a cross), but they also may include flowers, notes, poems, CDs, scarves, beer, candy, and other items of material culture that held meaning for the deceased and/or the bereaved. Roadside memorials, since they are a more individualistic form of monumentation and typically more temporary, allow for the expression of intersectional identity in a public format in ways that might not be as accepted in a formal location such as a cemetery.

While many shrines, particularly in the Southwestern US (e.g., Fig. 12.1), are relatively simple constructions focused on a cross and the mementos left by mourners (Collins & Rhine, 2003), some will highlight specific elements about how an individual lived and how they died. The phenomenon of *ghost bikes* is one

Fig. 12.1 Roadside memorial outside Albuquerque, NM. Note the accumulation of toys and the flowerpot at its base. Photo by the author, 2024

such type of roadside memorial. The first ghost bike appeared in St. Louis Missouri to memorialize the death of a cyclist. A bicycle shop owner had witnessed the collision and placed a mangled bike that had been painted stark white at the site of the crash (R. T. Dobler, 2011). Dobler (2010) notes that ghost bikes often occur at the site of fatal bicycle-automobile collisions. Ghost bikes are placed as roadside memorials, but also serve a purpose to larger communities to advocate for bicycle safety. These memorials are visible, and often painted white to be even more visible (Fig. 12.2). These particular types of memorials impart a significant amount of social information: the location of death, the type of death, and the avocation of the dead are all communicated by the placement of these memorials. As Dobler (2010, p. 38) notes, the presence of a ghost bike "converts the site of memorialization into a rallying point for the cyclist subculture, transforming the emotional energy of mourning into a force of protest." This type of memorial may be placed by fellow cyclists, not necessarily family members and so takes advantage of the visibility and sacredness of roadside shrines to pass along social messaging. Costantini (2017, 2019) examines these from a performative perspective, noting that the social messaging can be one that does not link to a specific cyclist, but to the larger social issues of bicycle safety. In this way, the ghost bikes are protests as much as they are memorials.

Fig. 12.2 Ghost bike memorial on the outskirts of Albuquerque, NM. Photo by the author, 2024

12.1.1 Spontaneous Shrines

Santino examined the role of spontaneous shrines from a folklore perspective (Santino, 2004). According to Santino, spontaneous shrines combine the personal and the public. In his work, he sees the creation of spontaneous shrines as a public display of personal grief, a form of ritual that allows for more individual participation since they are not sanctioned or organized by an overarching corporate entity such as a governmental body. One excellent example of this is the spontaneous shrine at the New York-New York Casino in Las Vegas, NV. The shrine began organically, with visitors who happened to be in Las Vegas on September 11, 2001 began to leave T-shirts on the hedge around the replica of the Statue of Liberty outside the New York themed casino and resort on the Las Vegas Strip. These T-shirts, depicting logos from Fire Departments, Police Departments, and so on from around the country were the clothes that individuals had packed for their vacation to wear. They were treasured items that spoke of the identity of the wearers. More than 5400 T-shirts were left by visitors (Ahmed, 2011).

The spontaneous shrine created by these T-shirts was intensely moving, and ultimately placed under UV-resistant glass in the same locations they'd been placed in 2002. In 2013, the display cases were removed as part of a redesign of the overall casino, but a commemorative plaque is still visible in front of the casino. In addition to T-shirts, letters, mementos, postcards, and personal items were also left. The New York-New York Casino donated all the objects accumulated between 2001 and 2012 to the UNLV Libraries Special Collections, and each item was cataloged and stored in 491 archival bins.[1] For example, Box 482 contains framed pictures and poems, a stuffed bear dressed as a firefighter, a dream catcher, and embroidered patches from fire departments and police departments from around the world (Brean, 2016). In 2011, 10 years after the 9/11 attacks, a plan began to use the T-shirts in a new memorial. UNLV's Library Dean, Patricia Iannuzzi worked with Troy Gillet, an alumnus of UNLV who submitted a proposal for a 32-foot high double helix made from about 3000 of the T-shirts to be placed in Lied Library on the UNLV campus titled "Common Threads". Quoted in the Las Vegas Sun, Gillet was interested in "people's motivation for bringing materials down to the New York-New York and why roughly 5,000 people decided to do this…. I wanted to make a statement on what ties us together as human beings, what people do in times of tragedy when they collectively have an unconscious need to reach out" (Ahmed, 2011). The shape of the piece in a double helix is specific, "a metaphor for what makes us human and what ties us together in times of tragedy" (Gillet quoted in Ahmed, 2011). For David Schwartz, the memorial and special collections gathered from the spontaneous shrine have tremendous power, both in terms of creating a collection for future research but also as a lasting memorial to the intense emotions surrounding 9/11. He notes that the current generation of college students have no memory of the 9/11

[1] The special collection can be searched here: https://special.library.unlv.edu/ark%3A/62930/f1zp5k

attacks and that the objects in the special collection have the power to connect them to this history (Brean, 2016).

Running through the meaning and importance of roadside memorials, ghost bikes, and spontaneous shrines is the seemingly senseless nature of the death(s) memorialized. We can see roadside memorials and spontaneous shrines as a new form of mortuary ritual, necessitated by changes in technology (increasing vehicular deaths with the use of automobiles, etc.). As Clark and Franzmann (2006, p. 582) state "[m]ourning rituals contrived in a previous era for a different grieving may seem inadequate and unable to construct meaning for a seemingly meaningless event. Something else is required. The construction and maintenance of a roadside memorial is one such new ritual. If we accept the general premise, that for those bereaved by sudden and violent death, such as road trauma, seek more meaningful ways to mourn, then we should search for those elements that empower the bereaved and struct and practice new or adapted mourning forms."

12.2 Tattoos as a Memorial

Here, I will focus on the concept of a memorial tattoo (Fig. 12.3), a tattoo which an individual gets as a reminder of the dead. According to a 2019 Ipsos poll of 1005 Americans interviewed online, 30% of American's have at least one tattoo, with age-related trends evident. Individuals under 55 are more than twice as likely to have at least one tattoo, with 40% of individuals between 18—34 and 36% of those between 34 and 54 with at least one tattoo (Ipsos, 2019). Later, in Chap. 16, I will focus on the preservation of tattoos after death that can be kept as a memento or personal relic. Getting a tattoo is a ritualized process, including the making of an appointment, deciding on a design, and witnessing the setup of the artist prior to beginning the tattoo. The needles are opened from their sterile pouches, and typically, the person getting tattooed will watch them being put into the tattoo machine. The artist will then look at the person about to be tattooed and ask "ready?". It is a process that is dictated by health codes but the systematic nature of getting a tattoo creates a ritualized atmosphere. No matter how many tattoos an individual has, there is always a little bit of anticipation. This is more likely to be true for individuals with multiple tattoos, you know what to expect from the process. Pain is exchanged for art, the power to dictate the appearance of your own body is purchased by not only the money spent on a tattoo but also the experience of the tattoo.

Grief and mourning are embodied experiences, and mourners experience them physically. One way to mitigate that or to embrace the physicality of grief may be getting a memorial tattoo. Davidson (2017) identifies five important aspects of memorial tattoos in what is described as "good grief". First, tattoos show continuing bonds between the living and the dead, drawing on continuing bonds and important images that held meaning to the dead. Second, tattoos are permanent markers of grief. As Davidson (2017, p. 35) notes "Our memorial tattoos, then are permanent embodied representations of love and loss. That our tattoos, like our loved ones,

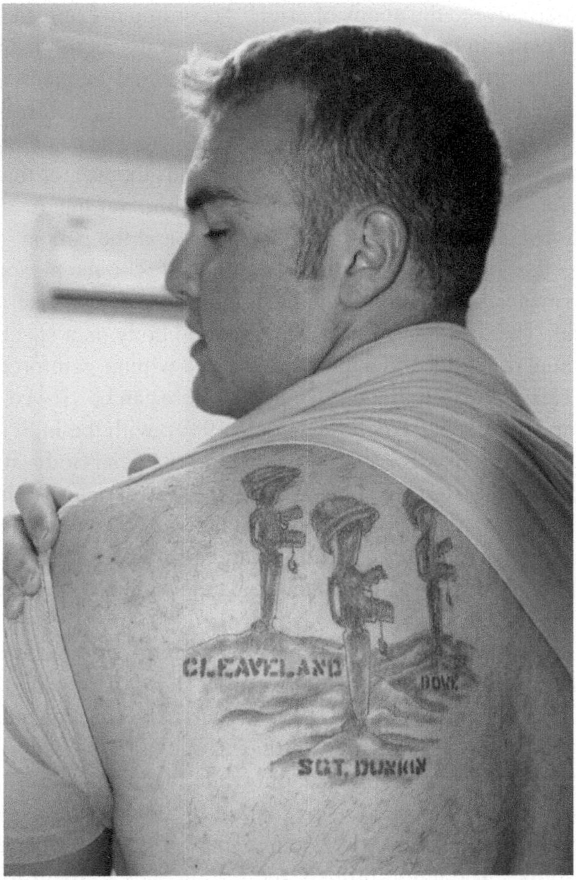

Fig. 12.3 Memorial tattoo on the back of Spc. Steven Baker (Photo taken September 2009). Photo attribution: Staff Sgt. Stephanie van Geete, Public domain, via Wikimedia Commons. https://upload.wikimedia.org/wikipedia/commons/9/93/US_Army_50997_Silent_Tribute.jpg

remain with us until our death is key to their import as memorials." Third, memorial tattoos show an adjustment to loss, a physical representation of the adjustment to a person's death and a way to incorporate them into their lives in meaningful ways. Fourth, memorial tattoos have the capacity to open dialogue on death and dying: "Tattoos invite casual conversation and once open, bearers of memorial tattoos use the opportunity to talk about their loved ones. This lifts stigma, helps to normalise death and assuage grief" (2017, p. 36). Finally, a tattoo marks *before* and *after*. Just as an individual might get a tattoo to memorialize college graduation of marriage, a memorial tattoo is a physical acknowledgement that things are forever different in their lived experience. Tattoos in general can be read as a language, symbolic representations that are used on the context of social and emotional bonding (Bergh, 2016), and "allow access to expression without the need of words" (Warnick & Toye, 2016, p. 134). Part of grammar of that language might be the role of pain involved with tattooing. Depending on the location of a tattoo on the body, it can be a very painful process. Cadell et al. (2018, p. e31) found that for those that experienced pain as part of the tattooing process, the physical pain was seen as an "accepted alternative to the agony of grief." According to Lanigan (2007, p. 42) the experience of pain can be a "cathartic component of the memorization process."

In terms of the Poetics model, the ritual of receiving a tattoo can be seen as incredibly meaningful. Memorial tattoos, in particular, are meaningful experiences. The memorialization is part of the process. The ritual nature of getting a tattoo (the setup, the process, the aftercare) reminds the wearer that their body is changed forever. This is typically done in a public setting (a shop), with others around. Conversation between the tattoo artist and the person getting the tattoo occurs (as well as others wandering through) about why the person is getting the tattoo, what it means, and so on. While the relationship can be an ephemeral one, it is meaningful. The artist, in forever changing the body, is a specialized practitioner, and the ritual of inking and wiping, inking and wiping, reinforces the change to the body. It is an expression of identity and therefore can be viewed in terms of the three bodies model (Scheper-Hughes & Lock, 1987), with the individual choices made through the agency of the wearer mitigated by the Social Body (which can be viewed through common designs of styles of tattooing—American Traditional, Japanese, Black and Grey, etc.). The Body Politic can be seen to control the process through the introduction of health codes. Tattoos are openly performative, especially those on visible parts of the body. When memorial tattoos are visible, they can be seen as part of the "embodied meaning making process in response to loss, whereby participants chose their memorial tattoos to create a sense of permanence in response to the demonstrated impermanence of their life story disrupted by bereavement"(Buckle & Corbin Dwyer, 2023, p. 4). On my left calf, I carry a memorial tattoo for my father, and the act of getting that tattoo was one of tremendous meaning. I chose the design carefully to symbolize freedom while focusing on bright bold colors that my father would have liked. As was reported by Buckle and Corbin Dwyer (2023, p. 7) for participants in their study, I often touch this tattoo when I think of my father. In their words: "The tattoo brought the person back to them in those moments and represented the constancy of the relationship." There is a concrete materiality to a memorial tattoo that is only brought about through the act of getting one. When this action of a tattoo is used to memorialize a friend or loved one, this becomes a constant act of memory, much in the way that tattooing stimulates the immune system (e.g., Lynn et al., 2020), memory of the dead is constantly stimulated for the wearer of a memorial tattoo. It may change the way others interact with the wearer, facilitating conversation and creating community.

In terms of the Poetics of the performance of tattoos, we can examine this in two different times: the tattooing process and the wearing of the tattoo. For Poetics to have an amplified meaning, there needs to be an audience. The audience may differ for these two different time periods, or it may be the same. It isn't uncommon to have friends and family members who knew the deceased present for the act of getting a memorial tattoo. Witnessing the tattooing process for those individuals may be meaningful. Tattooing studios or shops tend to be less-private spaces (though some will have private rooms); tattooing is effectively a public process. People coming in and other artists come by and witness the act, which can prompt discussion and remembrance. For the wearing of a tattoo, the audience includes the friends and family of the wearer and the deceased memorialized in the tattoo. The audience also includes those who did not know the deceased and who do not know the wearer; this is particularly true if the tattoo is in a visible location. It is a visible expression of connection to the dead, a statement by the wearer of their loss and their continued relationship to the memorialized individual. Similar to the manning of the Tomb of the Unknown at Arlington (discussed in detail in Chap. 13), the audience for the ritual of memorial tattoos is an amorphous one. It shifts depending on the part of the process examined and it is an ongoing one. There is no "off" switch to the performance, it is ongoing.

12.3 Virtual Memorials

Here we turn to the concept of grief and memorialized within an online space in the form of social media posts, virtual memorials and virtual cemeteries. Essentially, online spaces allow for a new public venue for the display of grief and memorialization. The online space is unique in that it provides both community and anonymity for posters. Kasket (2019) comments on the nature of social media to act as an archive of lived experience, particularly for those individuals growing up in a digital age; this archive allows for a more personal knowledge compared to curated photo album of individuals as we had in the past. She notes, "[w]hen someone dies two types of being cease to exist: the sentient, breathing physical being, and the being of that person's mind. If someone has generated a large amount of data and 'lived their lives online,' however, their hybrid digital being—a being composed of data but nevertheless resonant of both *res extensa* and *res cogitans*—has the potential to persist" (Kasket, 2019, p. 470). For Nansen et al. (2014), the persistence after death of an individual's data highlights what they term the "restless dead". For previous generations, the dead were conceptualized as 'at rest' or 'asleep', but the persistence of their data reconceptualizes individuals as somewhere between the living and the dead in terms of social interaction. In effect, the concept of social death becomes fuzzy, someone's individual quirks and personality are not forgotten: they are archived in the digital spaces of Facebook, Instagram, and X.

For Facebook users, instructions are given as to how to appoint a legacy contact to look after your profile if you die (Facebook.com, 2024b). The legacy contact has information on managing a deceased persons Facebook account as well (Facebook.com, 2024a). In general, once Facebook is informed that the account holder has passed away, the word "Remembering" is shown next to the person's name on their profile, but all the content a person shared remains visible. Depending on privacy settings that were established before death, people may post remembrances or condolences to the profile of the deceased. It is unclear what happens if you fail to designate a legacy contact, and the overall number of dead users on Facebook is unknown. Using a big data approach, Öhman and Watson (2019) attempted to estimate the number of dead users. Even if Facebook ceases to attract new users in 2018, at least 1.4 billion users will pass away before the year 2100. To come to this figure, they looked at projected mortality by age and nationality, projected population data for the twenty-first century by age and nationality, and current Facebook users for age group and nationality. Their goal in performing the study was to "bring in a more societal perspective and ask what the phenomenon of people dying on the internet means for us on a collective level" (Öhman email quoted by Greenspan, 2019). For Öhman and Watson (2019, p. 9), their data should be a wakeup call to develop infrastructure to properly curate and manage the large amount of data that will be present in the future, so that it doesn't become a casualty of corporate profit motive, saying "what makes data 'worth preserving' is ultimately their ability to directly or indirectly contribute to the company's profit." While the memorial pages still produce significant traffic on Facebook (along with the associated ad revenue), it is worthwhile to maintain them. They used the same methodology to look at the number of dead users on Instagram found that between 767 million and 4.3 billion Instagram users will die between 2019 and 2100 (Öhman & Watson, 2021).

Instagram freezes the profiles of dead users upon requests from loved ones, the account doesn't appear on the "explore" page, and no one can log into the account. Facebook memorialization is not limited to the archiving of individual profiles, though. Numerous groups exist that allow users to post individual remembrances and photos of tombstones of their loved ones. One such group, "Cemetery Enthusiasts" is a combination of users that are fascinated by tombstones and cemeteries *and* individuals who post photos of personally meaningful tombstones. Often, these posts become a place for communal grief and commiseration. One such post, from September 16, 2024 can be used as an illustration. The Post has the text saying "My oldest son! I learned a lot while raising him and I've learned even more since I lost him. I spend my lunch break with him every day! My heart will never be the same!!" and is accompanied by a photo of a headstone (Hall, 2024). The comments are supportive, with one saying "So sorry, it's the shittiest club you can belong to, sadly there are too many members" and another "I'm so sorry for the great loss you all have suffered. The headstone is beautiful. I'm sure he would've loved it". Commenters asked questions that began conversation about the meaning of the inscription, how her son had died, and so on.

For online spaces, Savin-Baden (2021) argues that new forms of grief can be performed. Not only Kasket's (2019) concept of digital persistence and Nansen et al.' (2014) understanding of the restless dead as discussed above, but also other forms. Media mourning, where mourning of public deaths such as school shootings occurs in spontaneous formats (a digital analog of the spontaneous shrines discussed above), can be associated with "bandwagon mourning" (Savin-Baden, 2021, p. 44) or "grief hyperjacking" (Abidin, 2022, p. 174). In media mourning, mourners who did not have a personal relationship with the deceased but disproportionately comment or post on individual or collective grief. For Abidin (2022, p. 174), this form of grief is a "shift from 'public grieving' to 'publicity grieving'". It may show a desire to join the supportive community created in those pages on the part of the commentators, or it may be a way to gain notoriety in the online space. Putnam (2020, p. 172) refers to this as virtual social capital, and notes that virtual communities "can support large, dense, yet fluid groups that cut across existing organizational and geographical boundaries, increasing the involvement of otherwise peripheral participants." The nature of the virtual space, with its niche communities allows for the creation of bonding social capital in a way that used to only be possible within physical spaces.

According to their website, the World Wide Cemetery (cemetery.org) is the oldest online cemetery site in the world ("Our Story," 2024). On their website, they state "The Internet forgets far more than we realize. Obituaries in newspapers and memorials on funeral sites or social media often disappear. A physical grave may be more permanent, but it may also be remote or difficult to visit." The World Wide Cemetery was designed as a way to leave "a lasting and accessible memory" by its designer, Mike Kibbee in 1995. A one-time fee is charged to create the memorial, and they will publish in English, French, German, and Italian. Memorial pages can be merged (such as for married individuals) if desired. QR codes can be purchased for placement on the physical tombstone if desired. Several other sites now exist to offer virtual cemeteries, complete with the ability to design a tomb and provide atmospheric weather for the visitors to the virtual memorial (Chayka, 2014). David Quiring, a monuments maker in Seattle, Washington, makes the "Living Headstone", where a QR code that can link to an online memorial or video of the deceased is attached to the surface of a physical marker. The addition, costing about $75 to the cost of a new monument and $150 to place on an existing tombstone, has been popular. In 2014, Quiring told a reporter for Newsweek that they'd distributed around 800 of the QR codes all around the world (Chayka, 2014).

Virtual memorials and social media archives force us to reexamine the composition of the body. If we continue to interact with the dead through their data, does the body become made of data? The virtual relationship we have with the dead is no less meaningful than that of any other relationship we have with the dead. In terms of mortuary performance, virtual interactions are ritualized, repetitive and performed for an audience: they meet the same standard used in 'traditional' memorials for giving closure to the bereaved and for creating group identity, particularly group identity focused on intersectional points (e.g., mothers who have lost children, people who have lost partners to suicide or cancer). These communities may exist completely online, without in-person meeting, but they still have the power of amplification mechanisms through the relatively formulaic nature of the interactions in an online public sphere. Öhman and Watson (2019, p. 9) argue that the social media archive created becomes "personal digital heritage" that becomes "part of our shared *cultural* digital heritage."

The persistence of online data also makes us define death differently, as Kasket (2019, p. 470) comments, "[i]f technology affords us the possibility to continue in our communities—not just having our names spoken, but also having our voices and images actively accessible—might this change our definitions and perceptions of death?... Will the people of my daughter's generation be somehow 'less dead' than my great-grandmother and her peers?". Do individuals who continue to be viewed online lack a social death? What happens when formats change or someone makes the choice to discontinue curation of these digital records?

12.4 Modern Material Culture: Material Remains as Modern Mortuary Performance

Material remains as mementos from funerals and memorial services have a long history. In the US, colonial-era through nineteenth century funerals were typically marked by the receiving of gloves as mementos, the creation of jewelry (sometimes made with the hair of the deceased), and other objects such as fans, and crepe bunting as a visible display of mourning (Bullock & McIntyre, 2012). By the twentieth century, prayer cards were a common item taken by mourners from a funeral service. By the end of the twentieth century, a new plethora of material culture was created for memorialization. Still popular are prayer cards and fans, but increasingly, items of everyday use are employed for memorialization such as T-Shirts and stickers applied to cars.

R.I.P. T-shirts (Rest In Peace T-Shirts, or memorial shirts), in particular, have a special place in mourning in the twenty-first century. These are shirts that typically will bear the image of a deceased person as well as some biographical information about them. Brooks addresses the history of memorial T-shirts, noting that there is little consensus about when and where they first appear in the US: sometime in the late 1980s or early 1990s, maybe from the east coast to the west coast, or somewhere in the south, all have been proposed in various popular sources. Generally, these are more commonly used as a mechanism for memorialization in African American communities. As Brooks (2018, p. 809) notes, "Memorial shirts are visuals that tell the story of a life, particularly a slain life. The shirts inform viewers of details about the person… including the reason the person is no longer alive. Moreover, the act of reading memorial shirts weds visual life writing and Black mourning practices and ritual traditions." Brooks (2018) sees memorial T-shirts as political statements as well, ways that marginalized groups perform both memorialization and identity. Bass-Krueger (2023, p. 402) refers to memorial T-Shirts as "walking memorials of sorts" that "allow death to come into the everyday discourse of Black communities while confronting larger society about its role in African American death." Kami Fletcher (2021) relates the experience of making and wearing R.I.P. T-shirts within her own family. Fletcher saw the creation of R.I.P. T-shirts for her relative Chill by Chill's mother Snooky as a public display of maternal grief. She also sees them as a form of "tangible condolence. Utilized this way, the shirts become powerful for persons who cannot be physically present when the bereaved gather to mourn and heal" (Fletcher, 2021, p. 215). Initially designed to be worn at funeral services and memorials, the shirts then become part of daily wear, "the material culture left behind that holds memories and allows for one more interaction with the dead" (Fletcher, 2021, p. 215). Seen as walking memorials, they can be worn as a way to keep the dead close and part of new activities.

After 9/11, R.I.P. T-shirts became more common in broader communities, but the political messaging continues to be foundational to their importance. In essence, RIP T-shirts are a way that marginalized communities sought to use the representation of the body of one person (the person memorialized) to interact with the Social Body, as a way to highlight the need for social change.

Another category of modern memorial is that of the car decal. These are often removable stickers that give the name of the deceased along with some demographic information about them. In their work, Cann (2014, p. 84) says that in putting up a decal, the "car essentially becomes a tombstone." Cann notes that the manufacturers of these decals noted that there is a high prevalence of memorials for young individuals, babies and children. So, while these may be as socially charged as RIP T-shirts, they needn't be. Because car decals or bumper stickers act as indicators of identity in a public sphere, the addition of a memorial decal is a public performance of grief and remembrance. Memorial decals may be used to memorialize individuals that the car owner didn't even know. Cann notes that memorial decals for Dale Earnhard Sr., who died during the Daytona 500 race in 2001, were common in the American South. Also common were memorial decals commemorating 9/11. The US is a car-obsessed country, and this may explain why the use of car decals as memorial space is less wide-spread in other countries. The memorial is temporary and removable. In some instances, the name of the deceased will be handwritten on a window in grease-paint only to wash away with the next rain storm or car wash. Others are custom made decals designed to be on the car long term. They identify the owner/driver of the car as a bereaved person in a quick, identifiable format: "in a culture dominated by fast-food and drive-through restaurants, texting and emoticons, car-decal memorials relay a message of grief" (Cann, 2014, p. 87). In public space, the car then becomes a movable shrine.

12.5 Decorating Graves

The offerings left at graves are the embodiments of actions conducted at the grave, the material remnants of ritual just as much as the remnants of feasting found in Roman cemeteries (e.g., Paige et al., 2019). We interpret these through ethnographic lenses as well as through the analysis of the material culture remnants of the activities performed at these sites. Oftentimes, these will involve every-day objects that have a different life history strictly by their incorporation in mortuary ritual. This incorporation of everyday, personal objects into mortuary performance reflects specific aspects of identity that the leavers of the objects wish to highlight. Vlach (1991) notes that visitors to graves in African American cemeteries are often leave glassware, metal containers, broken pitchers, cups and saucers, clocks, salt and pepper shakers, oyster shells, and mirrors. Visitors to Jewish cemeteries may leave pebbles. Visitors leave flowers, flags, and personal mementos at graves around the world. While not all of these (e.g., flowers) will leave material traces, many of these activities can be marked by their material remains.

Every object left is an intentional act producing a material record. While this doesn't necessarily impact the physical body, it does impact the mortuary assemblage. These visits and cemetery decorations are sometimes done to commemorate important events. In 2016, for example, the Chicago Cubs won the world series for the first time in 71 years. According to an story on National Public Radio's *All Things Considered*, the win was "somewhat bittersweet for many fans in Chicago, who remember parents, grandparents, spouses and other loved ones who didn't live long enough to see this day. So in cemeteries all across the Chicago area, fans are planting Cubs flags, pennants, flowers, balloons and little stuffed Cubby bears in front of the headstones of Cubs fans who are no longer with us." Celebrating the victory with the dead was a way of maintaining the relationship between the living and the dead, reaffirming the importance of an event through the leaving of a material element at the grave stone was a way to include the dead in those celebrations. This was also a way to remember the dead and to share that remembrance with others. Similar to this, also in 2016, the grave stone of Susan B. Anthony was covered with "I Voted" stickers, so much so that the tenders of the cemetery placed additional placards for people to put more stickers around the tombstone. In the first national election with a woman at the head of the ticket, women and men stood in line for hours to place their stickers on this tombstone to mark the occasion, and to honor someone who fought for female suffrage with their act of voting (Osterholtz, 2020).

Decorating cemeteries is a community-building activity. Decoration Day has a long history in the Southern US. This takes place on a Sunday in the Spring, usually coinciding with Memorial Day and involves the cleaning of the cemetery and placing of flowers graveside accompanied by singing and picnic dinners at the cemetery (Jabbour & Jabbour, 2010). This becomes a community event, a cultural event. Part of this event is the preparation and cleaning of cemetery, followed by mounding of clay or soil over the length of the grave, a holdover from the use of pine coffins that would deteriorate causing the grave to sink in as the body decomposed. Jabbour and Jabbour (2010, p. 25) note that while this tradition begins as a way to mitigate the collapse of graves, it remains "an annual ritual of renewed devotion," continuing the relationship between the living and the dead.

12.5.1 *Día de los Muertos (Day of the Dead)*

One specific category of burial decoration would be the celebration of the Día de los Muertos (Day of the Dead). Strictly, the Día de los Muertos is the Mexican celebration of All Souls' Day, usually observed on November 2. Increasingly, the Día de los Muertos celebrations encompass more than just this holiday, extending from October 31st through November 2 (Brandes, 2006), becoming the Días de los Muertos (*days* of the dead). Masses are said to honor the departed souls, but Mass may be the least important portion of the holiday from an identity perspective. More important is the decoration of cemetery spaces in bright bold colors celebrating death and the images central to it, including sugar skulls, skeletons, caskets and so on (Fig. 12.4). Community members clean, decorate and hold vigil in the cemetery spaces, decorating graves with flowers, candles, paper cutouts, and food such as

Fig. 12.4 Day of the Dead celebrations in the city cemetery in Leon Guanajuato. Photo © Tomas Castelazo, www.tomascastelazo.com/Wikimedia Commons. https://upload.wikimedia.org/wikipedia/commons/0/00/Day_of_the_dead_at_mexican_cemetery_3.jpg

tamales, *pan de muerto* (special breads), *pulque* (an agave-based drink), and sweets. This is essentially a celebration of the relationship between the living and the dead, rooted in a belief that the living owe some service to the dead (Marchi, 2013). These folk elements are local additions to the religious performance of Mass, and Brandes argues that it is these elements that have come to symbolize "Mexico and Mexicanness" (2006, p. 7), a key symbol of national identity. For example, the Nahua celebrate a day known as *el día de los accidentados* (the day of those who died an accidental death) or *el día de los matados* (the day of those who were murdered, commemorating a pre-contact category of the deceased where one's fate in the afterlife was dependent upon the way they died (Brandes, 1998, 2006). The celebrations are rooted in pre-contact harvest festivals, which were moved by Colonial powers to coincide with Catholic rites. Prior to Catholicization of the ritual, marigolds and raisins were placed on altars for the dead, this tradition is maintained in the Day of the Dead celebrations, with the creation of *ofrendas*, altars containing images of saints, the Virgin Mary, and devotional candles along with images of the deceased.

The holiday is also celebrated in the US, where it has come to symbolize ethnic identity, which may mean something different to those who celebrate, from Mexican identity, to others it may symbolize Latinx identity, and to others, an understanding of local community identity (Brandes, 1998, 2006). The holiday in the US gained more prominence in the 1970s, tied to social rights activists within the Chicano movement (Marchi, 2013). The holiday as celebrated in the US is seen by Marchi (2013) as a hybrid celebration, with a shifting from simple observances centered on cleaning the cemetery, leaving marigolds, and having a picnic in the cemetery to a more performative display with large skulls and skeletons throughout the cemetery. Marchi argues that the hybridization and performativity of US-based Day of the Dead displays can be seen as a mechanism for fostering a sense of identity amongst the Chicano movement, bringing together disparate Latinx groups in the US, and highlighting that identity over European ancestry. These celebrations are far more secular, with many seeing the religious elements such as Mass tied to a colonial religion imposed upon Indigenous groups. As Marchi (2013, p. 283) notes, contemporary Day of the Dead activities "were (and are) primarily meant to construct and perform cultural identity, rather than fulfill moral obligations to the dead." Brandes (1998) goes so far as to call the Day of the Dead a political event that expresses the complexity of Mexican-US relations.

12.6 What Do We Learn from Vernacular Memorials?

Mortuary performances tell us a lot about the living. In this section (Chaps. 10, 11, and 12), we can see the increasingly secular nature of mortuary performance as well an emphasis on the individual over society and a lack of consistent and formulaic mortuary performance. The majority of these changes have occurred in the latter half of the 20th and beginning of the twenty-first century. Are we beginning to have a new relationship with death? Increasing variability in what is considered fitting or appropriate in terms of body disposition methods suggest our cultural relationship with death and the dead is changing. Poetics provides a way to understand this relationship in fundamental ways. The rise of palliative care, funeral preplanning and living funerals speak to a desire by at least a portion of the population to open discussions surrounding the end of life, destigmatizing the topic. This may be linked to the ageing of the Baby Boomer generation. The Baby Boomers lived through times of social change and upheaval, they were socially engaged and broke barriers in social justice, gender roles, and how we interact with each other. Naturally, the generation that is responsible for Woodstock, shattering glass ceilings, and other social changes will have a different take on what it means to die, and how they wish to be remembered. Ellen McCarthy (2014, n.p.), in writing for the Washington Post puts it this way: "As Baby Boomers head toward retirement and the great hereafter they're thinking more about what will become of their remains. And what they're thinking is what they've thought during every phase of life: status quo? Buck that."

Modern forms of memorialization also reflect social change. In contrasting ways that we memorialize the dead (particularly in the US), we see a shift from culturally proscribed mourning and memorialization to one that better reflects the individuality and chosen communities of the dead and their bereaved. This chapter (Chap. 12) focused on vernacular memorials, those forms of memorialization that arose from individual experience and individual desire. This shows that individuals felt free to choose their own form of memorialization; this was not mediated by the state in an expression of the Body Politic. The Social Body, in the general acceptance of different forms of memorialization, acknowledges the primacy of the individual, and so the very variety of expression becomes the expression of the Social Body.

How has our relationship with death and dying changed? First, death is more personal. More and more Baby Boomers are looking to the future and planning for their own funerals. The reasons may be environmental or financial. Jay Castaño was interviewed by the Washington Post, and gave his reasons for green burial as both, saying "I want to be part of a tree, be part of a flower—go back to being part of the earth." Also, "I am a tightwad". For a growing number of Baby Boomers, funerals are seen as overly expensive while lacking a personal connection. Paying thousands of dollars for a coffin simply doesn't make sense to them. This change in perception will likely change the perception for the generations that follow, leading to new norms and new relationships with death. Individuals are also more empowered to make decisions about their own deaths and memorials. This coincides with a general decrease in religiosity in the US, opening up the possibilities for remembrance in a more personal format. Second, death and memorialization can be more informal. Green burials and ash spreading necessitates a change in how we look and act when these performances are conducted. Green burial typically involves the family in digging the graves, this is much better conducted in jeans than a suit or a dress. Scattering ashes in Disney World will likely be less obvious (remember this has to be covert) if dressed informally and acting casually. Finally, mortuary performances serve the function of a family reunion in many instances. Death and mourning are still family affairs but what that means is changing. Modern families are often separated by distance, and often mortuary performance can be the only chance to get together. So these become times for families and friends to gather that might not otherwise be able to take the time from work or their busy lives. Vacation elements are common.

12.7 Summary

This chapter has examined the changing nature of memorialization in the twenty-first century, focusing on vernacular (informal, personalized forms of remembrance outside traditional practices such as tombstones) memorials. Contemporary memorial practices such as roadside memorials, spontaneous shrines, memorial tattoos, and virtual memorials all reflect societal shifts towards individualized expressions of mourning and memory. Some forms, such as ghost bikes, blend grief and social activism, while memorial tattoos provide a deeply personal, permanent tribute. Virtual memorials and social media platforms blur the line between life and death, allowing ongoing interaction with the dead in a virtual, public space. This chapter situated these changes within broader societal shifts, such as a growing desire for individual agency in death rituals. It also touched on how traditional practices, such as Día de los Muertos, continue to evolve as expressions of cultural identity and community. Overall this chapter highlights how new forms of memorialization provide opportunities for connection, social identity, and public grieving in an increasingly individualistic society.

References

Abidin, C. (2022). Grief hyperjacking: Influencers,# ThoughtsAndPrayers, and the commodification of grief on Instagram. *The Information Society, 38*(3), 174–187.

Ahmed, A. (2011, September 7). 9/11 Memorial T-shirts fashioned into art at UNLV. *Las Vegas Sun*. Retrieved from https://lasvegassun.com/news/2011/sep/11/911-memorial-t-shirts-fashioned-art-unlv/

Bass-Krueger, M. (2023). Mourning dress in the west, 1800 until today. In V. Pouillard & V. Dubé-Senécal (Eds.), *The Routledge history of fashion and dress, 1800 to the present* (pp. 391–410). Routledge.

Bergh, L. (2016). Tattooing as memorial Pragmemes. In K. Allen, A. Capone, & I. Kecskes (Eds.), *Pragmemes and theories of language use* (pp. 585–597). Springer.

Brandes, S. (1998). The day of the dead, Halloween, and the quest for Mexican National Identity. *The Journal of American Folklore, 111*(442), 359–380. https://doi.org/10.2307/541045

Brandes, S. (2006). *Skulls to the living: Bread to the dead: Celebrations of death in Mexico and beyond*. Blackwell Publishing.

Brean, H. (2016, September 11). Spontaneous New York-New York 9/11 shrine lives on at UNLV as historical collection. *Las Vegas Review-Journal*. Retrieved from https://www.reviewjournal.com/local/local-las-vegas/spontaneous-new-york-new-york-911-shrine-lives-on-at-unlv-as-historical-collection/

Brooks, R. (2018). RIP shirts or shirts of the movement: Reading the death paraphernalia of Black lives. *Biography, 41*(4), 807–830.

Buckle, J. L., & Corbin Dwyer, S. (2023). Embodied meaning making: Memorial tattoos as a visual expression of grief. *Death Studies, 47*(1), 1–9.

Bullock, S. C., & McIntyre, S. (2012). The handsome tokens of a funeral: Glove-giving and the large funeral in eighteenth-century New England. *The William and Mary Quarterly, 69*(2), 305–346.

Cadell, S., Lambert, M. R., Macdonald, M. E., Davidson, D., O'Gorman, M., & Baljko, M. (2018). The pain of the tattoo was a relief: Advancing a theory of embodied pain in a study of memorial tattoos. *Journal of Pain and Symptom Management, 56*(6), e31.

Cann, C. K. (2014). *Virtual afterlives: Grieving the dead in the twenty-first century*. University Press of Kentucky.

Chayka, K. (2014, August 17). After death, don't mourn, digitize with sites like Eterni.me. *Newsweek Magazine*. Retrieved from https://www.newsweek.com/2014/08/29/after-death-dont-mourn-digitize-sites-eternime-264892.html.

Clark, J., & Franzmann, M. (2006). Authority from grief, presence and place in the making of roadside memorials. *Death Studies, 30*(6), 579–599.

Collins, C. O., & Rhine, C. D. (2003). Roadside memorials. *OMEGA-Journal of Death and Dying, 47*(3), 221–244.

Costantini, N. M. (2017). *The white bicycle: Performance, installation art, and activism in ghost bike memorials*. Ph.D. Dissertation. Louisiana State University and Agricultural & Mechanical College.

Costantini, N. M. (2019). Bikes and bodies: Ghost bike memorials as performances of mourning, warning, and protest. *Text and Performance Quarterly, 39*(1), 22–36. https://doi.org/10.1080/10462937.2019.1576919

Davidson, D. (2017). Art embodied: Tattoos as memorials. *Bereavement Care, 36*(1), 33–40.

Dobler, R. (2010). *Alternative memorials: Death and memory in contemporary America*. Ph.D. Dissertation. University of Oregon.

Dobler, R. T. (2011). Ghost bikes: Memorialization and protest on city streets. In M. Peter Jan & S.-C. Cristina (Eds.), *Grassroots memorials* (pp. 169–187). Berghahn Books.

Facebook.com. (2024a). *Managing a deceased person's account*. Retrieved from https://www.facebook.com/help/275013292838654

Facebook.com. (2024b). *What happens ot your Facebook account if you pass away*. Retrieved from https://www.facebook.com/help/103897939701143

Fletcher, K. (2021). Long live chill: Exploring grief, memorial, and ritual within African American R.I.P. T-shirt culture. In A. Thamann & K. M. Christodoulaki (Eds.), *Beyond the veil: Reflexive studies of death and dying* (pp. 209–234). Berghahn.

Greenspan, R. E. (2019, April 30). On Facebook, the dead will eventually outnumber the living. what does that mean for our histories? *Time Magazine*. Retrieved from https://time.com/5579737/facebook-dead-living/

Hall, S. (2024, September 16). My oldest son! I learned a lot while raising him and I've learned even more since I lost him. I spend my lunch break with him every day! My heart will never be the same!! [Facebook.com Social Media Post]. Retrieved from https://www.facebook.com/groups/165425904050656/permalink/1577361296190436

Ipsos. (2019). More Americans Have Tattoos today than seven years ago [Press release]. Retrieved from https://www.ipsos.com/en-us/news-polls/more-americans-have-tattoos-today

Jabbour, A., & Jabbour, K. S. (2010). *In the mountains: Traditions of cemetery decoration in teh southern Appalachians*. University of North Carolina Press.

Kasket, E. E. (2019). Mourning and memorialization on social media. In A. Attrill-Smith, C. Fullwood, M. Keep, & D. J. Kuss (Eds.), *The Oxford handbook of cyberpsychology* (pp. 467–484). Oxford University Press.

Lanigan, K. (2007). Memorial in ink. *Veterans of Foreign War Magazine, 95*, 42–43.

Lynn, C. D., Howells, M., Herdrich, D., Ioane, J., Hudson, D., & Fitiao, S. A. T. U. (2020). The evolutionary adaptation of body art: Tattooing as costly honest signaling of enhanced immune response in American Samoa. *American Journal of Human Biology, 32*(4), e23347.

Marchi, R. (2013). Hybridity and authenticity in US day of the dead celebrations. *The Journal of American Folklore, 126*, 272–301.

McCarthy, E. (2014, October 6). 'Green burials' are on the rise as Baby Boomers plan for their future, and funerals. *Washington Post*. Retrieved from https://www.washingtonpost.com/lifestyle/style/green-burials-are-on-the-rise-as-baby-boomers-plan-for-their-future-and-funerals/2014/10/06/d269cfbc-3eae-11e4-b03f-de718edeb92f_story.html

Nansen, B., Arnold, M., Gibbs, M., & Kohn, T. (2014). The restless dead in the digital cemetery. In C. M. Moreman & A. D. Lewis (Eds.), *Digital death: Mortality and beyond in the online age* (pp. 111–124). ABC-Clio.

Öhman, C. J., & Watson, D. (2019). Are the dead taking over Facebook? A big data approach to the future of death online. *Big Data & Society, 6*(1), 2053951719842540.

Öhman, C. J., & Watson, D. (2021). Are the dead taking over Instagram? A follow-up to Öhman & Watson (2019). In J. Cowls & J. Morley (Eds.), *The 2020 yearbook of the digital ethics lab* (pp. 5–21). Springer.

Osterholtz, A. J. (2020). Introduction: Processing and poetics, examining the model. In A. J. Osterholtz (Ed.), *The poetics of processing: Memory formation, cosmology, and the handling of the dead* (pp. 3–14). University Press of Colorado.

Our Story. (2024). Retrieved from https://cemetery.org/our-story/

Paige, J., Larson, K., Osterholtz, A. J., & Paraman, L. (2019). *Feasting with the dead: Preliminary analysis of faunal remains at the Put Dragulina Roman Cemetery*. Paper presented at the 84th Annual Meeting of the Society for American Archaeology, Albuquerque, NM.

Putnam, R. D. (2020). *Bowling alone: The collapse and revival of American community (Revised and Updated)*. Simon & Schuster Paperbacks.

Santino, J. (2004). Performative commemoratives, the personal, and the public: Spontaneous shrines, emergent ritual, and the field of folklore (AFS Presidential Plenary Address, 2003). *The Journal of American Folklore, 117*(466), 363–372.

Savin-Baden, M. (2021). *Digital afterlife and the spiritual realm*. Chapman and Hall/CRC.

Scheper-Hughes, N., & Lock, M. (1987). The mindful body: A prolegomenon to future work in medical anthropology. *Medical Anthropology Quarterly, 1*, 6–41.

Vlach, J. M. (1991). *By the work of their hands: Studies in afro-American Folklife*. University Press of Virginia.

Warnick, A., & Toye, L. (2016). Memorial tattoos as connection. In D. Davidson (Ed.), *The tattoo project: Commemorative tattoos, visual culture, and the digital archive* (pp. 134–135). Canadian Scholars Press.

Part IV
Bodies and Politics

Part IV focuses on the intersection of bodies and politics, focusing on the role of the dead to promote and negotiate political power and sovereignty. Together, these chapters highlight how the dead are not only repositories of memory and identity but also crucial players in the ongoing negotiations of social justice and historical reconciliation.

Chapter 13 examines the role of bodies in Nation Building activities. Key to this framing is the role of dead bodies in shaping cultural and national identity and political narrative. Here I delve into how the bodies of the dead can be powerful tools for the negotiation and performance of political power, social cohesion, and national identity. Chapter 14 looks at legislation and the shift in ethical approach within biological anthropology within the US with a special focus on the Native American Graves Protection and Repatriation Act (NAGPRA). In general, I chart the shift towards collaborative, community-engaged approaches to bioarchaeological research. In this chapter, the importance of ethical engagement with marginalized communities in bioarchaeology, focusing on the pivotal role of NAGPRA in reshaping the relationship between researchers and Indigenous peoples. It advocates for a collaborative approach to research that respects Indigenous sovereignty and addresses the legacy of exploitation in the field. Ultimately, the chapter contributes to the broader conversation about social justice, cultural identity, and the ethical responsibilities of bioarchaeologists in the twenty-first century. By emphasizing community engagement, it underscores the importance of creating a more just and equitable framework for the study of human remains.

Chapter 13
Nation Building with Bodies

Abstract This chapter examines the role of dead bodies in the construction of political narratives and national identities, exploring the intersection of memory, power, and social politics in the commemoration and manipulation of the deceased. Through case studies of memorials such as the Tomb of the Unknown Soldier at Arlington National Cemetery, the preservation of Communist leaders like Lenin, and the repatriation of Indigenous remains like Yagan's, this paper highlights the profound agency of dead bodies in shaping political discourse. Drawing on the concepts of *postmortem agency* (Crandall and Martin) and *dead body politics* (Verdery), the chapter analyzes how human remains, through their strategic placement, preservation, and memorialization, serve as conduits for national memory, territorial claims, and the negotiation of collective identities. The paper further explores how bodies of genocide victims—particularly in the context of the Rwandan genocide—become living archives that challenge historical narratives and amplify collective trauma. Through an exploration of the *necropolitics* involved in the memorialization of these bodies, the paper underscores how the dead continue to exert power over the living, shaping societal values, memory, and identity long after death.

The physical body is a malleable object, it can be used to promote nationalistic ideas and identity, it can be used as a tool to highlight the power of the Body Politic and to create national identity around the body of the leader. The phenomenological impact of interacting with the body heightens its effect. Memorials centering on the bodies of murdered individuals act both as a memorial to the massacre at Murambi but also serve as physical proof that the massacre took place. In this chapter, the use of the body will be examined from the perspective of the Body Politic and Social Body, and will focus on different uses of the body: for the promotion and performance of national identity and sacrifice (e.g., The Tomb of the Unknown Soldier at Arlington National Cemetery), the use of individual bodies and unidentified individuals to perform group identity, and the use of the bodies of dead leaders in post-Soviet eastern Europe, and the use of the body as a memorial in genocide. While these are different uses of the body, all center on the physical body as an important element in construction of identity, motivated by the Body Politic for political ends.

A. Osterholtz, *The Bodies Poetic*, Bioarchaeology and Social Theory,
https://doi.org/10.1007/978-3-031-85998-4_13

13.1 Tombs of the Unknown Soldier (Arlington National Cemetery)

Not everyone killed in conflicts can be identified, particularly from early conflicts before the introduction of modern biological anthropological methodology or the use of DNA. For these unknowns, creating a special monument to their sacrifice has always been a focus. The creation of a Tomb of the Unknown Soldier (Fig. 13.1) is a way to honor those who could not be identified or whose bodies were destroyed in war. As Poole (2009, p. 146) notes the individual buried within the Tomb of the Unknown Soldier "stood for all those lost," providing a place where mourning could occur, where families and friends could come to grieve. The Tomb of the Unknown is interesting for us in a Poetics sense, as it celebrates both individual and collective identity and allows families and friends to imagine the tomb is that of *their* individual fallen serviceman. It also allows the celebration of the collective identity of *soldier* or *American*. As Blaire and colleagues (2011, p. 460) express it, "Here is the hero of the Great War with no known identity except nationality." It highlights the intersectional identity element that is in common amongst all those who died in conflict.

The Tomb of the Unknown can be a place of reconciliation as well. At a ceremony at President Grant's Tomb in 1926, numerous military officials suggested

Fig. 13.1 Tomb of the Unknown Soldier with Tomb Guard at Arlington National Cemetery. Image attribution: DiscoA340, CC BY-SA 4.0 <https://creativecommons.org/licenses/by-sa/4.0>, via Wikimedia Commons. https://upload.wikimedia.org/wikipedia/commons/5/59/Tomb_of_the_Unknown_Soldier%2C_Arlington_National_Cemetery_%2820_September_2023%29_29.jpg

placing one Union and one Confederate together in a Tomb of the Unknown for the Civil War as a marker of unity between North and South ("Ask Common Grave for a Blue and Gray," 1926). It should be remembered, though, that this coincides with the rise of Jim Crow and the raising of Confederate monuments throughout the country, designed to glorify the Lost Cause.

Manipulation and monumentation of the body of the Unknown is a common Space for political meaning making. In their discussion of the Tombs of the Unknown in Europe, Hettling and Schölz (2019, p. 2/28) bring in the concept of the Body Politic, that the soldier has two different *bodies*, the private individual, mourned by friends and family and "the political role of the soldier (as a member of a political union, usually a nation). The private body was and remained perishable and could be mourned by relatives. In the political cult of the dead, the political body of the fallen soldier received a secular promise of permanence, and its death was ascribed political meaning… The idea of the political nation's continuity was linked to the promise to permanently remember the fallen. The private and political bodies can be respectively attributed to the gravesite and the memorial, each of which had different functions." The grave was a space for private mourning, but the memorial was a place of national unity. For Hettling and Schölz (2019), the tombs of the Unknown as a monument became important because of an expectation that every fallen soldier would be remembered by name. Prior to World War I, cenotaphs had been common for individuals who were missing in action or to commemorate mass graves of unidentified soldiers but it was not common to choose a single individual to stand in for all the unidentified and receive military honors. "His anonymity allowed the Unknown to be *a very particular someone* to a friend or family member, at least hypothetically (Blair et al., 2011, p. 459).

At Arlington, there are tombs of the Unknown Soldier for World War I, World War II, Korea, and Vietnam. It is unlikely that additional tombs of the unknown will be constructed, this will be discussed below in relation to the Vietnam Tomb of the Unknown Soldier. At the dedication of each of these monuments, the President of the United States has presided over the ceremony, acted as the ceremonial next of kin to the dead, and has accepted the interment flag for the burial. In doing so, the deceased ceases to be an individual but becomes emblematic of a larger concept, with the President (an office that in this case embodies the Nation) embracing the dead as kin. These flags are on display in the Memorial Display room at Arlington.

13.1.1 World War I

The tomb of the Unknown from World War I is one of Arlington's most revered monuments. The impetus for this monument can be traced to ceremonies conducted in Europe to honor unidentified soldiers in World War I. Marie Meloney, an editor of the *Delineator* Magazine wrote a letter to General March saying "It is the kind of thing which should have found birth in America… I want you to do the fine, big human thing that no one else in America has initiated… it is a big strong influence in the future… It brings patriotism home to men in a personal way" (Marie M. Meloney to Gen. Peyton C. March, Nov. 13, 1920, NARA RG 407, Box 563, File 293.8., quoted in Poole (2009, p. 146)). Providing a place of honor for the Unknowns was a way to celebrate the sacrifice of everyone, it showed to potential soldiers that sacrifice was a universal way to show love of country and that the country would honor that sacrifice, even in the absence of a name. Legislation established a special tomb to be constructed in Arlington National Cemetery "to bring home the body of an unknown American warrior who in himself represents no section, creed, or race in the late war and who typifies, moreover, the soul of America and the supreme sacrifice of her heroic dead" ("The Tomb of the Unknown Soldier," 2021).

The bodies of four unidentified soldiers were exhumed from cemeteries in France in October 1921. Loutenant Colonol G.V.S. Quackenbush supervised the recovery of these remains, and was specifically interested in making sure that no identifying elements be present in the remains. In his instructions, he noted that should "anything be found on the body or in the coffin which will tend to identify this particular body, an alternative body… will then be exhumed and similarly searched" (Poole, 2009, p. 149). These bodies were taken to the town hall of Châlons-sur-Marne (now Châlons-en-Champagne), where they were laid out. Major Robert P. Harbold of the Quartermasters Corp chose a sergeant of the Quartermaster Company of the 2nd Battallion, 50th Infantry, American Forces in Germany, to select the Unknown Soldier from amongst the four by placing a spray of white roses on one of the caskets. The three caskets of those not chosen were driven to the Meuse-Argonne Cemetery and buried in Graves 1, 2, and 3. The Unknown was then taken by caisson and rail to La Havre, France and then by the USS Olympia to Washington, D.C. Upon arrival there on November 9, 1921, the remains lay in state at the US Capitol Rotunda where approximately 90,000 visitors paid their respects on November 10. On November 11, the Unknown was placed on a horse-drawn caisson and carried through Washington DC to Arlington.

A state funeral ceremony was held in the newly constructed Memorial Amphitheater and the remains interred in the Tomb of the Unknown Soldier. President Warren G. Harding officiated at the ceremony and placed the Medal of Honor on the casket. The Tomb of the Unknown Soldier was initially a three-level marble slab. The lower two levels contained a rectangular opening through which the remains could be placed into the ground below. This simple tomb was added to with congressional appropriation of funds appropriated in 1929. The new tomb would consist of seven slabs of marble and include sculpture on all four sides. It is decorated with three wreaths on the north and south sides, the east (front) has three figures representing Peace, Valor, and Victory. The west features an inscription reading "Here rests in honored glory an American Soldier known but to God" ("The Tomb of the Unknown Soldier," 2021).

13.1.2 World War I and Korea

These two conflicts are discussed together and the Tombs of the Unknown were placed together. Following the conclusion of World War II in 1945, legistlation was introduced to create a tomb of the Unknown for World War II, and this was signed into law in 1946. This was to be installed by May 30, 1951. However, on June 25, 1950, North Korean troops crossed the 38th parallel and attacked South Korea (an ally of the US). Troops were dispatched and the Korean conflict (war) began. This war lasted 3 years, with positions gained and lost, and a generalized chaos that impacted how the US cared for the war dead. The unidentified dead from Korea were held in a mortuary near Yokohama, where they remained until 1956. These individuals were ultimately buried in the National Memorial Cemetery of the Pacific in Honolulu (known as the Punchbowl) in May of 1956. They were placed there alongside more than 2000 unknowns from World War II.

Construction was begun on two new Tombs of the Unknown to be buried alongside the Tomb of the Unknown from World War I to be interred on Memorial day, 1958. The World War II candidate was chosen from finalists selected from both the transatlantic and transpacific theaters. Thirteen sets of remains were exhumed from military cemeteries in North Africa and Europe and sent to Epinal, France where the final selection was made. A second candidate was selected from exhumed unknowns at the Fort McKinley American cemetery in the Philippines and from the Punchbowl in Hawaii. The Pacific candidate joined the Korean Unknown chosen from amongst those at the Punchbowl. All were sent to Guantanamo Bay, Cuba on May 15, 1958. The Unknown from World War II was chosen by Hospital Corpsman 1st Class William Charette. Charette was a Medal of Honor recipient for his service during the Korean war. He placed a wreath on one of the caskets, saluted and stepped back. The other Pacific candidate was taken eight miles out to sea, wrapped in sailcloth and given burial at sea. The Unknowns from World War II and Korea were taken together to Washington DC, following the same path taken by the Unknown for World War I. The caskets lay in state in the Capitol Rotunda for 3 days. They were then taken via matching black caissons draped in velvet and pulled by six gray horses. Upon entry to Arlington, the caisson with the Unknown from World War I pulled ahead of the caisson carrying the Unknown from Korea. Fighter jets and bombers flew overhead, each group flying a "missing man formation" (a formation where planes fly in a formation resembling a V; during the tribute, one of the planes breaks off and flies apart from the rest of the formation, symbolizing the loss of the individual being honored). President Dwight D. Eisenhower presided over the ceremonies that were also attended by numerous dignitaries and an associate justice of the Supreme Court. He placed the Medal of Honor on both caskets. The caskets were lowered into their crypts as Chaplains from Catholic, Protestant and Jewish faiths gave each gave a burial service. A twenty-one-gun salute was followed by the playing of Taps.

13.1.3 Vietnam

In 1973, the last American forces withdrew from Vietnam, and Congress authorized the creation of a Tomb of the Unknown for the conflict. This tomb remained empty until 1984. During this time, President Carter pardoned those who had evaded the draft, and President Reagan dedicated the Vietnam war memorial in 1982. Candidates had been selected to become the Unknown for the Vietnam war, but many held out hope that that these individuals could be identified through modern methods in forensic science. In effect, while there were many *unidentified* individuals from the war, there was doubt about whether those individuals were *unidentifiable*. This is an important distinction, and one that would render the tradition and ritual surrounding the Tomb of the Unknown moot. Of the final four candidates chosen, three were ultimately identified through dental and other medical records, leaving only one candidate which was designated as X-26. Several individuals had hopes that this individual, too, could be identified. Poole (2009, p. 245) essentially paints a picture of increasing political pressure to find an Unknown in hopes of "calming the old ghosts of Vietnam", despite the concerns of those in charge of identifications.

Despite these hopes and concerns, X-26 was selected for the Tomb, and in 1984, his remains were placed within the Tomb of the Unknown for the Vietnam conflict (Poole, 2009). His remains were transported from Hawaii to Washington, DC under a Marine honor guard, he lay in state in the Capitol rotunda, and was laid in the Tomb at Arlington. President Ronald Reagan gave a speech, where he in part said:

> The unknown soldier who has returned to us today and who we lay to rest is symbolic of all our missing sons. And we will present him the Congressional Medal of Honor, the highest military decoration we can bestow. About him, we may well wonder as others have: As a child, did he play on some street in a great American City? Did he work beside his father on a farm in America's heartland? Did he marry? Did he have children? Did he look expectantly to return to a bride? We will never know the answers to those questions about his life. We do know, though, why he died. He saw the horrors of war but bravely faced them, certain his won cause and the country's cause was a noble one, that he was fighting for human dignity, for free men everywhere. Today, we pause, to embrace him and all who served us so well in a war whose end offered no parades, no flags, and so little thanks… Let us, if must, debate the lessons learned at some other time. today we simply say with pride: Thank you, dear son, and may God cradle you in his loving arms. (transcript retrieved from https://www.upi.com/Archives/1984/05/28/Here-is-the-text-of-President-Reagans-remarks-at/9454454564800/)

The Vietnam conflict/ war was a contentious conflict that polarized the nation, to say the least. It was experienced in a multimedia fashion that hadn't been seen in prior conflicts, with color film broadcast on the evening news. A national draft was introduced, and so military service was not always sought by those who fought and those who died. Reagan's speech at the dedication of the Tomb of the Unknown for Vietnam is therefore a truly interesting piece of political theater. In his remarks, he addresses the social conflict surrounding the war, but then dismisses that the individual within the Tomb of the Unknown may have been drafted, did not desire to fight, and possibly died in a conflict he felt was unjust. In many ways, this speech embodies the role of the Tomb of the Unknown in the larger political sphere, one that highlights the role of country over individual and lauds those who die without understanding the individual circumstance under which they serve.

The unknown within the tomb was, however, believed to be First Lieutenant Michael J. Blassie. Blassie's airplane had been shot down near An Loc City in southern Vietnam, with witness statements saying that no parachute had been sighted, indicating the pilot had not ejected. A South Vietnamese reconnaissance team located the crash site months later and recovered human remains consisting of ribs, a pelvis, a humerus, military items such as part of a NOMEX flight suit, a raft, ammo pouch and part of a parachute. These and some personal items including Blassie's wallet and some cash that were delivered to a mortuary facility near Saigon 5 days after the recovery. His military ID was purported lost in transit between the crash site and the mortuary, leading to an identification of "Believed to Be", not a positive identification. Because the identification was not positive, the remains were stored at the mortuary in Vietnam for several years and ultimately made their way to the Central Identification Laboratory in Hawaii (CILHI). This was all prior to DNA testing, but the CILHI staff (not consisting of professional forensic anthropologists) found that the bones were inconsistent with Blassie in terms of stature and estimated age at death, believing the individual to be older and taller than Blassie. These findings placed the "Believed to Be" designation in doubt, and it was this that led to Blassie's remains being designated as X-26 (Wagner, 2013). By sheer happenstance, this set of remains was chosen as a finalist for the Vietnam Unknown and through the exclusion of the other candidates, became the Unknown.

Rumours about the identification of the individual within the Tomb of the Unknown began immediately and never subsided. Ultimately, Patricia Blassie, Michael Blassie's sister, joined the Air Force and was a Captain in 1994. Ted Sampley, a veteran and activist, contacted Patricia Blassie with what he felt was evidence that the individual in the Tomb of the Unknown was her brother. When the body had been prepared to send to Washington for the ceremony, the effects recovered with the remains had been placed with the remains. Patricia Blassie worked for the Air Force Office of Public Affairs at the Pentagon, and so approached the situation carefully. The Blassie family wasn't agreed about how to proceed, if Michael Blassie was indeed interred in the Tomb of the Unknown, it was a place of honor, and perhaps he should be left there. But as Patricia Blassie said in a CBS interview, "In the end, all you have is your name. When that's taken away, you're left with nothing. We just wanted Michael's name back" (Thomas, 2012), and they decided to go public with their beliefs.

On January 19, 1998, CBS News aired the story in which Patricia Blassie said "If it is Michael, he's not unknown. He's not identified, but he isn't *unidentifiable*," and demanded a DNA test of the remains within the Tomb. The remains were removed from the tomb on May 14, 1998 and subjected to testing at the Armed Forces Institute of Pathology (Wagner, 2013). While waiting for results, the remains "would occupy that liminal space between sacred and profane, simultaneously the object of national reverence as the honored Unknown, yet subject to scientific inquiry into the individual presumptive identity of 1stLt Michael J. Blassie. As with any rite of passage, ritual helped ensure a smooth transition from one realm to the next" (Wagner, 2013, p. 642). The osteological analysis was revisited (this time by forensic anthropologists) as well, with revised estimates for age at death that would be consistent with Michael Blassie. DNA testing gave a positive identification of the remains as Michael Blassie, and he was ultimately reburied in the Jefferson Barracks National Cemetery in St. Louis, MO, under a tombstone inscribed with his name.

After the removal of Blassie, the tomb was empty, and it remains so. It is technically now a cenotaph, not a tomb, since no remains are buried within it. A new inscription was placed on the tomb reading "Honoring and Keeping Faith with America's Missing Servicemen, 1958–1975" (Poole, 2009). The Medal of Honor that had been placed with Blassie within the tomb is on display in a glass case overlooking the amphitheater.

13.1.4 Unidentified Does Not Mean Unidentifiable

The Tomb of the Unknown for Vietnam is empty and is likely to remain so. Wagner (2013, p. 63) states that the story of the Tomb of the Unknown for Vietnam is "emblematic of a gradual reframing of the tangled relation between anonymous, collective, and individual memorializing at the heart of national projects commemorating wars and those who died fighting them. Whereas memorials of the past, including monuments to unknown soldiers, emphasized collective or anonymous groupings of war dead in articulating national identity, shifts in technology of identification… have enabled a heightened form of individuation. Naming each deceased, returning each set of remains to surviving families, no matter how partial or delayed, personalizes the ideals of sacrifice and honor embodied in the fallen soldier." What she terms the "technology of remembering" is tied to identification, something that comes about with increasing focus on individualization. In effect, she suggests that personalizing the narrative of heroism creates group identity in a powerful, but different mechanism than that used in creating collective identity through the veneration of an Unknown. Increases in technology, especially the rise of DNA-based identifications allows for greater individual identification, so it is unclear if any individuals are actually *unidentifiable* using the guidelines used at the Tombs of the Unknown Soldier.

In February of 1999, officials at the Pentagon concluded that there was little chance of any soldier who would be considered *unidentifiable*. DNA samples are now taken from all inducted members. A decision was made to keep the cenotaph within Arlington with the additions of a sign reading "The crypt is presently empty, but it serves as a tribute to all those who made the supreme sacrifice during the Vietnam Conflict" (Myers, 1999).

The Tombs of the Unknown only have power to act as a collective place of burial if there are indeed human remains housed within them. The intersectional identities that every person carries are stripped away in favor of a collective identity when a set of human remains is placed in the tomb. This symbolic act is a rite of passage on a societal level, allowing the nation to recognize and acknowledge a war. It is telling that the Tombs of the Unknown are only dedicated after the conclusion of the conflicts, and the speeches delivered at their dedications not only honor the dead but consecrate the ending of that conflict in real and meaningful ways.

13.1.5 The Tomb Guards

Soldiers who stand guard over the Tombs of the Unknown at Arlington National Cemetery are known as Tomb Guards. Prior to 1926, no military guards were posted at the Tombs. In 1926, the Army was given the honor of guarding the Tombs, and they continue to stand guard rain or shine, through snowstorms and hurricanes. On July 1, 1937, the size of the Guard was increased, and a 24-hour a day watch was begun. They are responsible for a constant vigil and preventing any desecration or disrespect directed towards the tombs ("Who are the Tomb Guards," 2021).

There are specific requirements for those who would serve in the Tomb Guards. There are three Reliefs, separated by height. The first Relief consists of individuals between 6′2″ and 6″4″ in height, the second between 6′ and 6′2″, and the third between 5′11″ and 6′. Each Relief consists of a Staff Sergeant and two teams consisting of an assistant Relief Commander and 4 Sentinels. Sentinels stand watch at the tombs, while the Relief Commander is responsible for conducting the Changing of the Guard as well as the welfare and morale of the Relief. Throughout their shifts, Sentinels march 21 steps across a black mat passing the grave markers of each of the Tombs of Unknown. They then turn 90-degrees and face east for 21 s, then north for 21 s and takes 21 steps north, repeating this process until relieved ("Who are the Tomb Guards," 2021).

The Tombs of the Unknown represent different things to different groups, it effectively elicits different responses from different intersectional identities. To the military, the Tombs are sacred spaces of obligation. Therefore, the changing of the Guard (which occurs three times a day), is a ritual that reinforces that sense of sacredness and obligation. Guards see themselves as "comrades-in-arms, keeping the unnamed soldiers company, especially in the dead of night, when the crowds have long departed" (Wagner & Matyók, 2018, p. 54). During the changing of the guard, all three sentinels, the two guards and their relief commander, move together with precision and coordination. We could draw this back to Foucault's mechanics of power (Foucault, 1977). Synchronized movements, gaits and posture, embody and perform military and state power. Within the changing of the guard is a display of arms. The relief commander inspects the uniforms of the sentinels and tests their rifles for spotless surfaces. The position of the rifle signals is functional, moved to the outside shoulder as the sentinels turn so that the weapon is always placed between the Tombs and any threat (Wagner & Matyók, 2018). Wagner and Matyók (2018, p. 56) see the sentinels as living, breathing embodiments of state power within the "curated space" of the Memorial Amphitheater. They go on to state that as "the designated protectors and purveyors of collective memory at the Tomb, the sentinels also instruct future generations about contemporary military ethos through the incorporating practices of their round-the-clock commemorative ceremony" (p. 57). In this way, the mortuary ceremony for the Unknowns is never-ending. It continues unabated regardless of anything else, sentinels stood during hurricanes, wars, around the clock.

This is a performance, conducted before an audience. The audience is expected to react and conduct themselves in specific ways, specifically silently observe while not drinking, smoking, or chewing gum. The audience is informed how they should respond and act by signage but also by the social cues of the other audience members, particularly those who come to visit the Tombs of the Unknown in commemoration or adoration. Through this, the audience absorbs the social messaging of the Tombs, even without understanding the full impact of the history of the monuments. McDonnell and Jackson (1999) anecdotally offer a visit to the Tombs for school children as a way to teach children abstract ideals of honor, service, and patriotism. The changing of the Guard at the Tombs of the Unknown has a common impact on the audience, regardless of age, with even children standing in silence and absorbing the social messaging (McDonnell & Jackson, 1999).

The performance of the guards continues after the last visitors have left the cemetery. Who is the focus of the performance when there is no one left to see it? In this case, the audience becomes the other guards, particularly the guards who are required to stay on site during overnight stays. Part of the impact of the Tomb Guards is the knowledge that the performance is ongoing and never stops, regardless of weather, regardless of any external political force. The structure of the Guard is consistent irrespective of administration or political affiliation. It stands as a social structure bigger than any one of the individuals participating in it. Additionally, the audience for after-hours performances can be imaged as those who even know that that the performance is ongoing. It is known and publicized that the Tomb Guards are perpetually on duty, this knowledge provides solace for the bereaved and amplifies the message. The audience for this may also be thought of as the dead, both in the tombs of the Unknown and in the cemetery surrounding them. It is a constant message to the dead that their sacrifice is honored regardless of the time of day or weather.

13.2 The Political Lives of Dead Bodies

The dead body (or a part of the dead body) has tremendous agency. Crandall and Martin (2014, p. 431) define postmortem agency as "the ability of dead bodies (in their new guises as objects, spirits, relics or other symbols, forms, or identities) to engage, influence, confine or structure the behavior of the living whether directly or indirectly." As non-human agents, bodies extend the agency not only of the living actors who manipulate the dead body but also of other dead people or social systems. The act of manipulation is a discursive act that communicates social identity and memory (Osterholtz 2020).

13.2.1 Authenticity, Statues, and the Life of Lenin's Corpse in Modern Russia

The unique power of the physical body can be used for nation building as can be seen in the above discussion of national memorials. The dead can also act as a conduit for social change or as a physical manifestation of political power or history. Katherine Verdery, in her work looking at the Political Lives of Dead Bodies (1999, 2004) primarily focuses on the bodies of Communist leaders in Eastern Europe and the use of their bodies in post-communist identity building. I will focus on these elements as well, and their importance with relation to space and place. Before delving into that, it is important to understand *why* the body can fill such a role. It is the unique physicality of the body as a reminder that can be touched, seen, and interacted with that provides a permanency to historical events. The dead body, if preserved, requires maintenance and a continued interaction with the body to keep it suitable as a national symbol. The continued maintenance of the body of the dead leader is itself an act that reinforces and acts as an amplification mechanism in a Poetics sense. This can be extended to statues of dead leaders, as Verdery notes that "statues are dead people cast in bronze" (1999, p. 5). While not everyone may be able to travel to see the preserved body of a dead leader, they can see the representation of that individual, faithfully reproduced in bronze. With the actual, physical body preserved there is a surety of representation. The properties of the body become transmitted to the bronze copies, and in that copying become amplified as messaging. This also explains why the destruction of these statues caries great meaning as well, destroying that original amplification of meaning and identity is a concrete and physical break with the past, creating something new and different.

The post-Soviet context that Verdery is working in is important to explore a bit. The changes in the late 1980s in the Soviet world led to changes in both economics and world view. Eastern Europe, in effect, was no longer sheltered from a western view of the world after the fall of Communist governments. These changes in worldview were analyzed through the lens of the dead bodies of Communist leaders by Verdery (1994, 2004). Political transformations become less of a technical process than a total shift in worldview. In presenting what she calls "dead body politics", Verdery (1999, p. 11) links the treatment of the dead to cosmologies and ways of relating the living to the dead in a new system. Why dead bodies? Bodies are material objects and therefore subject to "thereness". Corpses can be moved, displayed and located in strategic locations. This emphasis on landscape is in keeping with a Poetics approach, the bodies themselves mean different things in different places and their agency will have different impacts in different locations. For example, Lenin's body in Red Square signals specific ideas about state power but stored in a barn or private structure would lack the signaling ability. In Verdery's (1999, p. 14) words, "their 'thereness' undergirded the founding and continuity." We can also see the bodies of dead leaders as a physical representation of the political order, a physical manifestation of the Body Politic (along the lines of Scheper-Hughes & Lock, 1987). This isn't to say that the meaning of these bodies is static, it is in fact, remarkably dynamic. Yurchak (2015, p. 61) notes that Lenin's body has been perceived differently since it was first embalmed over 90 years ago. Over the past 25 years, the "body has been blamed for all the ills of the country and celebrated as a symbol of its successes."

Russian's perceptions of Lenin's body have shifted over the years. The Levada Center conducted surveys of Russian citizens. Their website (https://www.levada.ru/2017/04/19/vladimir-lenin/) presents summary data from surveys beginning in the 1990s and ending in 2017. The overall sample size for 2017 surveys was 1600 individuals from all throughout Russia. When asked about whether Lenin's body should be buried (and where) or left in place, longitudinal data is presented from surveys in 1997 through 2017. Options were presented to bury Lenin within a cemetery in St. Petersburg, to bury the body at the Kremlin wall, or to leave in the Mausoleum. An additional option of "I don't know/difficult to answer" was also given. In 1997, 37% felt it was appropriate to bury in St. Petersburg, with responses for this option hovering between 31% in 2008 and a low of 26% in 2017. The option to bury at the Kremlin Wall ranges from 11% in 2006 and a high of 32% in 2017. The option to leave the body in Mausoleum has a high of 44% in 2000 and a low of 25% in 2012. Finally, the "I don't know" response hovers for most year between 7 and 14% for all years presented except for 2012, which has the highest percentage of responses at 23%. For all of these responses, there is no particular visible trend, and the percentages of responses do not appear to vary with published economic rates of inflation or economic development (such as the economic data presented by Akindinova et al. (2016) and Cook et al. (2019)).

This may speak to the large-scale nature of the social changes that the end of communism brought; different age groups and different intersectional elements would experience these changes differently. Rural versus urban experiences would have differed significantly, and older individuals who experienced the Russian revolution or their children may have different interactions with the body of Lenin. Instead of being a stand-in for the Body Politic of the communistic state, at some point, the body of Lenin becomes a historical object. This is particularly true given the preservation efforts that have been ongoing since the initial embalming. Yurchak (2015) gives a detailed history of the preservation of the corpse, noting that as damage to the body was noted, the biological tissue was replaced with artificial materials. Recalling Sahlin's question of *does it have to be real to be real*, is the "authentic biomater" the important part of Lenin's body? Ultimately, if preservation efforts continue, the majority of the body will have been replaced with artificial materials. Yurchak (2015, p. 80) concludes "It is the figure of sovereign *Leninism*—the figure that is constructed as a dynamic *form* of Lenin's body and not its authentic biomater—that the Moscow Lab has been maintaining and improving for the past 90 years." History, and the role of Lenin is effectively being rewritten and kept alive through the maintenance of the body.

13.2.2 Balkan Bodies, Mass Graves, Landscape, and Identity

Central to much of Verdery's work is the idea of landscape. In no place is this more evident than in the role of burials within the Balkans, particularly the burial of average citizens. In recent history, the importance of burial can be traced to massacres committed by numerous groups during World War II. Croatian fascists (the Ustaša) and Serbian royalists (Četniks) were responsible for mass death, with victims thrown into caves, buried in shallow mass graves (e.g., Fig. 13.2), or left out to decompose. Ivan Goran Kovačić (1961) wrote "The Pit" in 1943. The poem is written from the perspective of a person thrown into a mass grave who is not yet dead and details the experiences of being surrounded by the dead and attempting to escape the pit. Kovačić himself was killed by Četnik forces later the same year, but the poem itself became a key piece of literature within Yugoslavian primary schools. A short excerpt is illustrative of the style and imagery evoked:

> From terror I was still—then they still too-
> Their weight decreased, a dead leg on my shoulder
> Dangled limp. They had pursued, but now
> Pursued no more!—My climbing had undone

Fig. 13.2 Mass grave of predominantly Croatian victims, killed after WW2, in the vicinity of Crngrob. Photo attribution: Ajznponar, CC BY-SA 4.0 <https://creativecommons.org/licenses/by-sa/4.0>, via Wikimedia Commons. https://upload.wikimedia.org/wikipedia/commons/e/e9/Crngrob%2C_mass_grave_01.jpg

The dead—I told myself.—That mangled noose
About your neck, a dead girl's locks have tangled!

Kovačić asks questions of how the dead and the activities leading to the mass graves will ultimately impact the region as well:

Is there a place where suffering and pain
Men suffer, and endure, but yet alive?
Is there a place where men forget again
And live with those who wronged them by their side?

Not uncommon to this region, there is an understanding during this conflict that contemporaneously fighting groups would again be living near each other. Kovačić questions, in the darkest hour, whether this is possible given the realities of the violence perpetrated by partisans on both sides. Concepts of Identity in the lands of the former Yugoslavia are incredibly complicated. Croat and Serb communities lived in close proximity to each other, often inhabiting the same spaces. One way that this was made possible was the active forgetting of massacres and massacre sites. With the rise of Josip Broz's (Tito's) regime, specific mention of massacres was forbidden, with abstractions such as "victims of fascism" and "domestic traitors" used to discuss the dead. According to Verdery (1999, p. 99), "The silencing of this grim subject meant that none of the murderers could be avenged (very important, in this area), and so neither the souls of the dead nor the minds of those close to them could rest in peace." With the weakening of the regime in the 1980s, people began to investigate and remove remains from known burial locations. The fact that villagers knew where to retrieve bones from speaks to an undercurrent of local knowledge, while these locations may not have been written down, they were passed as important inter-generational information that was important for local communities to maintain. This is effectively a form of culture-keeping that because it was focused on burial location and circumstances reinforces the importance of burial for group identity.

While there was an official forgetting of these acts, there was an active remembrance maintained within the local communities. Denich (1994, p. 367) refers to this as a moment when "nationalism burst from its hiding places and swept people by the hundreds of thousands into mass movements, then elections, then into confrontations that escalated, with astonishing speed, to civil war." The distance in time from the events as well as social changes made the re-remembering of these acts possible. In the former Yugoslav countries, as well, more recent conflicts likely put these acts into a different, historical context, leading to a desire to identify the locations and memorialize the dead. The primary fault lines in this new conflict generally developed along linguistic lines (Serb and Croat). Serbian and Croatian leaders revived nationalistic ideologies, but this is complicated in this area as Serbian and Croatian populations occupied overlapping territories, leading to areas in which most of the population was within the dominant nation and the creation of a marginalized minority population (Denich, 1994). Where Tito had steadfastly denied communist participation in genocide during World War II, identification of those that were killed by communist partisans in Croatia had a large role in the victory of

Framnjo Tudman's nationalist party in the first Croatian elections (Verdery, 1999, p. 101), when the "skeletons then served in the historical revisionism by which new nationalist histories emerged for newly emerging states." Viewed through this lens, the modern Croatian state is constructed on the foundation of the dead. The material remains of the dead, along with images of their retrieval and the monuments honoring them, form the basis for modern nation building. Televised funerals for bones collectivized and nationalized the dead. The anonymity of these individuals is necessary in this case, along similar lines as the importance of the anonymity of the individual within the Tomb of the Unknown Soldier: if the dead are anonymous they can serve as a stand-in for any fallen family member or community member. Because of this stand-in effect, the location of the burials of the dead become very important and are tied to specific land that now fell into new state boundaries, as Verdery (1999, pp. 101–102) summarizes, "graves laid out a geography of territorial claims and of personal commitment to those claims, for in these places 'our' dead were buried…. We might say that that these corpses assisted in reconfiguring space by etching new international borders into it with their dug graves."

Kovačić's "The Pit" and Kovačić himself takes on a new role in the post-Yugoslav countries. With the reframing of the violence of World War II, celebrations of "The Pit" and its author began to diverge. The political motivations of Kovačić's joining of partisan forces began to be minimized. Levanat-Peričić (2019) argues that the loss of ideology in reading the poem removes its historical context and changes the meaning significantly.

13.2.3 Politics, Repatriation, and Yagan

Another element of the political lives of dead bodies centers on the use of the body in postcolonial frameworks as an act of reparation. Later in this chapter, I will discuss the importance of the Native American Graves Protection and Repatriation Act, but here I would like to focus on repatriation and the use of the dead in non-US contexts. The dead and their location of burial can be an important act to begin to repair relationships. The repatriation of an individual named Yagan from the UK to Australia serves as an excellent example. Fforde (2004) provides the history for this individual. Yagan was a young Noongah man killed by white settlers in the area around Perth in Australia in 1833. Yagan had been actively involved with conflict with settlers and with attempts to discuss settler incursions onto Aboriginal lands. In 1833, a warrant was issued for his capture. He was shot by a white settler, and his body buried near the site of the murder. In August of 1833, a bounty of £30 was paid to James Keates for Yagan's head. The head was preserved via smoking and brought to England in September of 1833. The head was examined by phrenologists and ultimately presented to the Liverpool Royal institution where it may have been displayed in a case with other preserved heads to illustrate cranial variation. In 1894, the collections of the Liverpool Royal Institution were scattered and Yagan's head wound up at the Liverpool City Museum where it remained in the collections until the middle 1960s. In 1964, the Ethnology curator requested permission to destroy the skull, and it was decided to bury the preserved head in Everton Cemetery, in a grave that ultimately also housed stillborn children.

While Yagan's head was buried in England, his legacy in Australia grew. He was respected as a warrior and a protector of his people. He came from a prominent family and many Aboriginal individuals count lineal and cultural lines of descent from him and his family. For Aboriginal groups around Perth, his legacy has continued to grow, including the placement of a statue on Heirisson Island in 1984 after years of lobbying. The statue is a life-sized representation of the warrior and is located on Heirisson Island in the Swan River (Fig. 13.3). Yagan was also the special focus during the Kyana Corroboree in Perth in 1993. It was known that his head was removed from the grave and taken to England, and as early as the 1950s, there were calls to locate the head and reinter this with the body. It wasn't until 1993 that the disposition of the head was finally identified. In April 1994, the process of requesting the repatriation of Yagan's head began, but this process was complicated due to

Fig. 13.3 Yagan statue located on Heirisson Island, Perth, Western Austrialia. Photo attribution: Nachoman-au, CC BY-SA 3.0 <http://creativecommons.org/licenses/by-sa/3.0/>, via Wikimedia Commons. https://upload.wikimedia.org/wikipedia/commons/f/fd/Yagan_Statue_2005a.jpg

the presence of the stillbirths buried in the same cemetery plot. Permission for the disinterment of the stillbirths had to be sought from all of their families. An archaeological survey was conducted within the cemetery and alternative plan was developed to excavate in an adjacent plot and disturb the other interred individuals as little as possible.

Ken Colbung, a Noongah man who identified Yagan as an ancestor was prominent in promoting the legacy of Yagan and in attempting to recover the head from the UK. Colbung traveled to England and attempted to contact the families of the stillborn children to receive their blessing for the excavation, but ultimately left without success in July of 1997. Unbeknownst to Colbung (or anyone else), the excavation had been conducted on April 14, 1997 and a forensic analysis of the head was conducted the following day. On August 27, 1997, a Noongah delegation including Colbung arrived to collect Yagan's head. Colbung saw this trip as an opportunity to unite the Aboriginal community in Australia (Fforde, 2004). In September of 1997, a committee consisting of community members and elders as well as government representatives took on the task of locating the original burial site to reunite Yagan's head with his body. A week after the return of Yagan's head to Australia, the statue on Heirisson Island was beheaded with an angle grinder. The head was quickly replaced, but the replacement head was soon detached and stolen as well.

After numerous debates within the Noongah about the appropriate resting site. Yagan's head was buried in a traditional ceremony within the Swan Valley in July 2010. Though the burial site for his postcranial remains could not be identified, the site chosen for burial of Yagan's head was believed to be in the general area where he was killed. The reburial ceremony coincided with the opening of Yagan Memorial Park; the ceremony had over 300 participants and observers. Premier Colin Barnett is quoted in an Australian newspaper saying "I hope this corrects our history, balances our history and that Yagan will have his true place along with other leading Western Australians and he will be remembered and respected and provide some inspiration, particularly for young Aboriginal people, to succeed… and endeavour to go forward" (Lucev, 2010).

Yagan's legacy continues, the subject of contemporary art and discussion. In 2006, contemporary Aboriginal artist Julie Dowling completed Yagan (1833) and it hangs in the Art Gallery of Western Australia. Idle (2014, p. 111) compares this painting within the same permanent collection as one depicting a colonial settler, Mrs. Dance. Idle, in speaking of the painting of Yagan, notes "It disrupts the easy, unsophisticated understandings of colonial history depicted in Mrs. Dance, and draws attention to an Indigenous narrative at a time when cultural institutions and their (white) audiences were ready to hear and see them." Yagan has also been memorialized in the public space of Yagan Square in central Perth which opened in 2016. In particular, one piece of artwork within the space known as *Passenger* is designed to create a new body for Yagan. This "'new body' is conceived as a spatial history of passage" and is "collectivized to bear witness to the passing of generations of the unnoticed or excluded (history's refugees)" (Carter, 2016, p. 66). Yagan's physical remains will not be buried there, but the representation of the square as a replacement body calls to mind the use of statues (Verdery, 1999) or monuments as body substitutes (such as in cenotaphs, see also Basanti (2020)).

The story of Yagan, from a Poetics perspective is one of colonialism, the negotiation of Indigenous identity, and the resilience of that identity in the face of tremendous challenges. Yagan was a leader of his people and his death was mourned by many. His body was split into pieces and his head became an object of curiosity before being buried in an anonymous grave. Meanwhile, his Indigenous community had reconstituted his body in the form of a statue. This statue was seen as a replacement, a physical body for the dead and a visible object for the community to rally around. The destruction of the statue can be viewed in a similar fashion to that described by Verdery for the destruction of communist statues. David Martin (2007, p. 316), in discussing the destruction of Yagan's statue, speaks to the continuation of white settler violence and the power of representations to have "a life of their own, a statue which had stood untouched for well over a decade was made to recapitulate the original death of its once living likeness." Martin's argument is that destruction of statues is, in fact, a resanctification or rededication of them that can only be accomplished through the act of destruction.

In keeping with a Poetics perspective, acts committed in a public manner such as the destruction of public statues is both degenerative and regenerative (as I have argued elsewhere, e.g., Osterholtz, 2018b). The head of the statue was defaced multiple times, something that like the defacement of signs detailing the death of Emmett Till, become part of the story. For Martin (2007, p. 316), the headless statue "has had its 'sacredness' invigorated by the act of defacement." The act of repatriation of Yagan's head is a recognition of Indigenous sovereignty, and the protracted process of the repatriation speaks to the complicated interplay between governments, Indigenous groups (who didn't always agree on who should speak for them or where the head should be buried), and the legal framework. As Fford points out, one of the most complicating factors was the burial of the head in a grave for stillborn children. An order of exhumation was stalled due to the need to contact the families of those individuals to obtain their permission. If the head had remained in museum storage, the process for obtaining an export license would have been significantly truncated due to the fact that the identity of the individual was known.

13.3 Massacres, Genocide and Bodies

The role of the body in massacre and genocide is multifocal with respect to genocide. Bodies are used as political tools, and can also be used as memorials. The multifaceted role of bodies with respect to genocide is the work of Mbembé and Meintjes (2003) and their concept of *necropolitics*. Bodies themselves are killed, but this is not the only aspect of genocide that imparts meaning. It is the peri- and post-mortem destruction of the body, maiming and killing in a way that creates a specific type of taphonomic signature. This reduction and processing of the bodies is integral to the social messaging of the massacre or genocide itself. Systematic processing of the bodies can be seen as a Poetic ritual that when performed over and over, leads to identifiable traces such as those seen in extreme processing assemblages within the Southwestern US (e.g., Martin & Osterholtz, 2015, 2020; Osterholtz, 2010, 2012, 2013, 2018a, 2018b). The systematic nature of these acts argues for a performance that is consistent with a Poetics experience and can be related to identity formation/negotiation/renegotiation. But the historical context of this type of archaeological assemblage means that this must be placed within the context of *massacre* and not necessarily *genocide*. The difference here relates to the legal definitions at work. Massacres can be viewed as cultural systems with their own internal cultural logics that convey "powerful messages to witnesses and/or survivors about power dynamics" (Anderson & Martin, 2018, p. 1). Defining the term massacre is complicated, with Kuckelman and colleagues (Kuckelman et al., 2000) focusing on elements of demography, including non-combatants as victims. Anderson and Martin (2018) focus on the process of massacre as relational and dynamic power to convey social messages from the powerful to the rest of society. This is a collective understanding of massacre, but one that does not seek to understand or stipulate that there is a single reason for *why* massacres occur. Instead, they focus on the idea that massacres are situated within specific time and place and employ specific cultural logic that may take on different meanings when viewed outside that cultural system. Massacres tend to involve large-scale killing, but Anderson and Martin are careful not to designate a specific number of individuals who must be involved for the label to be appropriate. In this way, we can see massacre as an anthropological concept, not a legal one. Bioarchaeology contributes to the discussion of necropolitics in a few different ways: (1) bioarchaeological researchs provide information about the social identities of the dead through techniques such as osteobiography; (2) bioarchaeological analysis can elucidate the modalities of violence (such as perimortem versus antemortem lesions, compromised health status indicating subaltern status, trauma recidivism, intentional postemortem processing) to understand the totality of violent experience; and (3) bioarchaeologists can examine long temporal breadths, understanding more of the shifting nature of necropolitical processes (Geller, 2021).

Genocide, on the other hand, can be viewed as a legal term. The definition of genocide offered by the United Nations in the Convention on the Prevention and Punishment of the Crime of Genocide (CPPCG) is any of five "acts committed with intent to destroy, in whole or in part, a national, ethnical, racial or religious group". These acts consist of killing members of the group, causing serious mental or bodily harm, imposing living conditions intended to destroy the group, preventing births, and forcibly transferring children from the group. Victims are targeted because of real or perceived membership within a group. The CPPCG was enacted as a response to the Holocaust of World War II and was enacted as international law in 1951 with regular updates and readoptions through today (e.g., Human Rights Council, 2024). Genocide, as the CPPCG states, is based on *intent*. Intent is something that we often cannot know for past acts, so the attribution of genocide into antiquity is a fraught enterprise at best (Komar, 2008).

Numerous examples exist, sadly, of modern genocide in which we can identify the intent to destroy members of specific groups. Necropolitics become central to how we view the bodies produced as part of genocidal activities. According to Mbembé and Meintjes, "lifeless bodies are quickly reduced to the status of simple skeletons. Their morphology henceforth inscribes them in the register of undifferentiated generality: simple relics of an unburied pain, empty, meaningless corporealities, strange deposits plunged into cruel stupor." In viewing the bodies of victims as "other" or nonhuman things, we can see the ultimate use of the body as material culture. This builds on Foucault's (2004 [1976]) concept of biopower, the idea of dominion over body inscribed in modern state functions. Literally, states have the power to take life. Mbembé and Meintjes (2003, p. 12) draw this out to include the use of war as a way to achieve sovereignty, "Imagining politics as a form of war, we must ask: What place is given to life, death, and the human body (in particular the wounded or slain body)? How are they inscribed in the order of power?"

13.3.1 Memorializing Genocide with Bodies

The use of bodies of genocide victims or their belongings is well known. The use of these bodies or physical objects provides the physical link between the living and the dead, along the lines of mortuary monuments or cemeteries described by Buelow (2010). One hallmark of genocide tends to be the refusal of considerate or culturally appropriate burial for the victims. Memorials may form the only physical connection that can be touched and seen by survivors or witnesses.

The 1994 Rwandan genocide grew out of colonial conflict and an ill-advised attempt by Beligan colonial governments to provide majority rule, pitting Hutus against Tutsi and ultimately creating a power dynamic where neither group saw the other as as human as themselves. Hutu government soldiers disseminated propoganda via radio, and a large scale slaughter of Tutsis began. Soldiers had guns and grenades, but Hutu peasants who participated typically used everyday objects such as hoes and machetes to kill and dismember their neighbors. The massacres lasted for about 100 days and lead to the deaths of an estimated 800,000 individuals (Hohenhaus, 2013).

Memorials of the 1994 Rwandan genocide have become tourist sites, particularly the Gisozi Center in Kigali. Hohenhaus (2013, p. 142) sees visiting these as "the very darkest kind of dark tourism." He wraps these in concepts of place identity, which is crucial for understanding the unique role of the lcoations of the memorials in both "local 'sited'ness' and the greater 'national identity'." In particular, many memorials were made in the locations of mass killings, such as the Murambi school (Fig. 13.4). On April 16, 1994, approximately 65,000 Tutis ran to the school, seeking safety. After the victims were told to gather there, water was cut off and no food was given to them. They were then systematically attacked and ultimately overrun on April 21st. Thirty-four people are thought to have survived the massacre, with the rest dying at the school. Their bodies were preserved in lime and laid out on a wooden tables in a series of buildings (Boswell, 2017). According to Boswell (2017,

Fig. 13.4 "Deep gashes delivered by the killers are visible in the skulls that fill one room at the Murambi School". Aftermath of Rwandan Genocide. Photo taken during the official visit of US Rep. Frank Wolf, Public domain, via Wikimedia Commons. https://upload.wikimedia.org/wikipedia/commons/d/db/Rwandan_Genocide_Murambi_skulls.jpg

p. 84), memorials such as Murambi are a way in which one "explores the relationship between materiality, testimony, the imagination and knowedge, finding new forms of meaning and the interplay between these equally important reference points." Memorials such as Murambi force the viewer to become a witness to the genocide, complete with the phenomenological elements.

According to Boswell (2017, p. 85), "tourists and other visitors to Murambi cannot avoid the fact that the schoolrooms are still permeated by the nauseating stench of bodily decomposition. If one looks closely, one can start to tentatively decipher the evidence of bodily violence without any specialist training. Broken bones, fractured children's skulls, rictus of corpses that still seem to be yelling: we read these wounds and gestures with our senses and our emotions. What is being evoked is not precise historical knowledge." This is not sterile history presented, it is designed to evoke a visceral reaction. At its base, the memorial at Murambi is physical proof of the massacre and the treatment of the bodies. In this way, they form a "corporeal historiography" (Major, 2015, p. 177). The bodies themseves are used as the archive documenting the genocide. As Major (2015, pp. 177–178) puts it, "The affective ability of human bones to demand attention, and the sense of ongoing and disruptive prsence that they exude, has become the means through which a collective, politically amenable, identity is being consolidated." The use of the body as memorial in Rwanda builds on the unique physicality of human remains (their 'thereness') and the phenomenological nature of interacting with the decomposing body. Visiting the memorials forces the viewer to become a witness to the genocide, the phenomenological elements are effective amplifcation mechanisms for the performance of the memorial.

13.4 Summary

Modern nation states have unique abilities to use the bodies of leaders (as in the work of Verdery), soldiers, and citizens for nation-building. Bodies come to symbolize power, whether that is the power of the state as evidenced in the post-mortem usage of dead leaders in the former Soviet Union or the power of the mob to perform genocide as in Rwanda. Bodies can also become a focal point for important elements of national identity, such as sacrifice and spreading ideals around the world, such as the use of the body in Arlington National Cemetery in the US. Individuals such as Yagan can become rallying points for the performance of identity in contested spaces, but the physical remains are an important element and ground the conversation surrounding repatriation, structural violence, and group identity. The breadth of uses of the body highlights the malleability of the body and its ability to embody ideas of both the Body Politic and the Social Body.

References

Akindinova, N., Kuzminov, Y., & Yasin, E. (2016). Russia's economy: Before the long transition. *Russian Journal of Economics, 2*, 219–245.

Anderson, C. P., & Martin, D. L. (2018). Introduction. In C. P. Anderson & D. L. Martin (Eds.), *Massacres: Bioarchaeology and forensic anthropology approaches* (pp. 1–11). University of Florida Press.

Ask Common Grave for a Blue and Gray. (1926, April 26). *New York Times*, 11. Retrieved from https://www.nytimes.com/1926/04/26/archives/ask-common-grave-for-a-blue-and-gray-veterans-suggest-that-unity-of.html?searchResultPosition=1

Basanti, D. S. (2020). Poetics of the house: Changing realities of body and person in Aksumite mortuary practices. In A. J. Osterholtz (Ed.), *The poetics of processing: Memory formation, cosmology, and the handling of the dead* (pp. 103–122). University Press of Colorado.

Blair, C., Balthrop, V., & Michel, N. (2011). The arguments of the tombs of the unknown: Relationality and national legitimation. *Argumentation, 25*(4), 449–468. https://doi.org/10.1007/s10503-011-9216-9

Boswell, M. (2017). Reading genocide memorial sites in Rwanda: Eurocentrism, sensory secondary witnessing and shame. *Memoires en jeu, 3*, 80–87.

Buelow, D. A. (2010). *Peripheral memory: New York's forgotten landscape*. Massachusetts Institute of Technology.

Carter, P. (2016). Ghosting: Putting the volume into screen memory. In P. D. Marshall, G. D'Cruz, S. McDonald, & K. Lee (Eds.), *Contemporary publics: Shifting boundaries in new media, technology and culture* (pp. 61–76). Palgrave Macmillan.

Cook, L., Holm-Hansen, J., Kivinen, M., & Kuhnle, S. (2019). Introduction: The Russian welfare state in a time of economic stagnation. *Russian Politics, 4*, 291–301.

Council, U. N. H. R. (2024). Resolution adoopted byt eh human Rights Council on 3 April 2024. Retrieved from https://documents.un.org/doc/undoc/gen/g24/059/80/pdf/g2405980.pdf: https://documents.un.org/doc/undoc/gen/g24/059/80/pdf/g2405980.pdf

Crandall, J. J., & Martin, D. L. (2014). The bioarchaeology of postmortem agency: Integrating archaeological theory with human skeletal remains. *Cambridge Archaeological Journal, 24*(03), 429–435.

Denich, B. (1994). Dismembering Yugoslavia: Nationalist ideologies and the symbolic revival of genocide. *American Ethnologist, 21*(2), 367–390.

Fforde, C. (2004). Yagan. In C. Fforde, J. Hubert, & P. Turnbull (Eds.), *The dead and their possessions: Repatriation in principle, policy, and practice* (pp. 229–241). Routledge.

Foucault, M. (1977). *Discipline and punish: The birth of the prison* (A. Sheridan, Trans.). Vintage Books.

Foucault, M. (2004 [1976]). *Society must be defended. Lectures at the Collège de France, 1975–76* (D. Macey, Trans. M. Bertani & A. Fontana Eds.). Picador.

Geller, P. L. (2021). *Theorizing bioarchaeology*. Springer.

Hettling, M., & Schölz, T. (2019). Bereavement and mourning. In U. Daniel, P. Gatrell, O. Janz, H. Jones, J. Keene, A. Kramer, & B. Nasson (Eds.), *1914–1918 International Encylcopedia of the First Wrold War [Encyclopedia]* (pp. n,p.). https://doi.org/10.15463/ie1418.11401

Hohenhaus, P. (2013). Commemorating and commodifying the Rwandan genocide: Memorial sites in a politically difficult context. In L. White & E. Frew (Eds.), *Dark tourism and place identity: Managing and interpreting dark places* (pp. 142–155). Routledge.

Idle, J. (2014). Yagan, Mrs dance and whiteness. In V. Castejon, A. Cole, O. Haag, & K. Hughes (Eds.), *Ngapartji Ngapartji. In turn in turn: Ego-histoire, Europe and indigenous Australia* (pp. 109–124). ANU Press.

Komar, D. (2008). Patterns of mortuary practice associated with genocide: Implications for archaeolgoical research. *Current Anthropology, 49*(1), 123–133.

Kovačić, I. G. (1961). *Jama (The pit)*. Matica Hrvatska.

Kuckelman, K. A., Lightfoot, R. R., & Martin, D. L. (2000). Changing patterns of violence in the Northern San Juan region. *Kiva, 66*(1), 147–165.

Levanat-Peričić, M. (2019). Forgetting strategies in (re) structuring collective memory of Goran's poem the pit. *Poznańskie Studia Slawistyczne, 16*, 159–174.

Lucev, A. (2010, July 10). *Yagan's head reburied in Swan Valley*. The Western Australian. Retrieved from https://archive.wikiwix.com/cache/index2.php?rev_t=20110823030033&url=http%3A%2Fau.news.yahoo.com%2Fthewest%2Fa%2F-%2Fbreaking%2F7563246%2Fnyoongar-warrior-yagans-head-reburied%2F#federation=archive.wikiwix.com&tab=url.

Major, L. (2015). Unearthing, untangling and re-articulating genocide corpses in Rwanda: Déterrer, démêler et réarticuler les corps du génocide au Rwanda. *Critical African Studies, 7*(2), 164–181.

Martin, D. L. (2007). Of monuments and masks: Historiography in the time of curiosity's ruin. *Postcolonial Studies, 10*(3), 311–320.

Martin, D., & Osterholtz, A. J. (2015). *The poetics of processing: Deviant performances and memory making in the ancient southwest*. Paper presented at the 114th American Anthropological Association Annual Meeting, Denver, CO.

Martin, D. L., & Osterholtz, A. J. (2020). The poetics of corpse fragmentation and processing in the ancient southwest. In A. J. Osterholtz (Ed.), *The poetics of processing. Memory formation, cosmology, and the handling of the dead* (pp. 85–102). University Press of Colorado.

Mbembé, A., & Meintjes, L. (2003). Necropolitics. *Public culture, 15*(1), 11–40.

McDonnell, M., & Jackson, J. (1999). The close up capital experience: Middle schoolers travel to Washington, D.C. *Clearing House, 72*(6), 385–391. https://doi.org/10.1080/00098659909599429

Myers, S. L. (1999, February 25). Arlington yields to science: No more 'unknowns' likey. *New York Times*, pp. 1, 24.

Osterholtz, A. J. (2010). *Hobbling and torment at sacred ridge: Perimortem damage to the feet as a mechanism of social and physical control*. Paper presented at the Paleopathology Association Meetings, Albuquerque, NM. Poster retrieved from

Osterholtz, A. J. (2012). The social role of hobbling and torture: Violence in the prehistoric southwest. *International Journal of Paleopathology, 2*(2–3), 148–155. https://doi.org/10.1016/j.ijpp.2012.09.011

Osterholtz, A. J. (2013). Hobbling and torture as performative violence: An example from the prehistoric southwest. *Kiva, 78*(2), 123–144.

Osterholtz, A. J. (2018a). Each one the same: Performance, demography, and violence at sacred ridge. In C. P. Anderson & D. L. Martin (Eds.), *Massacres: Bioarcheaology and forensic anthropology approaches* (pp. 82–93). University of Florida Press.

Osterholtz, A. J. (2018b). Interpreting and reinterpreting sacred ridge: Placing extreme processing in a larger context. *Kiva, 84*(4), 461–479.

Osterholtz, A. J. (2020). Introduction: Processing and poetics, examining the model. In A. J. Osterholtz (Ed.), *The poetics of processing: Memory formation, cosmology, and the handling of the dead* (pp. 3–14). University Press of Colorado.

Poole, R. M. (2009). *On hallowed ground: The story of Arlington National Cemetery*. The Story of Arlington National Cemetery.

Scheper-Hughes, N., & Lock, M. M. (1987). The mindful body: A prolegomenon to future work in medical anthropology. *Medical Anthropology Quarterly, 1*(1), 6–41. Retrieved from http://www.jstor.org/stable/648769

The Tomb of the Unknown Soldier. (2021). Retrieved from https://www.arlingtoncemetery.mil/Explore/Tomb-of-the-Unknown-Soldier

Thomas, B. (2012, November 8). Last soldier buried in tomb of the unknowns wasn't unknown. *Washgton Post Magazine*. Retrieved from https://www.washingtonpost.com/lifestyle/magazine/last-soldier-buried-in-tomb-of-the-unknowns-wasnt-unknown/2012/11/06/5da3e7d6-0bdd-11e2-a310-2363842b7057_story.html

Verdery, K. (1999). *The political lives of dead bodies: Reburial and Postsocialist change*. Columbia University Press.

Verdery, K. (2004). Dead bodies animate the study of politics. In A. C. G. M. Robben (Ed.), *Death, mourning, and burial: A cross-cultural reader* (pp. 303–310). Blackwell Publishing.

Wagner, S. (2013). The making and unmaking of an unknown soldier. *Social Studies of Science, 43*(5), 631–656. Retrieved from http://www.jstor.org/stable/43284199

Wagner, S., & Matyók, T. (2018). Monumental change: The shifting politics of obligation at the tomb of the unknowns. *History & Memory, 30*(1), 40–74.

Who are the Tomb Guards. (2021). Retrieved from https://tombguard.org/guards/who-are-the-tomb-guards

Yurchak, A. (2015). Form versus matter: Miraculous relics and Lenin's Scientiic body. In O. Hakola, S. Heinämaa, & S. Pihlström (Eds.), *Death and mortality: From individual to communal perspectives* (pp. 61–81). Helsinki Collegium for Advanced Studies.

Chapter 14
NAGPRA, Sovereignty, and the Ancestors

Abstract This chapter engages with the evolving ethical landscape of bioarchaeology primarily in the US with a focus on the Native American Graves Protection and Repatriation Act (NAGPRA). Within the chapter, the shift from early practices of human remains research to a collaborative, community engaged approach. Special focus is given to the work of Commission for the Ethical Treatment of Human Remains of the American Anthropological Association (AAA), whose ongoing work continues to address the legacy of exploitation in biological anthropology. Drawing on the Commission's final report, this chapter explores key concepts such as accountability, transparency, and the ongoing need for ethical engagement with marginalized communities. It emphasizes the critical importance of *sovereignty* over ancestral remains, with a focus on federal repatriation legislation, and argues for a collaborative approach to research that centers the voices and interests of Indigenous and other affected groups. The chapter highlights how these changes contribute to a broader conversation about social justice, identity, and the ethical obligations of bioarchaeologists in the twenty-first century.

The approach taken in this volume has emphasized the materiality of the dead. Human remains elicit a different response than other elements of material culture. Depending on the community of practice and descent communities feeling, human remains may be more culturally sensitive than other types of material culture. They are therefore subject to additional levels of protection and ethical concern, particularly in the Americas. As my experience with Indigenous human remains has been focused in the US, I will focus on the Native American Graves Protection and Repatriation Act of 1990 (NAGPRA) that provides a minimal standard for ethical action on the part of scientists who wish to study human remains. This is not to say that important conversations occurring in Canada and the rest of the Americas as well as other areas impacted by settler colonialism. Repatriation of Indigenous remains have become central to issues of sovereignty. In Chap. 13, the example of Yagan was used to explore similar issues in Australia. That case study was presented separately but could easily be included here. In organizing the chapters for this volume, I chose to present this chapter as one focused on the US and the legacy of

A. Osterholtz, *The Bodies Poetic*, Bioarchaeology and Social Theory,
https://doi.org/10.1007/978-3-031-85998-4_14

NAGPRA. The issues surrounding human remains form the basis for a modern can become a focal point for land claims, rights of access, and agency.

14.1 The Native American Graves Protection and Repatriation Act (NAGPRA)

The Native American Graves Protection and Repatriation Act of 1990 (NAGPRA) (25 USC 32) and the associated regulations (43 CFR 10) set guidelines for the responsibilities of federal agencies and institutions that receive federal funds (such as museums, universities, state agencies, and local governments) with respect to the return of Native American human remains, funerary objects, sacred objects, and objects of cultural patrimony to lineal descendants, culturally-affiliated Tribes and Native Hawaiian organizations. Regulations relating to implementation of NAGPRA have continued to evolve, with the most recent updates published online August 15, 2024. These regulations loosen the definition of cultural affiliation to one of a "reasonable connection between human remains or cultural items and an Indian Tribe or Native Hawaiian organization based on a relationship of shared group identity. Cultural affiliation may be clearly identified by the information available or reasonably by the geographical location or acquisition history of the human remains or cultural items".

Items subject to repatriation are now defined as "a funerary object, sacred object, or object of cultural patrimony according to the Native American traditional knowledge of a lineal descendant, Indian Tribe, or Native Hawaiian organization" (Title 43, Subtitle A, §10.2). This broadening of the scope of NAGPRA allows for the repatriation of items found outside of a burial context as well as expanding the idea of to whom those objects or remains should be returned. Prior interpretations of NAGPRA set a distinction between items that were culturally affiliated through lineal descent and those that were considered too old or without firm cultural ties. The introduction of *historical* geographical locations and Indigenous knowledge broadens the scope of repatriation and prompts additional interactions between anthropologists and Indigenous groups in ways that will likely lead to fruitful collaboration. A reliance of geographical location has long been a critique of NAGPRA. Dongoske (1996, p. 289) notes that "the regulations fail to recognize individual tribal histories and the development of Indian Reservations." The history of forced removals, development of Reservations, and the mobile nature of non-agricultural groups in the past make the use of current geographical location problematic to say the least.

The purpose of this section of this volume is not to give a step-by-step guide for how to implement NAGPRA but to examine the larger framework of how we interact with NAGPRA as practitioners in bioarchaeology and how we can view NAGPRA as a framework for interacting with lineal and social descendants in our work. NAGPRA is a legal framework that sets out to formalize responsibilities and requirements for reporting and creates a system by which cultural affiliation can be identified. The new regulations acknowledge the importance of Indigenous knowledge and allow for this Indigenous knowledge to used as a claim for the disposition of human remains and culturally important objects. Watkins (2004) notes, however, that the system in place (i.e., the role of the NAGPRA Review Committee), particularly with respect to human remains or objects collected long ago and lacking a lot of cultural context, allows for multiple claims to be filed and potentially pits Indigenous groups against one another. Since disposition and the ultimate use of the objects and/or human remains is at the discretion of the successful claimant, completing claims may be made by groups with very different ideas. In the case of Spiro Mounds (Oklahoma), two Indigenous groups (the Caddo and the Wichita) can make claims for cultural affiliation to the objects removed from graves at Spiro. In this case, museums were able to retain and display objects while the cultural affiliation is being discussed between the Caddo and Wichita (Watkins, 2004). Watkins ties the calculatedly neutral position of museums to continued structural violence of the initial excavations.

14.1.1 Framing Our Relationship to the Dead: ~~Who owns the dead?~~ Who Is Responsible to the Dead?

Framing our approach to NAGPRA is important. NAGPRA establishes a legal framework of responsibility for the limited analysis and repatriation of currently held ancestors within publicly funded institutions such as museums and academic institutions. Is this a relationship of ownership? In recent years, a small but vocal minority has argued that the limitation of analysis of Indigenous remains hampers their academic freedom, typified most recently by Weiss and Springer's (2020) *Repatriation and erasing the past*. In this book, and subsequent social media posts, Weiss argued that her academic freedom was being infringed upon by the restrictions of NAGPRA, and that NAGPRA preferenced Indigenous religious concerns over scientific exploration. This is not the viewpoint held by the majority of bioarchaeologists, nor is this a constructive viewpoint. Weiss was moved out of a curatorial role over an Indigenous collection, and ultimately resigned her tenured position. The publication of *Repatriation and erasing the past* had an immediate reaction on the part of both Indigenous scholars and bioarchaeologists. Halcrow et al. (2021) highlight the false dichotomy between what Weiss and Springer term "objective science" and Indigenous knowledge. In 2024, Weiss published *On the Warpath: My Battles With Indians, Pretendians, and Woke Warriors*. This volume is an autobiographical work that, based on the book description, argues that the approach to NAGPRA taken by universities in particular infringes on her intellectual activities.

The contention between Weiss and the vast majority of bioarchaeologists and archaeologists really centers on the concept of ownership. Weiss is essentially arguing that the ancestors under her care are in fact objects that can be owned and studied. The modern conception Indigenous remains is one of *responsibility to*, not *ownership of*. As Supernant (2021, p. 271) notes regarding collaborative relationships between anthropologists and Indigenous groups, "this shift in practices emphasizes changing relationships of power in projects to where anthropologists working with ancestral remains and objects are working by, with, and for Indigenous peoples, rather than on them." When reframed into a position of responsibility to the dead and their lineal and social descendants, the ethical issues become clearer, and our role is to attempt to undo damage that has been done in the past through analysis (including destructive analysis) of Indignous remains prior to the NAGPRA framework. Permissions for analysis were rarely saught, and the wishes of stakeholders were not followed.

Indigenous remains were set into a category of "other", and not afforded the same protections as the remains of European American groups (which were protected by desecration of graves statutes). The discipline of physical anthropology was founded upon an assumption of access and authority over the "other." Blakey (2022, p. 8) describes early practicioners such as Aleš Hrdlička: "their science (a subjective theoretical choice) was poor (racial cleaving against evolutionary continuity) and questionable intentions reflected a self-serving and deeply subjective vantage of whiteness." This view is echoed by participants in Listening Sessions conducted by the Commission for the Ethical Treatment of Human Remains of the AAA (the Commission). In the final report of the Commission, Agarwal et al. (2024, p. 25) state, "The clear message from participants, which is legislated through NAGPRA, was that institutions do not *own* Ancestors and cultural materials, and that all decisions and actions should flow from that reality. In other words, if institutions don't have the right of ownership, then they don't have the right to display, to conduct research, etc."

NAGPRA (as with any legislation revolving around repatriation) is, at its heart, simply a legal framework that acknowledges Indigenous sovereignty, and shifts the focus to responsibility, what Blakey (2022, p. 9) refers to as the First Principal, "to acknowedge that these ancestral remains do not belong to us but most legitimately fall under the stewardship of their social descendants." The framing here is important, that we (as anthropologists) owe the dead respectful treatment. Our role is to assist social descendants in their duties to the dead, however those stakeholders are defined.

I received my BA in Anthorpology in 1999, 3 years after the implementation of NAGPRA, and this has been the operating paradigm for my entire career. The act was passed in 1991 with institutions given 5 years to produce inventories prior to the implimentation of the legistlation in 1996. As we move forward, it will become increasingly rare for bioarchaeologists to have been trained outside of a NAGPRA framework. Those of us brought up within a NAGPRA framework, in effect, have a different relationship with both Indigenous groups and their ancestors. Halcrow and colleagues acknowledge that the ideas espoused by Weiss are indicative of a mindset and operating viewpoint that is not held by the majority of practicing bioarchaeologists today.

Halcrow et al. (2021, p. 213) frame their discussion in terms of violence, noting that "*Repatriation and Erasing the Past* is a racist book and is violent towards Indigenous people." The framing of structural violence is one also used by Buikstra et al. (2022), acknowledging the foundational elements of biological anthropology rooted in racism and colonialism while building on these towards decolonization. Decolonization involves the active participation and engagement with stakeholders at every stage of research, including whether research should be conducted.

NAGPRA doesn't mean the end of bioarchaeological research, as Weiss suggests. In fact, robust community engagement leads to better questions asked and a more grounded focus for research efforts. In their ongoing research into a Genizaro cemetery in New Mexico, Pamela Stone and Debra Martin have developed a model for community engagement that actively encorporates lineal and cultural descendants of those buried within a community cemetery. The research questions they're asking have been developed in collaboration with the stakeholders. In a recent conference presentation, Stone (2024, p. 960) stated:

> My goal here is to underscore the need to rethink traditionally colonial research questions and modes of inquiry, and to revision bioarchaeological practices that are community engaged and intertwined with Indigenous/descendant knowledge of their ancestors. A community-engaged bioarchaeological project begins by bringing in Indigenous people and/or descendants from the very beginning of the project, centering their questions and bringing forward their knowledge of the past provides grounding and inclusion. Shifting away from solely the analytical goals of bioarchaeologists to those of the community creates a shared vision of the past that is co-created by invested descendants and bioarchaologists.

This inclusion of additional voices prompts new questions and new avenues to explore about how the dead continue to influence the living. My own research in Croatia is community engaged as well. Recently, my Croatian colleagues and I collaborated on a conference presentation (Osterholtz et al., 2024) about the ethical nature of bioarchaeological research on individuals buried within Croatian medieval cemeteries. In our conversations, my Croatian colleagues found the use of lineal descent terms such as "ancestor" to be inappropriate, preferring the phrase "social descendants." While they do not see themselves as the lineal descendants of those individuals, they do believe that ethical research conducted in a collaborative manner is appropriate and feel that as culturally affiliated historical societies and local museums, an increased knowledge of lived experience in the past adds to the overall understanding of history within their regions.

The specific examples of the Historic Belen Bioarchaeology Project and my own work in Croatia highlight the importance of community engaged research. NAGPRA should always be seen as foundational for our discipline going forward, but it forms only the minimum requirements for engagement with stakeholders. Additional work is needed to develop robust models for community engagement, such as that used in the Historic Belen Bioarchaeology Project or the clientage model developed by Blakey and colleagues for work on the African Burial Ground in New York City (Blakey, 2010, 2022; Blakey & Rankin-Hill, 2004).

As discussed in Chap. 5, White, Colonial power structure within the US limited the ways in which African American identity could be formed and performed. The same processes occurred with respect to Indigenous and Native American groups. Social bonds and traditional cultural practices were disrupted and broken through forced migration and mass mortality events perpetrated upon these groups. The celebration of ancestors and the rituals through which they are honored remain foundational to identity in modern Indigenous and Native American groups. NAGPRA is one mechanism to codify and acknowledge this mortuary and commemorative space. The increasing importance of community engaged projects is a further reframing of the repsonsibility of anthropologists to work towards rightin these historical wrongs.

14.1.2 Who Is an Ancestor? Stakeholder? Affiliated Group?

In looking through the abstracts for the 89th annual meeting of the Society for American Archaeology from 2024, the term "stakeholder" or "stakeholders" appears 89 times, "Ancestor" or "Ancestors" appears 218 times, at the same conference in 2023, "stakeholder" or "stakeholders" appears 62 times in the abstracts and "Ancestor" or "Ancestors" appears 136 times. In 2022, "stakeholder" or "stakeholders" appears 91 times in the abstract and "Ancestor" or "Ancestors" appears 43 times. This shows a general trend towards a preference for using terms in our research that acknowledge the relationship between living Indigenous groups and their ancestors that for so long have formed the basis for research in North America. But what is an ancestor? Is this limited to individual of lineal descent? Watkins (2004) points out that while lineal descent is preferenced in NAGPRA claims, it is not actually defined in the statute. Is this strictly a genetic relationship? Such relationships have usually been preferenced in repatriation claims around the world, as noted by Fforde (2004) in relation to the return of Yagan's head to Australia. As a known individual, repatriation would likely have been less complicated than an unknown and unnamed skull. If descent is strictly identified based on what Watkins (2004, p. 152) describes as the ability of an individual to "trance ancestry, directly and without interruption, by means of a traditional kinship system or by the American colonial law system of descendancy," adopted relatives will be left out unless that adoption can be somehow documented. In extending the descent relationship into a time before written records exist, these important relationships are likely not to be recognized.

The term "stakeholder" is broader and not contingent upon a direct line of descent. This may include culturally affiliated Indigenous groups and community members who feel a kinship (however defined) to the dead.

The terminology used to describe those connected to the dead will vary significantly. As mentioned above, my Croatian colleagues do not find the term "ancestor" to be appropriate, but numerous Indigenous groups within the US find the term "ancestor" to be most representative of their relationship to the dead. The take-home message here is that there is no one "right" way to describe the relationship between the living and the dead. It is a continually evolving relationship, and so the terminology we use to describe it will necessarily be situational and must remain flexible.

Most modern bioarchaeologists have been open about the origins of the discipline and feel a genuine desire to redress the approaches taken by those who came before us. Central to this was the development of the Commission by the American Anthropological Association under the leadership of Michael Blakey. The Commission was charged with reviewing and assessing legislative, policy and professional guidelines/standards relating to the study and storage of human remains (Agarwal et al., 2024). They also conducted numerous listening sessions all around the world. In their final report, the Commission members noted that there is no single fix for the legacy of biological anthropology. Instead, they "recognize that accountability, transparency, cooperation, and ethical anthropological practices must be ongoing, relational, and dynamic" (Agarwal et al., 2024, p. 8). As the Commission rightly notes, structural violence and exploitation involving marginalized groups (and often the remains of their Ancestors) continues in the form of gentrification, imperialism, mining, and war. Larger issues of social justice are central to this work and must enter around the voices of marginalized groups. One of the avenues for the promotion of identity is sovereignty over the remains of their Ancestors, helped by federal legislation that mandates repatriation. When permissions are granted for research to occur, the approach *must* be one of collaboration, not consultation. This framing centers Stakeholders in an important way, and makes sure that when research is conducted, it serves the needs and interests of those stakeholders. It is important to recognize the power that physical human remains have as well as the associated artifacts for the establishment, negotiation, and performance of sovereignty for Indigenous groups. Indigenous groups have used the legal frameworks developed by the Body Politic (i.e., the US government) to reclaim the bodies of their Ancestors. Legal frameworks have largely had the effect of developing mechanisms to control Indigenous bodies and limit Indigenous agency. These same legal frameworks, through the process of repatriation, can be seen as an expression of group identity and agency in the 21st century.

14.2 Summary

This chapter has examined the ethical, legal, and cultural requirements and implications of repatriation of the remains Indigenous individuals within the US, focusing on the Native American Graves Protection and Repatriation Act (NAGPRA) of 1990. NAGPRA sets guidelines for the return of Native American ancestors as well as funerary objects and sacred items to culturally affiliated tribes. This chapter examines debates over the responsibility of the living to the dead, arguing that this should be one of the responsibilities of researchers to the dead and their descendants, with a focus on collaboration and community engaged work. When reframed in terms of responsibility and community-lead and -engaged research, bioarchaeology can be an amazing tool for understanding lived experience in the past as well as working towards the goals and needs of culturally affiliated communities.

References

Agarwal, S., Blakey, M. L., Champney, T. H., de la Cova, C., Davis, J. L., Martin, D. L., Weik, T. et al. (2024). *The commission for the Ethical Treatment of Human Remains: Final Report.* Retrieved from https://www.americananthro.org/wp-content/uploads/tcethr-report-2024-06.pdf?_gl=1*vos7oo*_ga*NTQyNTQxODYuMTcyOTE4NzI3NA..*_ga_NHV0Y97DC9*MTcyOTE4NzI3NC4xLjAuMTcyOTE4NzI3NC42MC4wLjA

Blakey, M. L. (2010). African burial ground project: Paradigm for cooperation? *Museum International, 62*, 61–68.

Blakey, M. L. (2022). Walking the ancestors home: On the road to an ethical human biology. *Anthropology Now, 14*(1–2), 1–20.

Blakey, M. L., & Rankin-Hill, L. M. (Eds.). (2004). *Skeletal biology final report: The New York African burial ground, Vol. 1 and 2.* The African Burial Ground Project, Howard University for the United States General Services Administration, Northeastern and Cribbean Region.

Buikstra, J. E., DeWitte, S. N., Agarwal, S. C., Baker, B. J., Bartelink, E. J., Berger, E., Brickley, M. B., et al. (2022). Twenty-first century bioarchaeology: Taking stock and moving forward. *American Journal of Biological Anthropology, 178*, 54–114.

Dongoske, K. E. (1996). The native American graves protection and repatriation act: A new beginning, not the end, for osteological analysis – A Hopi perspective. *American Indian Quarterly, 20*(2), 287–297.

Fforde, C. (2004). Yagan. In C. Fforde, J. Hubert, & P. Turnbull (Eds.), *The dead and their possessions: Repatriation in principle, policy, and practice* (pp. 229–241). Routledge.

Halcrow, S., Aranui, A., Halmhofer, S., Heppner, A., Johnson, N., Killgrove, K., & Schug, G. R. (2021). Moving beyond Weiss and Springer's repatriation and erasing the past: Indigenous values, relationships, and research. *International Journal of Cultural Property, 28*(2), 211–220.

Osterholtz, A. J., Novak, M., & Valent, I. (2024). *Approval and Agency with Skeletal Collections: Lessons from Northern Croatia.* Paper presented at the 2024 American Anthropological Association Meeting, Tampa, FL.

Stone, P. K. (2024). *Community-engaged bioarchaeology: Decolonizing research.* Paper presented at the 89th Annual Meeting of the Society for American Archaeology New Orleans, LA.

Supernant, K. (2021). Decolonizing bioarchaeology? Moving beyond collaborative practice. In C. H. Meloche, L. Spake, & K. L. Nichols (Eds.), *Working with and for ancestors: Collaboration in the care and study of ancestral remains* (pp. 268–280). Routledge.

Watkins, J. (2004). Artefactual awareness: Spiro mounds, grave goods and politics. In C. Fforde, J. Hubert, & P. Turnbull (Eds.), *The dead and their possessions: Repatriation in principle, policy, and practice* (pp. 149–159). Routledge.

Weiss, E., & Springer, J. W. (2020). *Repatriation and erasing the past.* University Press of Florida.

Weiss, E. (2024). On the Warpath: My Battles With Indians, Pretendians, and Woke Warriors. Washington, D.C., Academica Press.

Part V
Bodies and Material Culture

Part V focuses on the postmortem agency of human remains and human tissues, particularly with the use of these human tissues in art, education and as relics. Together, these chapters highlight the complex and multifaceted ways that human remains continue to shape cultural identity, political narratives, and spiritual practices across time and space.

Chapter 15 explores the use of human cadavers in medical education, with a focus on the procurement of these bodies and the changing ethical standards in medical education surrounding human remains. Chapter 16 examines relics and reliquaries. Personal, religious, monarchic, nationalistic, and scientific relics are all examined. Through examples from religious practice, nationalism, science, and personal mourning, the power of relics to shape collective identity, social cohesion, and cultural continuity is highlighted.

Chapter 15
The Body in Education, Art, and Religion

Abstract The phenomenological elements of the body allow for its use in art, medicine, and as a teaching tool. In this chapter, the role of the body as material used for culturally and scientifically important aspects will be addressed, beginning with the role of the body in educational spaces. The role of medicinal aspects from use of cadavers as teaching tools as well as the role of mummia or medicinal cannibalism will also be addressed, focusing on the use of the physical body as a medicine beginning in the Middle Ages. Finally, the ritual role of anthropophagy (cannibalism) will be approached, from both a religious and compassionate viewpoint. In examining anthropophagy, both compassionate mortuary cannibalism and religious cannibalism can be examined as a form of communion, a ritual feast that seeks to promote group cohesion. The human body is uniquely able to fill this need, and without the human body as food in these cases, the ritual feast would not have the same meaning.

The phenomenological elements of the body allow for its use in art, medicine, and as a teaching tool. In this chapter, the role of the body as material used for culturally and scientifically important aspects will be addressed, beginning with the role of the body in educational spaces. The role of medicinal aspects from use of cadavers as teaching tools as well as the role of mummia or medicinal cannibalism will also be addressed, focusing on the use of the physical body as a medicine beginning in the Middle Ages. Finally, the ritual role of anthropophagy (cannibalism) will be approached, from both a religious and compassionate viewpoint. In examining anthropophagy, both compassionate mortuary cannibalism and religious cannibalism can be examined as a form of communion, a ritual feast that seeks to promote group cohesion. The human body is uniquely able to fill this need, and without the human body as food in these cases, the ritual feast would not have the same meaning.

A. Osterholtz, *The Bodies Poetic*, Bioarchaeology and Social Theory,
https://doi.org/10.1007/978-3-031-85998-4_15

15.1 The Body as Education

The use of the body as a teaching tool is long-standing, beginning in a modern sense with the clandestine dissections of the Middle Ages and continuing through to the modern use of cadavers and skeletons in medical and anthropological training. In this section, I will focus on the use the use of the body in dissection settings as well as the individuals who were more likely to wind up in anatomical collections and as medical cadavers.

15.1.1 Egyptomania and Mummy Unwrapping Parties

While the use of mummies as medicine and raw materials (discussed later in this chapter) was common throughout the Renaissance and Early Modern period, Egypt and Egyptian mummies as art and scientific objects really came into focus following Napoleon's occupation of Egypt in 1798. In this occupation, Napoleon famously brought over 160 scientists and artists to document the monuments, geography, flora and fauna (McCouat, 2019). By the 1830s, a trade in Egyptian mummies as scientific objects was flourishing as diplomats, tourists, and upper-crust Europeans vied to create scientific cabinets of curiosity in their homes (precursors to modern museums and a show of sophistication and education). Egyptomania swept around the world by the late nineteenth and early twentieth centuries. Newly publicized discoveries that could be seen via moving pictures such as the opening and emptying of King Tutankhamen's tomb in the 1920s exposed a public to a view of the ancient world that many had never seen or had contact with before (Fritze, 2021). This new media allowed anyone visiting a movie theater (with news reels played prior to the feature) to experience Ancient Egypt. The excavation of Tutankhamen's tomb was financed by George Herbert, 5th Earl of Carnarvon, who discovered Egypt and Egyptology during a convalescence from an injury. He began to finance excavations, ultimately gaining a concession to work within the Valley of the Kings and employing Howard Carter as an excavator. Tutankhamen's tomb was opened in November of 1922, and the process of carrying out the objects was fully documented with motion picture cameras. Carnarvon died on March 19, 1923 from blood poisoning from an infected razor cut.

With Lord Carnarvon's death soon after the excavation of Tutankhamen's tomb, fascination with not only the archaeology but the possibility of curses and the occult fed the flames of Egyptomania (Fryxell, 2017). Adding the mystical element made Egyptian history and contemporaneous exploration topics of popular attention. Increased interest also came from published travelogues, the books published by tourists to Egypt, some with extracts serialized in local newspapers (Baber, 2016; Fritze, 2021). This brought about changes in architecture, includingtomb and memorial design in modern cemeteries (e.g., Fig. 15.1). Known as Egyptian Revival Style, this focuses on the form of pyramids, mastabas, and obelisks as mortuary monuments. Broman (2001, p. 33) notes that pyramids in particular "contain the

Fig. 15.1 An Egyptian Revival Style tomb in Allegheny Cemetery, Pittsburgh, PA. Photo credit: Cbaile19, CC0, via Wikimedia Commons. https://upload.wikimedia.org/wikipedia/commons/f/f3/Hartley-Given_Mausoleum%2C_Allegheny_Cemetery%2C_2015-06-19%2C_01.jpg

aura of mystery that ancient Egypt represents and are most closely associated with burial and death." Particularly in the US, Giguere (2009) argues that Egyptian Revival Style monuments evoked a sense of timelessness, and were used as a way to proclaim that the US was based on an ancient past. It was also a way to perform class status in American cemeteries, since the monuments were relatively expensive.

Mummy unwrapping parties brought the exotic into the home or lecture hall, combining spectacle and scientific investigation into ancient life. Sheppard (2012) details the unwrapping of two mummies by Margaret Alice Murray in 1908 in the Chemistry Theatre at Manchester University. Murray was a trained Egyptologist who studied under Sir Flinders Petrie.

Murray, in performing the unwrapping in a university setting, was being overtly scientific in approach, but the spectacle was an important element of engagement with the general public. As Sheppard (2012, p. 526) notes, such activities "changed the sites where archaeological knowledge was created, thus allowing the public to participate in the 'exotic' archaeological fieldwork and the scholarly process of science." Murray was the first woman known to publicly unwrap a mummy in England, which likely added to the overall spectacle in a time when women were rarely involved in university lecturing. Mummy unwrappings were seen as scientific endeavors, as opposed to a simpler hunt for jewels and gold known to be within the mummy wrappings. By the late nineteenth century, mummies were typically seen as scientific specimens with scientific aims of understanding more about mortuary customs including embalming methods. The unwrapped mummy was then often placed on display in museums as an educational display.

Early mummy unwrappings occurred in homes, as early as the eighteenth century when Dr. John Hadley produced an early publication about a mummy unwrapping in his home in front of an audience made up of numerous surgeons and interested onlookers. The stated goal was to understand Egyptian mummification techniques (Aufderheide, 2003). Ultimately, this spectacle shifted from the drawing room of salon of private homes to lecture halls such as the unwrappings performed by Murray in 1908.

15.1.2 Museum Displays: Fetishizing Death and the Other?

Is it acceptable to display human remains in museums? There is no easy or simple answer to this question, just a general response of "it depends." It depends on the wishes of the Stakeholders and descent communities. In the case of the Body Worlds exhibits, the prominent display of consent forms is an effort to allay concerns about the display of the bodies. European museums are generally more accepting of the display of burials within their displays, particularly those from prehistoric, Roman, and/or Medieval peoples that lived in the area that the museum was built (Tarle et al., 2020). Within countries characterized by settler colonialism, such as the US, Canada, Australia, and New Zealand, there is a growing concern over the display of human remains within museum settings, particularly the display of human remains from Indigenous contexts. It is generally not acceptable to display human remains from Indigenous contexts; even grave goods are usually removed from display. The American Anthropological Association (AAA)'s Commission for the Ethical Treatment of Human Remains stresses a collaborative approach (Agarwal et al., 2024), and it is one that should also be followed in decisions regarding museum displays, not just analysis (as discussed in Chapter 14). Growing concern over the display of *all* human remains in US museums led to the Mütter Museum in Philadelphia pulling all of their online exhibits as of 2023 until a policy can be put into place. As their website states, "we have concluded that planning the future MUST include examination of questions such as collection provenance, consent, and what constitutes respectful display of human remains, especially at a time when these standards are evolving rapidly" ("Mütter Matters," 2024). This viewpoint is not universal, though. Ciliberti et al. (2020) note that human remains, when presented in a respectful manner have a unique ability to promote understanding of different cultural perspectives and practices and support inclusion.

For some Indigenous groups such as the Xaltocan of Mexico, the display of human remains is integral to their understanding of who they are and how they honor their dead (Overholtzer & Argueta, 2018). In their work, Overholtzer and Argueta worked with descendants of the pre-Aztec city state of Xaltocan to design museum displays that prominently feature human remains. According to Overholtzer and Argueta, the collaborative design of museum displays has been a way for the descendants to reclaim their own history. Central to this is the collaboration from the beginning of the project between the archaeologists and the Indigenous groups. "When residents were asked what should be shown in the museum, the answer was unequivocally bones, and all of them" (Overholtzer & Argueta, 2018, p. 516). The display of human remains in this context was acceptable as a performance of sovereignty as well as being in line with Mesoamerican beliefs about the dead. Human remains are ritually charged objects and teaching tools for the descendants of the Xaltocan. Through interaction with human remains, children "learn that when death occurs, you must be with others, thus bringing people together" (Overholtzer & Argueta, 2018, p. 521), and that "engagements with local human remains exist in a realm of familiarity, not fear" (Overholtzer & Argueta, 2018, p. 522). The case study of the Xaltocan is instructive because it shows us that there is no one-size-fits-all approach to the display of human remains. Displaying human remains from Indigenous groups that do not approve is a denial of agency and sovereignty. But when Indigenous groups wish to have the remains displayed, a denial of that desire due to blanket bans of any display of human remains is also a denial of agency and sovereignty.

Riederer and Bueno-López (2014, p. 366) note that display of human remains should be approached with caution, and "should be not to satisfy simple human curiosity but to improve scientific education and communication, and in doing so to justify the propose of use of body donations." Modern anatomical displays of contemporary human remains, such as Body Worlds utilize technology and preservation methods (such as plastination) that provide long-term preservation and the ability to pose and position the bodies in essentially any position desired by those exhibiting them.

Plastination is the process developed by Dr. Gunther von Hagens. Riederer (2014) refers to bodies preserved using plastination as "plastinates", further removing them from their original human condition. The Body Worlds website gives a lay-person's understanding of the process:

1. Fixation and dissection: preservative solutions are pumped through the arteries of the body to kill bacteria and prevent decomposition. After fixation, the skin, fatty and connective tissues are removed.
2. The body is placed in an acetone bath where water and soluble fats are dissolved. Under freezing conditions, acetone draws out and replaces water within the cells of the body.
3. The body is placed into a bath of liquid polymer, such as silicone rubber, polyester, or epoxy resin. A vacuum is created, leading to the boiling of acetone, as the acetone vaporizes and leaves cells, the liquid polymer is drawn in.
4. After vacuum impregnation, the body is flexible and can be positioned as desired. Anatomical structures are properly aligned and fixed according to the needs of the exhibit.
5. When the positioning is complete, the specimen is cured or hardened. The method of hardening depends on the polymer used, and may be accomplished with gas, light or heat ("Plastination Technique," 2020).

Overall, the plastination process may take approximately 1500 working hours to complete, but once complete, the plastinate will not decay and will not lose color or texture. Riederer (2014) examined plastinates that had been created 15-years prior and they maintained their true-to-life appearance. This is important for their long-term use as education tools. The Body Worlds display has garnered significant praise around the world since the traveling exhibitions began travelling. I remember seeing one such exhibit in the early 2000s in Salt Lake City. I went to the exhibit with a friend and her then 12-year-old son. I pointed out the position of the *serratus anterior* muscle on the first plastinate that we saw and then we found the *serratus anterior* muscle on each of the rest of the plastinates throughout the exhibit. It was an excellent way to teach a young, interested student how that particular muscle moved, extended, and contracted, given the variety of positioning that was present in the exhibit. He did the same for his class group when they went to the exhibit later.

The structuring of the exhibit was designed to see the body as universal, elements of individuality were stripped from the bodies: skin, hair, and body fat removed, they were muscle, bone, tendon, and nerves on display. The plastinates were displayed as art objects, full of motion and energy, but not really seeming human, more like a representation of human-ness (my impression, at any rate). It was not until the last room of the exhibit that the humanity of those on display was returned, the final room displayed consent forms for those on display. This experience is one that I think of often because it was such an effective tool for teaching, just like the cadaver-based Anatomy and Physiology labs that I taught for pre-med and pre-nursing students while I was a graduate student. I used the idea of experiential and phenomenological learning to teach muscle actions, using the cadavers in those labs as teaching tools and not thinking of their lives prior to the donation of their bodies. They were very much an interactive textbook and instructor. This concept is explored in detail below with respect to cadavers used in medical education.

15.1.3 Display as Punishment

The display of the body can also serve as a deterrent, or a reinforcement of 'proper' social behavior. In this way, the body is education, and so is presented within this portion of the chapter. There is an element of public shaming present to the display of the dead in some contexts. Sugg (2016) argues that during the seventeenth century, "corpses of criminals were left on display as a deterrent." This practice was common into the eighteenth century. With the Murder Act of 1752, all convicted murderers in England were to be either dissected or gibbetted. In England and Wales, more than 1000 convicted murderers were thus punished between 1740 and 1800 along with another hundred or so gibbeted for other offences (King, 2017, p. 6). As noted elsewhere in this volume, early modern states used the post-mortem display of convicted murders as an additional form of punishment beyond simple execution. For example, gibbetting (placing the corpse in a cage) in a central and highly visible location, was a common post-execution punishment in England. This was done to create a "lasting warning and example" (King, 2017, p. 17) to others. Even after these practices were banned in England, they were used on colonial enslaved persons. Following a slave rebellion in 1736 on Antigua, authorities burned

58 rebels alive, broke 5 individuals on the wheel, and 6 were reported to have been in gibbeted and starved to death (King, 2017). After death, their corpses were mutilated, their heads cut off and affixed to poles, their bodies burned.

Post-execution punishments to the body had numerous impacts to the community. First, the body was removed from a standard and culturally proscribed mortuary program. The body could not be buried in a community cemetery, but was instead added to a medical school's anatomy collection, put on display in a public display (in the case of gibbetting) or was burned, thereby denying the deceased's family access to the same mourning rituals. Second, there was a public performance of the punishment, or the knowledge of one occurring. From a Poetic perspective, we can view these public performances of death as important for conveying social information from the Body Politic utilizing the individual bodies of the condemned as the language and grammar of the message.

Once the skeleton of the condemned was articulated, it could be placed on display in anatomical theaters or museums. The remains of Jakob Karrer von Getweiler are also a useful example. In May of 1543, Karrer was executed via beheading in Basel for the crime of attacking his legal wife with a knife after she discovered he was a bigamist. The body was granted to the Flemish anatomist Andreas Vesalius for dissection, which was performed in front of an audience. He used these remains as the basis of illustrations for his *De humani corpus fabrica libri septem (Seven books on the structure of the human body*, Fig. 15.2) (1555). Within the chapter beginnings of book 1, the process of macerating the remains is depicted within the opening letter of the chapter, and the process is described within Chapter 39 of book 1. The skeleton has been stored at the Institute of Anatomy since the dissection and maceration. It was restored in 1985 and is currently still on display (Fig. 15.3). The depictions of the dissected body of Karrer have cultural sticking power, with the depiction of the body semi-dissected, lacking a mandible and showing deep muscles of the arms and legs (Vesalius, 1555, p. 230) in the TV show *Bones*. The episode *The Corpse in the Canopy* (Season 8 Episode 12) utilizes this imagery in the corpse found hanging in the canopy of a bed.

The activities of William Burke and William Hare are a well-known example. Burke and Hare began to supply dissecting rooms in Edinburgh in the 1820s, beginning with grave robbing. Ultimately, they turned to murder to obtain fresher corpses, and are known to have killed at least 16 people by 1828 (Rosner, 2009). To save his own post-conviction body, Hare turned on Burke. Burke was hanged, and his body was then anatomized and rearticulated. It is still on display at the University of Edinburgh Anatomy Museum (Fig. 15.4; University of Edinburgh, 2024). The ultimate disposition of William Hare is unknown, his death and burial remain shrouded in mystery.

15.2 Dissection and Anatomization

The body has been a focus for medical education for several hundred years, but the process by which medical cadavers have been procured has changed a lot over time. Legal frameworks in the United States mirrored those of England with respect to anatomization. In 1647, a Massachusetts resolution allowed the anatomization and

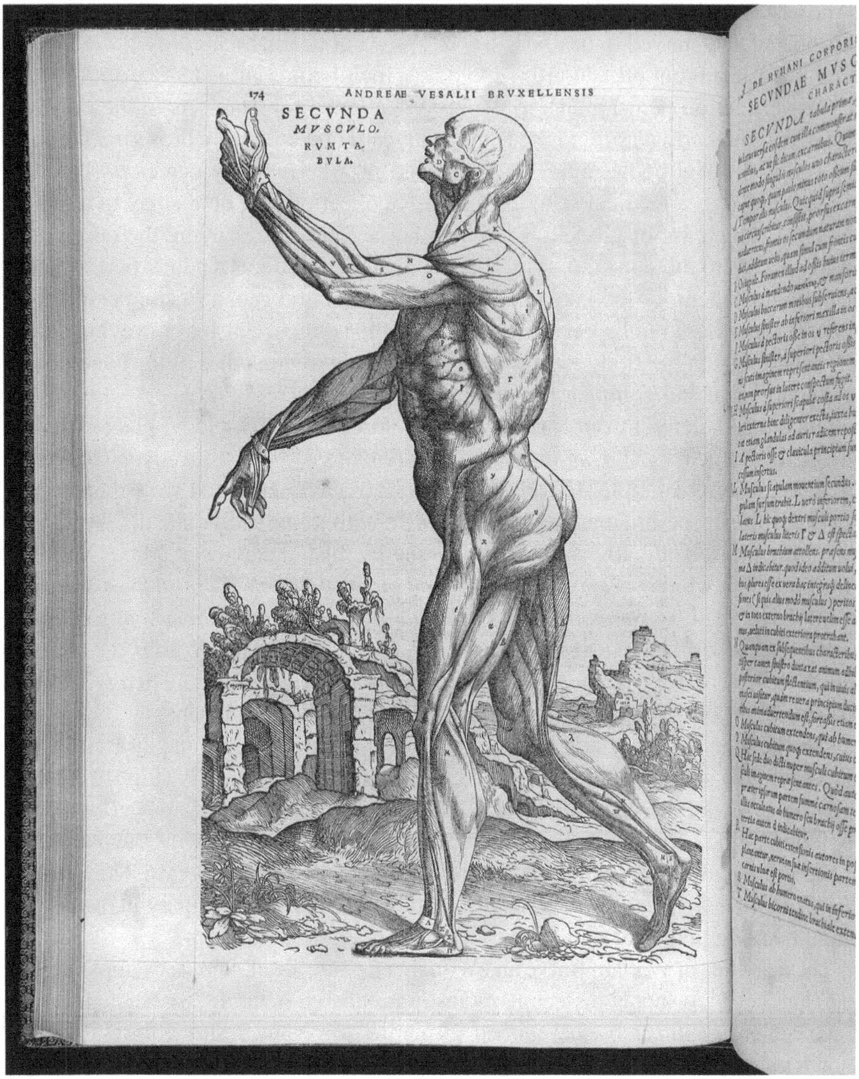

Fig. 15.2 L0031072 A. Vesalius, De humani corporis fabrica libri septum. Credit: Wellcome Library, London. Wellcome Images images@wellcome.ac.uk http://wellcomeimages.org Second muscle man. De humani corporis fabrica libri septem Andreas Vesalius Published: 1543. Copyrighted work available under Creative Commons Attribution only licence CC BY 4.0 http://creativecommons.org/licenses/by/4.0/

dissection of a single individual every 4 years. This was obviously not going to suffice to train physicians, and they were successful in lobbying state governments for increasing access to executed criminals. By the eighteenth century, Massachusetts, New York and New Jersey all had legalized the dissection and anatomization of executed individuals. But even this did not provide enough bodies for dissection. Demand quickly outstripped supply, leading to grave robbing on massive scales.

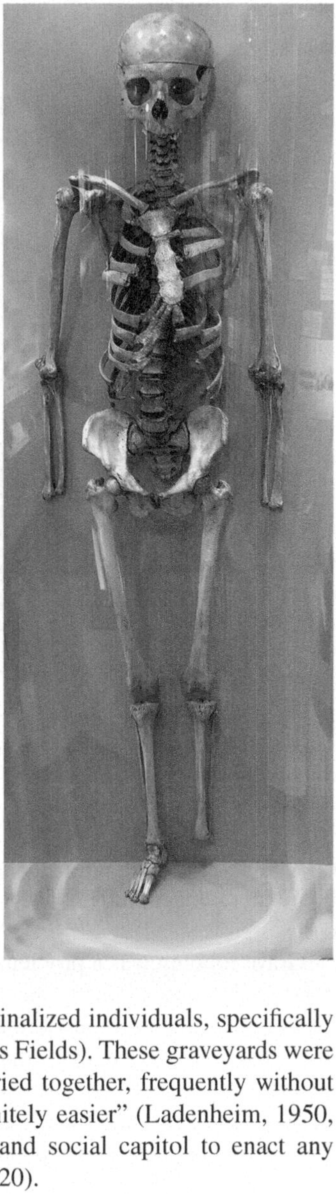

Fig. 15.3 Skeleton of Jacob Karrer anatomized and articulated by Vesalius in 1543, on display in the Anatomical Museum of Basel. Photo attribution: Amada44, CC BY-SA 3.0 <https://creativecommons.org/licenses/by-sa/3.0>, via Wikimedia Commons. https://upload.wikimedia.org/wikipedia/commons/a/a5/Skelett_im_Anatomischen_Museum_Basel_-_4675.jpg

Grave robbers tended to target cemeteries of marginalized individuals, specifically African American and pauper's cemeteries (Potter's Fields). These graveyards were targeted "chiefly because large numbers were buried together, frequently without coffins, making the procurement of cadavers infinitely easier" (Ladenheim, 1950, p. 26). These groups also lacked the economic and social capitol to enact any changes to these processes (Hodge & Nystrom, 2020).

Dissection of convicted and executed individuals was relatively common practice in early modern western European societies. King (2017) notes that dissection was the most popular form of post-execution punishment (as discussed above). In his analysis, he looked at 29 different publications of pamphlets available to the English public between 1700 and 1752, he found that 8 of the publications recommend this particular form of punishment. Dissection, either public or private, was popular with the populace and the medical establishment who used these

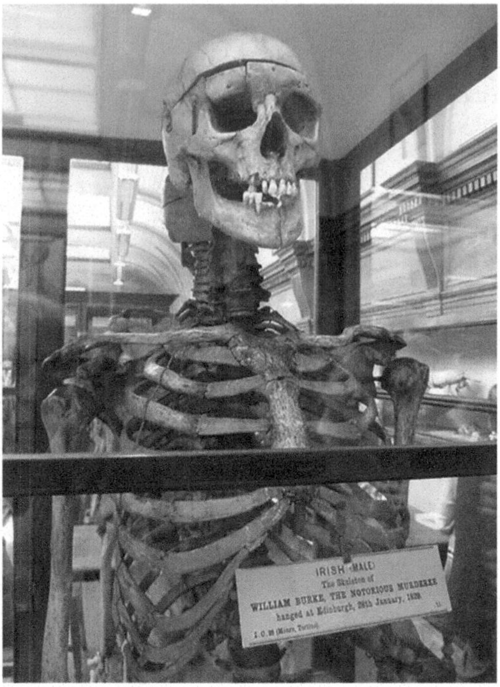

Fig. 15.4 The Skeleton of William Burke on display at the University of Edinburgh Anatomy Museum. Photo attribution: Kim Traynor/The skeleton of William Burke. https://upload.wikimedia.org/wikipedia/commons/6/68/The_skeleton_of_William_Burke_%28geograph_2621217%29.jpg

dissections as training for medical students (Figs. 15.5 and 15.6). For some time, the bodies of executed individuals was the only legal source of cadavers and demand for corpses grew steadily.

Grave robbing was a problem on both sides of the Atlantic. The medical school at Edinburgh was a famous one, and they required significant numbers of bodies for dissection. Students were often employed in the procurement of bodies for dissection. Students would read death announcements and wait for an opportunity to take a fresh body. Mac Gregor (1884, p. 29) recounts one such story, noting how the surgical students waited while a woman mourned over a freshly dug grave before recovering the body. They apparently did not wait quite long enough, as they were leaving by boat, "they again observed the woman running backward and forward in the churchyard with her arms waving, apparently acting under the most intense excitement. Her agonizing cries quickened the use of the oars, and hurriedly they left the heart-rending scene behind them." There was a general feeling amongst the students that the needs they had to learn anatomy outweighed the desire of the bereaved to have their relatives left in peace. This feeling has not completely disappeared. Humphries (2014, p. 71), writing as a fifth-year medical student, acknowledges that the acts of the resurrectionists were "gruesome", but goes on to say "yet the bodies obtained in this way allowed them to expand their knowledge of the

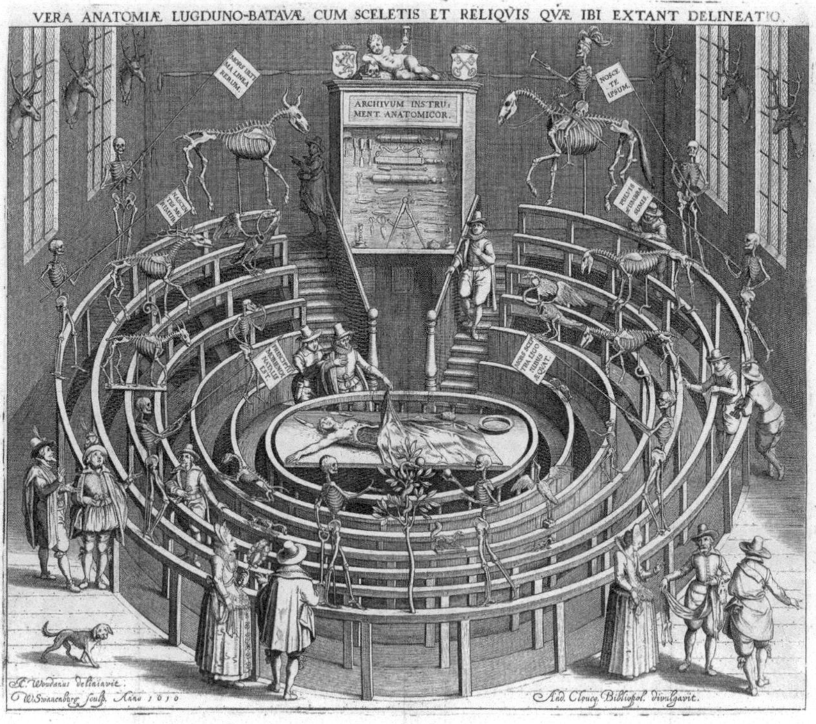

Fig. 15.5 Anatomical Theater of Leiden University. Engraving by Willem Swanengurgh based on a drawing by Jan Cornelisz circa 1610. Image attribution: Willem Isaacsz Swanenburg, Public domain, via Wikimedia Commons. https://upload.wikimedia.org/wikipedia/commons/f/ff/Anatomical_theatre_Leiden.jpg

human body, improve their surgical skills and make many anatomical discoveries, providing invaluable information on underlying disease processes." Faculty members at the medical school were complicit in these illegal acts, never asking where the bodies came from.

Other individuals also began to specialize in grave robbing as a profession (known as resurrectionists). Wiliam Burke and William Hare are a good example of this (discussed above in greater detail). In the US, funeral homes, hospitals and urban slums were also sources for fresh cadavers. Blakely (1997, p. 5) describes the procurement of cadavers as "necessary, opportunistic, and discriminatory". Within the US, dissection of African Americans occurred more often than others. Access to cadavers for dissection was a recruiting point for some Southern schools. Savitt (1982, p. 339) quotes an advertisement for the Medical College of South Carolina in 1831 noted the availability of African American cadavers to carry out "proper dissections… without offending any individuals." From this, it is clear that the bodies that mattered were those of the white citizens, not the families of those whose bodies would be dissected.

Fig. 15.6 Skeletons on display at the rebuilt Anatomical Theatre in Leiden, now a popular museum dedicated to Anatomy and Physiology and the history of medical education. The provenance of these skeletons on display is not listed in the museum. Photo by the author, 2023

15.2.1 Cadaveric Processing and Post-dissection Rituals

Bodies used for study and dissection began to be preserved via embalming around the same time that funerary embalming (discussed in Chap. 6) became more common. The goal of medical embalming is different than that in funerary embalming. Funerary embalming in the latter twentieth and twenty-first centuries is designed to provide short-term preservation of the remains, long enough for visitations and open-casket services but not designed for permanent or long-term preservation (Mitford, 2000). Medical embalming, on the other hand, is designed to maintain life-like colors and textures as long as possible to facilitate the use of the cadaver as a teaching tool. It is illustrative that here, the terminology shifts from that of being a body (a person) to a cadaver (an object).

Medical literature is consistent in the use of the term cadaver. The body has become a teaching tool and serves a very different purpose than it did during life. Bilge and Celik (2017, p. 1263) detail a study in which they tested several different embalming fluids on donated cadavers. "We aimed to determine a suitable, simple and cost-effective embalming method that preserves color, texture, pliability and flexibility of the tissues for a long time without a disturbing smell or mucosal irritation. The embalmed cadavers were expected to be durable against environmental effects, utilizable for multiple and repetitive surgical trainings and instrumentations." Study with cadavers is seen as a bedrock foundation for medical training, and improvements in preservation of tissues to maintain life-like color and texture are important for training physicians to understand what *normal* and *abnormal* feel and look like. But in perhaps no other medium is the concept of the body as material culture more obvious. Throughout their article, Bilge and Celik (2017) present the cadavers as commercial objects, noting the cost-effective nature and non-odor of their methodology. The focus is very much on what these objects (these cadavers) can provide, without an acknowledgement of their humanity. These bodies have transitioned completely into material culture that is adaptable, changeable and discardable by the living. Ciliberti et al. (2016, p. 266), in fact, argue that human cadavers should be preferentially used for medical research to protect animal rights, concluding that society "accepts that body donation offers a source of health for everybody." The body, converted into a cadaver, is a tool for the betterment of society through increased knowledge and training.

While viewing the cadaver as an object, professors of anatomy often also use the cadaver to teach bed-side manner to the students, effectively using the dead to teach students how to interact with the living. Students are taught to see the cadaver as a patient. Ghosh (2017) refers to the cadaver as the first teacher a student has in anatomy, and notes that the "cadaver is a complex, socially constructed entity, laden with all possible attributes, from frivolous to tragic to sinister to religious. However, in the context of an anatomy dissection room, it is a teacher whose function is not to symbolize mortality but to help us learn" (Ghosh, 2017, p. 130). The importance of cadavers is such that they regularly become elements of design within the schools of medicine (e.g., Fig. 15.7). Medical students are surrounded by reminders of the gifts given by the donors, and so the body becomes fundamental to their future careers.

Disposal of the remains after dissection or use in medical training followed a different path from a typical traditional funeral. They experience a different death history than a body that is disposed of through normative mortuary activity. In August of 1989, human bones were discovered in the basement of the original Medical College of Georgia building in Augusta. The remains were commingled, representing everything from fragments to whole limbs to whole individuals. They had seemingly been tossed on the dirt floor and capped with quicklime to reduce the smell of decomposition. Remains included African Americans, European Americans, males and females, and ranged in age from fetal to older adults. Mixed in with the remains were other artifacts, including ceramic jugs, scalpels, pipettes, syringes, test tubes, microscope slides, coins, buttons, fabric, and so on. In other words, bones and body parts were disposed of with other waste and trash (Blakely, 1997). Disposal in basements was not uncommon. Blakely and Harrington (1997) cite examples of bones found in the basements of medical schools in London as well as in the US. Remains analyzed by Hodge and Nystrom (2020) from Holden Chapel at Harvard University was also recovered from the basement. These remains were also recovered with animal remains, ceramics, scientific glassware, and architectural debris (Hodge, 2013; Nystrom, 2016).

Fig. 15.7 Cadaver frieze at the School of Medicine, Tulane University. Image attribution: Tulane Public Relations, CC BY 2.0 <https://creativecommons.org/licenses/by/2.0>, via Wikimedia Commons. https://upload.wikimedia.org/wikipedia/commons/0/02/School_of_Medicine_%283630515886%29.jpg

Modern donated bodies are follow various mortuary pathways following their use as teaching tools. The University of Minnesota's Policy regarding the use and disposal of donated bodies states that donated remains may only be held as long as studies are ongoing and must be disposed of immediately after the completion of approved studies. The policy specifically states that researchers "must comply with the time frame in which the human anatomical donation can be studied, cremated, and the cremated remains returned to next-of-kin/authorizing person" (University of Minnesota, 2016). This rather sterile wording of a university operating policy doesn't reflect the memorialization that often accompanies the cremation and return of those remains.

Cadavers are converted through the post-dissection rituals at medical schools back into bodies of formerly living people. "Proper cremation/burial of the remains of human bodies after anatomical dissection is an act of bidding farewell to a teacher who has enlightened the students with the knowledge of human anatomy and has selflessly contributed towards the welfare of the society" (Ghosh, 2017, p. 131). Funeral services should form a cornerstone of anatomical study, without them "the entire exercise of cadaveric dissection would become too mechanical and appear as an act driven by the rigors of medical sciences. Funeral services are an opportunity to reflect on the humanistic side of the *person* [my emphasis]" (Ghosh, 2017, p. 132). Commonly, after study, remains are cremated and returned to the family (e.g., Jones et al., 2014; Minnesota, 2016; Sakai, 2008), but ceremonies are also very common. At Tzuchi University Medical School in Taiwan, students wrap the body (cadaver), write a letter addressed to their 'silent mentor' and participate with family members in funeral rites (Kanter, 2010; Tseng & Lin, 2016). In Thailand, students are led by Buddhist monks to the site of final cremation (Prakash et al., 2007).

Medical schools commonly hold rituals of remembrance, centering on the relationship that has developed between the living students and their dead instructors, now once again seen as embodied beings. In a study of memorial ceremonies at Anatomy programs, Jones et al. (2014) conducted a survey to illustrate the extent that these ceremonies exist and their purposes. Of the 88 allopathic medical schools that responded to their request for information, 95.5% confirmed the presence of yearly memorial services associated with donation programs. Only 9.5% of the respondents used the term "cadaver" in the title of their ceremonies (Jones et al., 2014), suggesting that the transition from cadaver back to person was an implicitly important element of the ceremonies. The majority of the ceremonies were at least co-organized by medical students and faculty. Family members of donors were involved in the planning of the ceremonies in 16.7% of the cases. The majority of these ceremonies had been going on yearly for over 20 years (at the time of the survey), which reflects changes in medical school pedagogy in the 1970s and 1980s for a more humanistic approach (Jones et al., 2014). Pawlina et al. (2011, p. 139) describe the "Convocation of Thanks", the annual ceremony commemorating the gift of body donation to the Mayo Clinic Bequest program in the Department of Anatomy, College of Medicine at the Mayo Clinic during this ceremony, students, donors, and their families interact and students "develop a richer understanding of the last days of life that the donors experienced. Students' recognition and expression of gratitude for the sacrifices of donors and their families can enhance an emerging sense of professional responsibility."

The donated body acts as a transformative object important for the creation of the a new intersectional identity of "Doctor." The ceremony, like many others in the U.S. incorporate both donor's family members, students, and faculty and include music, shared meals, artistic exhibitions, and floral arrangements. When the 26th "Convocation of Thanks" was held in 2010, 350 guests attended. Students reflected publicly on their anatomy experience through a variety of media, including music, poetry, ballet, and multimedia presentation (Pawlina et al., 2011). The first names of the donors are also read aloud. It is this renaming of the dead that solidifies their re-personhood, and completes the ritual transfer of responsibility to students for the care of future patients. Some medical schools erect monuments dedicated by the students to the donors and their families. The monument at Keele University School of Medicine contains the inscription "these benevolent individuals and their families have given so much for the benefit of medical education and research" (Fig. 15.8).

In effect, researchers that argue for a formal disposal of the person turned cadaver turned person again, are arguing for closure to this process. In donating their bodies, a transition occurs transforming the body into a cadaver/teacher. When this role is complete, an additional formalized ritual is needed to return the humanity to the object and provide closure to the instruction of the pupils. This can easily be viewed

Fig. 15.8 Monument at Keele University School of Medicine dedicated to those who donated their bodies to the Anatomy Facility at the Keele University School of Medicine and their families. Photo attribution: Keele Cemetery memorial to body donors by Jonathan Hutchins, CC BY-SA 2.0 <https://creativecommons.org/licenses/by-sa/2.0>, via Wikimedia Commons. https://upload.wikimedia.org/wikipedia/commons/c/c5/Keele_Cemetery_memorial_to_body_donors_-_geograph.org.uk_-_5277709.jpg

within the Hertz/Van Gennep rites of passage model as well as through Geller's Death History model of interaction with the dead. Different forms of memorialization occur with these different types of disposal, from burial within a university owned plot within a local cemetery (if family members do not wish the return of the remains) (Ramsey-Stewart et al., 2010) to a completely typical disposal after return to the families.

Another common method of processing and divergence from the typical life history of a body in mortuary processes was anatomization and ultimate inclusion in anatomical collections. In their recent work, Hodge and Nystrom (2020) focus on the Poetics of anatomization and medical education in early American medical schools. They note the social disparity between those who studied and those who were studied, with medical students entering an elite social status while learning from the bodies of the less-advantaged. The higher social status of medical students was reinforced by the educational space, "Any structure is inherently exclusive, more so within a closed university setting" (Hodge & Nystrom, 2020, p. 198). The dissection of individuals was seen as more important than the wishes of the bereaved, since the training of physicians was seen as a tremendous social good. The personhood of the dissected was negated. The space this was engaged in took on a different meaning and allowed for different relationships to be negotiated. As Hodge and Nystrom (2020, p. 204) note, "spaces of body processing—including anatomical theaters, examination rooms, disposal features, and cemeteries—were experienced as sites of experimental discourse. They permitted practitioners to engage in transgressive—and even illegal—behaviors in a protracted environment (openness about body snatching, dissection, callous treatment of the dead, handling unclothed corpses, examination of removed organs, disposal of body parts with common trash, etc.)."

Fear of resurrectionists led to new innovations in cemetery design, including the introduction of mortsafes in burials themselves as well as the introduction of high walls, iron gates and watch houses within the cemeteries (Humphries, 2014). Mortsafes consisted of iron cages placed over newly buried graves, these could be temporary or permanent. In Scotland, 'mort stones' were also used, these large stone slabs were lowered using a tackle and paced over the grave within the graveyard (Humphries, 2014). Coffin making also changed for those who could afford it, iron coffins became more common for the well-to-do, and other anti-grave robbing devices were also employed.

15.2.2 The Use of Unclaimed Bodies in Medical Education

Bodies for dissection and anatomization also drew upon individuals with less social power than those who would dissect them, from alms, poorhouses and asylums. These individuals were all the recipients of significant social support, the common thinking went, and so "if an individual died while in the almshouse, they should repay their debt to society and their body should be used to benefit medical science and society at large" (Nystrom, 2016, p. 14). This viewpoint is reflected in state laws that allowed for the unclaimed dead to be used for medical education. Comer (2022, p. 822) argues that laws allowing for the use of unclaimed bodies are "government sanctioned body snatching as it provided the bodies of the poor without their consent prior to their burial for anatomical dissection, rather than after burial as occurred through grave robbing."

Caplan and DeCamp (2019, p. 360) conducted a survey of medical schools within the US. Of the 89 responses, 11 institutions (12.4%) reported the "possible" use of unclaimed bodies for educational purposes. Shupe et al. (2023), focusing on Texas, reported that 6 of the 14 schools (42.9%) who responded for their survey either directly procured unclaimed bodies or received transferred unclaimed cadavers from other universities. This marks an increase in the number of unclaimed dead used for medical education between the years 2017 and 2021.

There is an increasing understanding of the ethical murkiness of using unclaimed bodies for medical education. In 2016, a series of articles in the New York Times focused on Hart Island, the Potter's Field in New York City (Bernstein, 2016a, 2016b, 2016c, 2016d). In these articles, Bernstein highlighted the continued use of a law that dated to the 1850s allowing for bodies unclaimed for 48 hours after death (PBH § 4211) (2016d) and offered suggestions on how to avoid this from happening (Bernstein, 2016a). Later that same year, Governor Andrew Cuomo signed a bill into law to outlaw the use of unclaimed dead in medical education (Bernstein, 2016c). Medical schools withdrew an objection to the new bill, saying that they would increase their donation program to overcome any shortfall of cadavers for medical education (Bernstein, 2016b). Families of the dead who cannot afford burial may be more likely to agree to donation, according to concerns expressed by advocates (Bernstein, 2016c). Is this really informed consent? Or consent born out of necessity?

According to a statement on the American Medical Association (AMA)'s website (2024, p. n.p.), there are several ethical concerns associated with the use of the unclaimed dead in medical education. First is the lack of informed consent. Second, "by definition, unclaimed bodies come from vulnerable populations of people: functionally, those who are poor and/or have little-to-no social capital. Economic justice is also connected to structural racism; marginalized racial/ethnic groups are disproportionately impoverished, making the use of unclaimed bodies an issue of systemic racism, and one that was also associated with slavery." Finally, there is a risk of "moral injury" to the learners in medical school (AMA, 2024), particularly to students who may not be aware that the cadavers who become their "first patient" were unclaimed bodies (Peeler, 2020).

For Peeler (2020, p. 2002), finding out the cadaver used was an unclaimed person was a lesson in structural violence: "For me, it was a first lesion in the uncomfortable truth that people with less power and fewer advocates can be vulnerable to exploitation by the medical profession. The lives of people who become unclaimed bodies are lived on the margins because of their poverty or aloneness. I saw these future unclaimed bodies during my clinical rotations throughout the rest of medical school. I cared for them.... Each time I saw patients like these, I thought of what could happen to them after death. And more important, I thought of the ways in which the medical system could deny them the ability to control their own future while they were alive." Modern medical students have shown increasing awareness of the ethical issues surrounding this issue, as shown through the work of Matheson et al. (2023). They surveyed 212 medical students and found that while 97% were comfortable dissecting willing donors, only 31.1% were comfortable dissecting the unclaimed dead. The conclude by saying "medical school represents a critical time in the professional development of trainees, and development practices which align with the moral code of local institutions and stakeholders is crucial" (Matheson et al., 2023, p. 12). Essentially in training future physicians to care for their patients, they must have their "first patient" consent to be used as a teaching tool.

15.3 Historical Documented Collections (HDCs) and Anthropological Research

The concern over consent has impacted the study of bioarchaeology in important ways as well, particularly surrounding the analysis of individuals from non-Indigenous groups. De la Cova and colleagues (2024) note a dearth of ethical standards for research utilizing historical documented collections (HDCs) within the US (see also de la Cova, 2019, 2020a; Nystrom, 2014; Hodge & Nystrom, 2020; Williams & Ross, 2021). These collections are result of procurement of indigent individuals or individuals who remained unclaimed after their deaths. These collections, with their known age at death, adult stature, and known sex, formed the basis for the development of a good deal of the methodology used to estimate age at death and sex in both forensic anthropology and bioarchaeology.

These individual bodies have been treated as though they were "neutral bodies of evidence, rendering individuals within them scientific objects without human identities exempt from the ethical and legal issues surrounding other collections of deceased individuals (e.g., Native and Indigenous individuals)" (de la Cova et al., 2024, p. 2 of 10). But these bodies are not neutral, nor are they exempt from ethical concern. De la Cova (2020b, p. 213) looked at HDCs as the embodiment of structural violence, arguing that "the bodies of those who make up anatomical collections were utilized as social tools to reinforce and cement social stratification in the late nineteenth- and twentieth-century American society." Taking a Poetics approach, de la Cova (2020b, p. 227) argues that through postmortem processing, "the anatomist exerts power and control over the powerless poor, ultimately determining their postmortem fate."

de la Cova and others organized a workshop titled "Ethical Futures for Curation, Research, and Teaching in Biological anthropology in November of 2021 (de la Cova et al., 2024). There, 46 researchers (including academics, students, and museum professionals) participated in panels and discussions surrounding the use of HDCs within research and teaching. Central to these discussions was the rights of the deceased individuals and the responsibilities of researchers to the dead and their descendants (and culturally affiliated communities). Rehumanization or reindividualization was advocated for by numerous participants, arguing for partial de-anonymization and the use of first names of the deceased, but some participants also argued that anonymization may protect personal identity.

Issues of consent were extensively discussed, including the need for the institution of a review-based system similar to an Institutional Review Board (IRB) approach utilized in the social sciences. This review board would review proposals for proposed outcomes and potential harms from the proposed research. The workshop also highlighted the need for continued improvements in the teaching of ethical frameworks within biological anthropology. Importantly, the language used by the workshop participants and in the culminating journal article (de la Cova et al., 2024) focuses on language that affirms personhood of the individuals within HDCs, utilizing "deceased individuals", "individuals", and "people" instead of "human remains", "specimens", or "samples" (de la Cova et al., 2024, p. 8 of 10). They also state that "[o]ther terms and phrases such as 'responsibly' or 'ethically sourced' skeletons, which borrows from descriptors for commercial products such as coffee and sugar" are harmful.

HDCs currently occupy contested space within biological anthropology. There is a tension evident between the desire to honor the individuals that make up the HDCs, and proponents of their continued use for research argue that research utilizing these collections "highlights the embodied effects of structural and race-based violence" (Zuckerman et al., 2021, p. 39; see also Nystrom, 2014; Hodges & Nystrom, 2020). In this view, as Zuckerman et al. (2021, p. 39) state, "anthropology can illuminate past lived experiences that are otherwise invisible or inscrutable."

Following the Ethical Futures workshop, the Smithsonian Institution placed restrictions on research and temporarily halted the acquisition of additional individuals into their curated spaces. Similar policies have been taken by other major American museums, including the American Museum of Natural History (New York, NY), the Cleveland Museum of Natural History (Cleveland, OH), and the Peabody museum of Archaeology and Ethnology at Harvard University (Cambridge, MA). As de la Cova and colleagues (2024, p. 1 of 10) note, "[t]oday the field of biological anthropology is in an uncomfortable position, which is perhaps exactly where it needs to be. Discomfort, after all, is a catalyst for change." Between the Ethical Futures workshop and the AAA's Commission for the Ethical Treatment of Human Remains, the field of biological anthropology sits at a point of tension, but one that is informed by emerging models of community engagement and the consent of descent communities for research (however those descent communities are defined).

15.4 The Body as Medicine: Medicinal Cannibalism

Medicinal cannibalism was practiced extensively through Europe from at least the sixteenth to eighteenth centuries (Gordon-Grube, 1988). As Himmelman (1997, p. 184) notes, "Cannibalism is *never* just about the consumption of the human body, but neither is it ever *merely* symbolic. It operates in the ambiguous intermediacy between the two. The body 'embodies' that which cannot be symbolized—it *is* the liminal space that has always divided rational man from the world, the spiritual from the material, nature from culture.... In eating human flesh man's world conflates, converges on a single, indeterminate point where all things connect, lending to the body a significance that far exceeds individual personhood." In this way, "Medical attention *to* the body is necessitated by its physical deterioration; medical use *of* the body is made possible by its spiritual, occult properties" (p. 185). Being made of the body, human tissue has the ability to heal the body.

15.4.1 Mumia or Mummy

Mumia or Mummy is the substance created through processing the mummified body, normally a powder that could be incorporated into salves, medicines, and paint. Liquids may also be produced through other forms of processing as well. The development and use of mummy (also known as mumia or mumiya) and its ultimate cessation as a curative is a history of medicinal ethics and outlook: who has the control over the body? Do we honor the body as an embodied individual or is this more appropriately raw material that can and should be used to treat the living?

Dannenfeldt (1985) traces this back to the use of bitumen in the Fertile Crescent under the name of *mumiya*. Pliny the Elder wrote of the use of bitumen in the treatment of cataracts, leprosy, gout, toothaches, coughs, dysentery, and wound healing. Bitumen was used as treatment for abscesses, fractures, concussions, paralysis, epilepsy, vertigo, ulcers, and disorders of the liver and spleen according to the Arab physician Avicenna (also cited in Dannenfeldt (1985)). Bitumen was believed to be an important element in body preservation in Egypt, particularly due to the dark color of Egyptian mummies. Definitely by Ptolemaic times, bitumen was a common additive for embalming the bodies of the dead. Ultimately, the name of the substance comes to signify the process, and the mummified body begins to embody the healing properties of bitumen. Bitumen in its natural form was common enough in some areas, but as popularity as a medicine increased, supply of natural bitumen could not keep up. Visitors to Egypt were able, however, to purchase mummies that had been preserved with bitumen. In effect, the preserved bodies became a source of bitumen that was in greater abundance than naturally occurring bitumen. Donenfeld (1985, p. 167) details a case in 1424 of authorities in Cairo arresting individuals who had collected a large number of cadavers. They ultimately confessed to processing the bodies, boiling them and collecting the oils which rose to the surface of the pots. The oil was sold to Europeans as medicinal mummy (giving the source of Antonie I. Silvestre de Sacy, 1826–1827 *Chestromathie Arabe*).

By the sixteenth century, the meaning of the medicinal term *mummy* had morphed into what Tomé Pires ([1516] 1944), a Portuguese apothecary described as "an exudation from corpses, obtained in this fashion: when the man dies they remove his intestines, lungs, etc., and throw in myrrh and aloes; then they sew up the corpse and put it in a sepulchre with holes. This mixture with the moisture of the body drips out and is collected. This liquor is called mummy" (quoted in Dannenfeldt (1985, pp. 165–166)). Occurring only a short time later, there is an additional shift from the liquid derived from decomposition or processing or the use of the entire body as medicine. Again, the bodies of Egyptians were readily available and well-preserved. Illicit digging of graves and tombs provided well-preserved bodies that could be processed and exported to Europe. By the end of the sixteenth century, there is a well-trod path of exportation of Egyptian mummies to Europe, and again demand outstripped supply for authentic mummies. By the mid-sixteenth century, the manufacture of fraudulent mummies was identified (Dannenfeldt, 1985). Citing Ambros Paré's *Oeuvres completes* (1840–1841), Dannenfeldt (1985) details the story of one merchant encountering Guy de la Fontain, physician to the King of Navarre in Alexandria. Encountering a large number of bodies, the physician asked where they had come from and inquired about the ancient embalming methods that had preserved them. The merchant instead informed the physician that he had produced them himself. The bodies were referred to in the text as "mumia," noting a transition from person to body to substance. This text is also interesting in that it blames a specifically Jewish merchant, a way of othering this population and saying that it could not have been a Christian who had committed this act which was presented not only as a kind of desecration but also as a kind of fraud. While Christians in Europe were happy to consume mummy as a curative, they still seem to have had misgivings about the way it was produced, and so ascribing the creation of fraudulent mummy to a non-Christian group was a way to other Jewish populations, a form of structural violence in itself.

Ultimately, while authentic ancient Egyptian mummies were preferred, the use of any properly prepared body as mummy was still medicinally important and powerful. During the Renaissance, mumia from emblamed virgins was deemed to be more medically effective and therefore sold at a higher price. While the importance of naturally occurring bitumen was still identified, mumia was sold at approximately twice the cost at Augsburg (Dannenfeldt, 1985), suggesting that there was a synergistic effect of using mummy that had been treated with bitumen as opposed to simply using bitumen.

By the early modern period in Europe, mummy or mumia could have been made from either mummified remains or locally prepared fresh corpses. Noble (2011, p. 2) notes that this ties to a commonly held notion of the times that "the human body contains a mysterious healing power that is transmitted in ingested matter such as mummy," a concept that for Noble typifies the time period. She ties this to the idea of the eucharist consumption during communion, a tie to "the *anima* that has never really been present, except as an already multitemporal trace" (Noble, 2011, p. 3).

Noble (2011) ties this Early Modern practice of medicinal cannibalism to the exchange and consumption of bodies and body parts, linking anatomization and medical cannibalism as two aspects of the same "medical market relations" (p. 4), with medical cannibalism based on medical practice that depended on "violated, manipulated, fragmented, processed, circulated, marketed, and consumed human body, made possible by a culture that witnessed public displays of violent executions and bloody anatomies" (p.8). It is, in effect, a natural extension of manipulation of the body for social ends.

Le Breton (2017, p. 233) argues that the use of the body and body parts as medicine can be drawn to concepts of identity, during the Early Modern period, "Individualism had not yet made the body into a self-contained property, an articulation of an individual's singularity. In the communitarian regime, the body belonged to the species, to the group.... Human remains were regarded as a precious resource by those who knew how to transform them into remedies... consumed internally in the form of elixirs, syrups, pills, tablets, pastes, and dried powders." The taste is described in James' (1747, p. 512) *Pharmacopoeia* as "acrid and bitter" (as cited in

Gordon-Grube (1988)). The father of toxicology, Paracelsus was a known proponent of the use of mumia in medicine in the sixteenth century. According to Crone (2004, p. 65), Paracelsus' use of this stems from both a belief in the medicinal value as well as a the metaphysical value. He quotes Paracelsus as saying "mumia is the man himself, the mumia is the balsam which heals the wound: mastic gums, glaze will not give a morsal of flesh; but they can protect the working of Nature so as to assist it." The views expressed by Paracelsus, then, reflected a larger medical knowledge that since mumia was made of man, it would be useful in healing man. But there is an added element here, "Mumia is… an immaterial force or energy that strives to protect the person and to correct ay sickness or injury" (Crone, 2004, p. 65). The life-force of the mummy can be instrumental in boosting the life force for the living patient.

Universal love of mummy as a curative was not complete, but interestingly, their criticisms often touched not only on the concept of medicinal cannibalism but also on the social qualities of those converted into mummy. Essentially, physicians such as Fuchs in his *Paradoxorum medicinae* of 1566–1567, argue that the social quality of the individual that is converted into mummy has an effect. While mumia from bodies recovered from tombs is efficacious, "certain imposters produce a rival of it, the gory matter of cadavers received evidently from the gallows or from the torture wheel, spotted with the feces of corpses in place of aloes and myrrh, which they sell to others as the real thing" (quoted in Dannenfeldt (1985, p. 176)). Fuchs is in effect arguing that it is unpalatable enough to be consuming human flesh, but even more unpalatable to be consuming the flesh of someone who was publicly hanged. Physicians such as Ambrose Paré, while prescribing mumia earlier in his career, began to see it as ineffective, his concern was not in ethical considerations of consuming human materials, arguing that if it were effective, there could be an argument for continued use of mumia, but since it is not effective, it should not be used. By the early eighteenth century, medical understandings were shifting within Europe and the use of mumia, while still listed in numerous medical texts, was not commonly used in practice.

15.4.2 Other Early Modern Uses: Salves

There is certainly an element of magic or magical thinking involved with at least some of the uses of human remains in medicine. From at least the Renaissance, powdered skull was often used to combat epilepsy and other diseases of the head, a kind of sympathetic magic. Human fat was applied externally to treat rheumatism, nervous complaints, gout, breast cancer, aches and pains. Dried and powdered blood was sprinkled into wounds or stuffed up nostrils as a way to staunch bleeding. In some cases, *usnea* was also added to these curatives. Usnea is identified as materials or substances that grew on human remains after exposure to weather, such as moss or maggots (Ziller Camenietzki, 2001). As Sugg (2016) notes, at least some of these curatives, particularly the introduction of powdered substances into open wounds to stop blood loss, would likely have been at least partially efficacious. Powdered substances may stimulate coagulation.

As the seventeenth century progressed, Paracelsian concept of mummy as human materials became more common in Europe, partly due to a decrease in availability in Egyptian mummy and partly due to a general rise in the type of metaphysical medical understanding that he professed. This allowed for the increased use of human remains as curatives outside the realm of what had been considered mummy earlier. Though the terminology is the same, the meaning at this time does not necessitate a firm relationship with either Egypt or Bitumen.

A wound salve in use at the end of the sixteenth century is a good example of this. Given to Emperor Maximilian by Baptista Porta, the recipe called for poultice made up of "moss from a dead man's skull, human fat, mummy, human blood, linseed oil, turpentine, and bole-armenick" (Crone, 2004, p. 141). It was important for the usnea (moss) to be gathered during a crescent moon (Ziller Camenietzki, 2001). This salve was applied not to the wound, but to the weapon that produced the wound. The weapon was then covered with a white cloth and kept in a temperate place. Bleeding of the wound would be treated with the patients' urine and covered with a white cloth. If very deep it could be stitched. If this series of ritualized actions occurred, the wound would be healed. As Ziller Camenietzki (2001, p. 85) notes, "it was not possible to ignore the distinction between natural and artificial motion. Natural motion had intrinsic, internal causes. If the curing of a wound were artificial, this would necessarily depend on the action of an artifice, an agent capable f bringing about the cure. With the weapon salve, an immaterial substance would be responsible for the transport of a curative virtue of the potion that had already been applied to the weapon, until the specific wound had been cured."

At the heart of the development of medicine in the Early Modern period is a changing understanding of the human body and the human soul. With increasing anatomical study and knowledge, the way the body works was becoming less of a mystery. An increased interest in understanding *why* a treatment might work was becoming more prominent. Libavius, a fellow physician argued that the weapons salve essentially made little sense. He instead argued that "the actual wound cures itself naturally. These ceremonies are magical or illusory or even faked by chance to encourage the patient, who then recovers faith and directs the virtues naturally by imagination to the affected spot, cures himself fully" (quoted in Ziller Camenietzki (2001, p. 85)). He argued that the effect of cure was either natural, brought about by the patient's natural healing, or demonic, with the salve "a trick and irony of the devil. We conclude that if there is anything of a cure in the unguent, it is false or magic, or it cures by itself... or it is operated by the devil" (quoted in Ziller Camenietzki (2001, p. 86)).

The use of the body as medicine has a very long history, and what is defined as *mummy* or *mumia* has changed as well. The study of the medicinal consumption of human remains can be effectively seen as a shift in how the body is viewed and constructed socially and the role that it can play in healing and medicine. If the body is seen as individualized, as begins to happen in the early modern period, then the use of mummy as a curative goes out of style. If the body is seen as dividual (as described elsewhere in this volume), then the body can more easily be manipulated to meet social ends. We also see in this section that the social qualities of the body used to create mummy have an impact on how it is seen. Mummy derived from those identified as ancient Egyptian nobles was more highly prized than that derived from executed criminals or lower class Egyptians. As Sugg (2016, p. 59) notes, "for most early-modern physicians, the key question about mummy was not, 'Should you eat people?' but rather: 'What sort of person should you eat?'."

15.5 The Body as Art: "Pharaohs as Paint and Paper"

Above, the medicinal consumption of mummy was explored. But mummies, particularly those purportedly from Egypt, came in wrappings. The production of other substances from the mummies and their wrappings also occurred. The oil paints Mummy Brown and Caput Mortem (Death's Head) were thought to be made from either ground mummy or burned mummy wrappings. As McCouat (2019, p. n.p.) notes "given that Europeans happily ate, drank and rubbed mummy on themselves, you should not now be surprised that they also painted with it." The pigment identified as mummy brown, Caput Mortuum, or Egyptian brown, was a rich brown pigment made from ground mummy mixed with white pitch and myrrh. One colourman, or creator of pigments, wrote in 1915 that one Egyptian mummy furnished material sufficient to satisfy his customers' demands for about 20 years (McCouat, 2019). The pigment had its greatest popularity in the mid-18th to 19th centuries. French Artist Martin Drölling was rumored to have created his Mummy brown from the remains of French kings disinterred from the abbey of St. Denis in Paris (McCouat, 2019). According to Field's Chromatography (Field, 2007), mummy as a pigment fell out of favor with artists, partly due to its composition but also its capacity to interact poorly with other pigments. This color was available into the twentieth century, with Roberts of London (a well-known pigment producer) stocking the color through the 1930s. The color Egyptian brown is still made, but is composed of quartzite, goethite and hematite (McCouat, 2019).

15.6 The Body as Sustenance: Anthropophagy

Anthropophagy (cannibalism) can be defined as the consumption of human flesh by other humans. A general definition is important here because the analysis of cannibalistic activities should be viewed without judgement. Burley (2016, p. 490) draws a distinction between "normative or institutionalized cannibalism, which is part of a cultural system of way of life, and non-normative forms that are out of step with the standard values of the culture in which they occur." Survival cannibalism would fall into the second category, as would cannibalism that occurs as part of criminal murders (e.g., Jeffrey Dahmer's crimes). Acts such as communion and compassionate cannibalism would fall into the first category. *Exocannibalism* refers to the consumption of human flesh from members outside the consumer's social group, *endocannibalism* refers to the consumption of a flesh from a member of the own consumer's group. The important element for culturally proscribed cannibalism is that the act fulfills a cultural need and carries great meaning.

15.6.1 Communion: Ritual Consumption of Human Tissue

Within the Catholic church, transubstantiation transforms the communion wafer into the body of Christ and wine into the blood of Christ. During the act of communion, this wafer is consumed by the faithful. Many cultural groups have prohibitions against cannibalism, and it is perhaps this proscription that gives the ritual act of cannibalism such tremendous meaning. As Pan (2022, p. 870) notes, there is an ethical tension when "we generally think that cannibalism (eating human flesh) is wrong, yet Christianity (one of the major world religions) actively encourages and celebrates this story in which Christians eat the flesh of their God." As Journet notes (1974, p. 735), this sacrifice of Jesus is seen as purifying and as an act of solidarity, "by the personal eating of the victim, to signify that one was offering himself together with the victim." The transubstantiation was codified in the Thirteenth session of the Council of Trent (1551), a refutation of the reformation (Fastiggi, 2020; McCue, 1968). The act of communion forms the center of religious experience for the Catholic faith and is represented in art, religious texts, and is represented within church architecture (e.g., Fig. 15.9). Communion takes place at the main altar in the church and is therefore the focus of religious performance, a performance focused on the transubstantiated human body that is to be consumed by the faithful.

Christians are, of course, not the only group to use cannibalism as ritual. Indigenous groups within precolonial North America practiced ritualized cannibalism of war captives, accounts of these activities were recorded by missionaries and the material traces of such activities have been found archaeologically (e.g., Glencross, 2014). Cevasco (2016) notes the similarities between the vessels used by missionaries during communion and by Indigenous groups during ritualized feasting on human flesh. The particulars of such rituals are of less importance here and varied with time, location, and cultural group. What is important for the current discussion is that the ritualized consumption of human flesh was an important social activity and was performed in a highly ritualized fashion designed to reinforce social identity and hierarchy.

Williamson (2007) details missionary and settler accounts of Iroquoian ritualized cannibalism, noting that the rituals were formulaic, and the dispersal of human remains followed a recognized pattern. In no way was this gustatory or voluntary cannibalism, the consumption of human flesh was very much a ritual practice that sought to reinforce group identity: war captives and outsiders were those who were

Fig. 15.9 Stained glass window featuring the elements of the Eucharist from St. Michael the Archangel Parish, Findlay, Ohio. Image attribution: Nheyob, cropped by Tahc, CC BY-SA 4.0 <https://creativecommons.org/licenses/by-sa/4.0>, via Wikimedia Commons. https://upload.wikimedia.org/wikipedia/commons/9/97/St_Michael_the_Archangel%2C_Findlay%2C_OH_-_bread_and_wine_crop_1.jpg

sacrificed and eaten. Cevasco (2016, p. 559) takes an expanded view of what communion means, defining communion as "the ritualistic consumption of the human body as an expression of community." For Indigenous groups, communion was a way of "showing community belonging in the midst of conflict." A similar pattern of exocannibalism was practiced by the Wari' of western Brazil (Conklin, 2004). In Wari' exocannibalism, the bodies of captured enemies were abused, treated disrespectfully, and the flesh roasted and eaten off the bone. We can still view this as a communion, since it was a public ritual performed as an expression of community.

The role of cannibalism within extreme processing in the precontact American Southwest has long been debated. Extreme processing assemblages are those with evidence for burning, tool marks, dismemberment, and commingling within the assemblages. Debra Martin and I have previously written on extreme processing and Poetics extensively (e.g., Martin & Osterholtz, 2020; Osterholtz, 2010, 2012, 2013, 2014a, 2014b, 2016a, 2016b, 2018, 2020; Osterholtz & Martin, 2017). My purpose here is not to relitigate whether or not cannibalism occurred as part of extreme processing, the bones from these assemblages exhibit all of the markers that Turner and Turner set out in their checklist for identifying cannibalism (e.g., C G. Turner, II, 1993; C G. Turner, II & Turner, 1992; Christy G. Turner, II & Turner, 1999). But as Glencross (2014) and Williamson (2007) both note, the osteological signature that Turner and Turner identify as cannibalistic can also be found on remains that have extensive postmortem processing. To paraphrase what I wrote in 2018, cut marks on a cranium simply tell you that a cranium was cut; it doesn't tell you why, or what happened to the excised tissue. I have also argued in the past that whether or not cannibalism took place is not the most important issue when examining extremely processed remains. In fact, the time and energy spent in systematic processing was likely the goal of the enterprise itself, not necessarily the ultimate consumption or nonconsumption of flesh. The systematic nature of the processing argues that there was a "right" way to process human remains, suggesting that this ritual activity held significant social importance Osterholtz (2014a).

Why are we so fascinated with whether or not cannibalism took place? Cevasco (2016) notes that many human groups have a general prohibition against the consumption of human flesh, and Pan (2022) discusses the ethics of eating human flesh, drawing attention to the dichotomy that we place between human meat and the flesh of other animals. One of the case studies that Pan discusses is that of a man who served his amputated foot to friends in the form of tacos (Mufson, 2018). The Reddit user who goes by the name IncrediblyShinyShart (Shiny) was in a motorcycle accident which necessitated the amputation of a foot and lower leg. He requested to keep the amputated limb, to which the doctor assented. Three weeks after the accident, Shiny invited friends to a brunch at which the main course was fajita tacos made from his severed limb. In the Vice Article, Shiny details the process of finding a friend to serve as the chef and of letting friends know what they would be eating. Ultimately, 10 agreed to the meal. The leg was marinated overnight and sauteed with onions, peppers, salt, pepper and lime juice. The Vice article (Mufson, 2018) includes a link to the recipe provided by the chef for those interested (noting that it is acceptable to substitute a different type of protein). In the interview with Mufson, Shiny describes the taste as "super beefy" and "tough and chewy". All participants had been fully informed of what they were eating, and for Shiny, the act was a bonding experience and a way to provide some closure on the accident that caused the amputation. He concludes that eating the foot "was a funny and weird and interesting way to move forward." I remember discussing this case with students back in 2019 in terms of ethics and consent. There was an excellent discussion about whether this meat would be considered vegan, since consent of the 'donor' was obtained and the participants had full knowledge of what they were eating. All of us discussing the case were moderately uncomfortable having the conversation, but it was a useful exercise in thinking of our cultural biases regarding cannibalism.

We rarely, in our modern lives, put things like communion into a cannibalism framework. But when we do, we can begin to recognize the cultural significance of such activities. It is precisely because it is uncomfortable and transgresses lines of cultural appropriateness that acts like cannibalism take on tremendous meaning. They are culturally expensive, and at the barest level, damage the community by removing a member. In Poetics terms, they are both destructive and constructive, and the act itself is an amplification mechanism to promote and solidify group identity. In analyzing the case of the leg tacos, it is important to note that the meal was presented as a communal feast, which Pan would identify as a communion. It brought those individuals closer together through a culturally transgressive act that could only be brought about through the sacrifice of Shiny's limb. In general, it was a positive experience, though, if uncomfortable for all participants involved.

15.6.2 *Consuming Grief: Compassionate Cannibalism*

The consumption of human flesh as part of mortuary ritual has been extensively studied. Conklin (1995) discusses funerary cannibalism in contrast to the exocannibalism performed by the Wari'. While the bodies of captured enemies are treated with conspicuous disrespect, the bodies of the dead are never left alone, are held by the living and cared for. The elaboration of funerary ritual tended to be related to social status or age, with elders receiving a more elaborate funerary ritual than infants. A ritual firewood bundle was created of roofbeams taken from each house in the dead person's village. This fire is lit, the body is dismembered and internal organs removed to be roasted. Body parts considered inedible (e.g., hair, genitals, intestines, nails) were burned. Mortuary specialists sever the head, remove the brain, dismember the body and place the body parts on a roasting rack. Throughout this, grief is visible in the community. The most vocal displays of grief came with dismemberment, as this act represented the "most radical alteration of the corpse and the mourners' relations to it, a graphic severing of the attachments represented in the body" (Conklin, 1995, p. 81). The closest relatives to the dead person do not consume the corpse, this would be considered autocannibalism and was believed to be fatal. Affines and other non-kin were responsible for consuming the corpse, this may include a dead person's spouse's siblings, a spouse's parents, the parents' siblings, and so on. The consumption of the corpse was an obligation, refusal to do so was seen as an insult to the dead person's family, a rupture of social practice. Flesh not eaten as part of the funeral feast was cremated in the pyre. In this we can see that the point of the consumption was not to increase caloric intake or for gustatory reasons. Cultural rules dictated who would be allowed to eat and who would not, and so the formulaic nature of the feast adds to the cultural meaning of the activity. Consuming the flesh of one affine's reinforced community bonds, reaffirming a commitment to each other and the overall social system. In this way, Pan (2022) would also view this as a communion. It is separated here as a specific form of communion that operates within a mortuary space. This particular type of communion can only occur with the death of a community member.

In her work with Wari', she describes the response of the Wari' to the anthropological fascination with cannibalism. In her words, "Elders were bemused and at times rather irritated by anthropologists' singular obsession with the eating of bodies, for they insisted that cannibalism cannot be understood apart from the entire complex of mortuary rites and mourning behaviors aimed at reshaping emotional and spiritual relations between the living and the dead" (Conklin, 1995, p. 76). There is an understanding on the part of the Wari' elders of the importance of this particular mortuary custom. According to Conklin (1995, p. 84), "In fulfilling mortuary obligations, including disposal of the corpse, Wari' families linked by marriage affirm continuing commitments that transcend the lifetime of any individual member." As a public display of love, affection, and community identity, mortuary anthropophagy serves a function that nothing else could within these communities. Discontinuation of this ritual leads to a rupture of the social structure.

15.7 Summary

This chapter has focused on manipulation and cultural uses of the body, from educational materials to artistic materials to substances consumed as food and/or medicine. In all instances, the materiality of the body is paramount, without the specific material elements of the body, these uses would not be possible. Our fascination with Egypt and the mummies that were found there leading to changes in mortuary monumentation in the US, and the co-optation of the architectural elements to proclaim a deeper past than our own national history would allow as well as a mechanism to proclaim wealth and status. The elements of Egyptian architecture that are highlighted in cemeteries from the nineteenth century onward are those related to tombs and the bodies discovered within them. These bodies, mummies, became a point of fascination and served as a way to gain and keep the interest of scholars and laypersons alike during unwrapping parties. Central to our love of Egypt is our understanding of the central role of the Dead within Egyptian culture and history. The use of cadavers as teaching tools is foundational to a modern approach to medical education. In learning from the dead, medical students develop the skills to save lives. A growing understanding of the personhood of the individuals who become cadavers has lead to the creation of new rituals surrounding the disposition of cadavers when they are converted once again into bodies after their use as teaching tools. Central to our understanding of the central role of communion is an understanding of the unique role of the body to cement social relationships, within communities (in the case of mortuary cannibalism) or within the larger community (through communion).

References

Agarwal, S., Blakey, M. L., Champney, T. H., de la Cova, C., Davis, J. L., Martin, D. L., Weik, T. et al. (2024). *The commission for the Ethical Treatment of Human Remains: Final Report*. Retrieved from https://www.americananthro.org/wp-content/uploads/tcethr-report-2024-06.pdf?_gl=1*vos7oo*_ga*NTQyNTQxODYuMTcyOTE4NzI3NA..*_ga_NHV0Y97DC9*MTcyOTE4NzI3NC4xLjAuMTcyOTE4NzI3NC42MC4wLjA

American Medical Association (2024). *Ethics of unclaimed bodies in cadaveric dissection in medical education* [Press release]. Retrieved from https://www.ama-assn.org/system/files/issue-brief-unclaimed-bodies-medical-education.pdf

Aufderheide, A. C. (2003). *The scientific study of mummies*. Cambridge University Press.

Baber, T. T. (2016). Ancient corpses as curiosities: Mummy mania in the age of early travel. *Journal of Ancient Egyptian Interconnections, 8*, 60–93.

Bernstein, N. (2016a, May 15). How to Avoid the Fate of a Common Grave. *New York Times*. Retrieved from https://www.nytimes.com/2016/05/15/nyregion/how-to-avoid-the-fate-of-a-common-grave.html

Bernstein, N. (2016b). New York City's Medical Schools will stop using unclaimed bodies. *New York Times*. Retrieved from https://www.nytimes.com/2016/08/11/nyregion/new-york-citys-medical-schools-will-stop-using-unclaimed-bodies.html

Bernstein, N. (2016c). New York State bans use of unclaimed dead as cadavers without consent. *New York Times*. Retrieved from https://www.nytimes.com/2016/08/20/nyregion/law-bans-the-use-of-unclaimed-dead-as-cadavers-without-consent.html

Bernstein, N. (2016d, May 15). Unearthing the Secrets of New York's Mass Graves. *New York Times*. Retrieved from http://www.nytimes.com/interactive/2016/05/15/nyregion/new-york-mass-graves-hart-island.html

Bilge, O., & Celik, S. (2017). Cadaver embalming fluid for surgical training courses: Modified Larssen solution. *Surgical and Radiologic Anatomy, 39*(11), 1263–1272. https://doi.org/10.1007/s00276-017-1865-4

Blakely, R. L. (1997). A clandestine past: Discovery at the Medical College of Georgia and Theoretical Foundations. In R. L. Blakely & J. M. Harrington (Eds.), *Bones in the basement: Postmortem racism in nineteenth-century medical training* (pp. 3–27). Smithsonian Institution Press.

Blakely, R. L., & Harrington, J. M. (1997). Grave consequences: The opportunistic procurement of cadavers at the medical College of Georgia. In R. L. Blakely & J. M. Harrington (Eds.), *Bones in teh basement: Postmortem racism in nineteenth-century medical training* (pp. 162–183). Smithsonian Institution Press.

Broman, E. (2001). Egyptian revival funerary art in green-wood cemetery. *Markers: The Annual Journal of the Association for Gravestone Studies, 18*, 30–67.

Burley, M. (2016). Eating human beings: Varieties of cannibalism and the heterogeneity of human life. *Philosophy, 91*, 483–501.

Caplan, I., & DeCamp, M. (2019). Of discomfort and disagreement: Unclaimed bodies in anatomy Laboratories at United States Medical Schools. *Anatomical Sciences Education, 12*(4), 360–369. https://doi.org/10.1002/ase.1853

Cevasco, C. (2016). This is my body: Communion and cannibalism in colonial New England and New France. *The New England Quarterly, 89*(4), 556–586.

Ciliberti, R., Fulcheri, E., Petralia, P., & Siri, A. (2020). Sharing ethics of displaying human remains in museums. *Medicina Historica, 4*, 1–8.

Ciliberti, R., Martini, M., Boonsignore, A., & Penco, S. (2016). Break with tradition: Donating cadavers for scientific purposes and reducing the use of sentient beings. *Annali dell'Istituto Superiore di Sanità, 52*, 261–268.

Comer, A. R. (2022). The evolving ethics of anatomy: Dissecting an unethical past in order to prepare for a future of ethical anatomical practice. *The Anatomical Record, 305*(4), 818–826. https://doi.org/10.1002/ar.24868

Conklin, B. A. (1995). "Thus are our bodies, thus was our custom": Mortuary cannibalism in an Amazonian society. *American Ethnologist, 22*(1), 75–101.

Conklin, B. A. (2004). Hunting the ancestors: Death and alliance in Wari' cannibalism. In A. C. G. M. Robben (Ed.), *Death, mourning, and burial: A cross-cultural reader*.

Crone, H. (2004). *Paracelsus, the man who defied medicine: His real contribution to medicine and science*. The Albarello Press.

Dannenfeldt, K. H. (1985). Egyptian Mumia: The sixteenth century experience and debate. *The Sixteenth Century Journal, 16*(Summer), 163–180.

de la Cova, C. (2019). Marginalized bodies and the construction of the Robert J. Terry anatomical skeletal collection: A promised land lost. Bioarchaeology of marginalized people. M. Mant and A. J. Holland. London, Academic Press: 133–155.

de la Cova, C. (2020a). Making silenced voices speak: Restoring neglected and ignored identities in anatomical collections. In C. M. Cheverko, J. R. Prince-Buitenhuys, & M. Hubbe (Eds.), *Theretical approaches in bioarcaheology* (pp. 150–169). Routledge.

de la Cova, C. (2020b). Processing the destitute and deviant dead: Inequality, dissection, politics, and the structurally violent legalization of social marginalization in American anatomical collection. In A. J. Osterholtz (Ed.), *The poetics of processing: Memory formation, cosmology, and the handling of the dead* (pp. 212–234). University Press of Colorado.

de la Cova, C., Hofman, C. A., Marklein, K. E., Sholts, S. B., Watkins, R., Magrogan, P., & Zuckerman, M. K. (2024). Ethical futures in biological anthropology: Research, teaching, community engagement, and curation involving deceased individuals. *American Journal of Biological Anthropology, 185*(2), e24980. https://doi.org/10.1002/ajpa.24980

Fastiggi, R. (2020). The contributions of the Council of Trent to the Catholic reformation. *Perichoresis, 18*(6), 3–20.

Field, G. (2007). Field's chromatography; or treatise on colours and pigments as used by ARtists. An entirely new and practical edition; revised, rewritten, and brought down to the present time. In T. Salter (Ed.). Retrieved from http://www.gutenberg.org/files/20915/20915-h/20915-h.htm

Fritze, R. H. (2021). *Egyptomania: A history of fascination, obsession and fantasy*. Reaktion Books.

Fryxell, A. (2017). Tutankhamen, Egyptomania, and temporal enchantment in interwar Britain. *Twentieth Century British History, 28*(4), 516–542.

Ghosh, S. K. (2017). Paying respect to human cadavers: We owe this to the first teacher in anatomy. *Annals of Anatomy - Anatomischer Anzeiger, 211*, 129–134. https://doi.org/10.1016/j.aanat.2017.02.004

Giguere, J. M. (2009). *"The dead shall be raised": The Egyptian revival and 19th century American commemorative culture*. The University of Maine.

Glencross, B. A. (2014). Into the kettle: The analysis of commingled remains from Southern Ontario. In A. J. Osterholtz, K. M. Baustian, & D. L. Martin (Eds.), *Commingled and disarticulated human remains: Working toward improved, theory, method, and data* (pp. 67–84). Springer.

Gordon-Grube, K. (1988). Anthropophagy in post-renaissance Europe: The traditional of medicinal cannibalism. *American Anthropologist, 90*, 405–409.

Himmelman, P. K. (1997). The medicinal body: An analysis of medicinal cannibalism in Europe, 1300-1700. *Dialectical Anthropology, 22*(2), 183–203.

Hodge, C. J. (2013). Non-bodies of knowledge: Anatomized remains from the Holden Chapel Collection, Harvard University. *Journal of Social Archaeology, 13*(1), 122–149.

Hodge, C. J., & Nystrom, K. C. (2020). Dissection as social process: Anatomical settings in the nineteenth-century United States. In A. J. Osterholtz (Ed.), *The poetics of processing: Memory formation, cosmology, and the handling of the dead* (pp. 189–211). University Press of Colorado.

Humphries, E. (2014). Murder, mortsafes and Moir: A medical student looks at anatomy teaching in Aberdeen. *Journal of the Royal College of Physicians of Edinburgh, 44*, 67–71.

James, R. (1747). *Pharmacopoeia universalis: or, A new universal English dispensatory. Containing, I. An account of all the natural and artificial implements and instruments of pharmacy, together with the processes and operations, whereby changes are induced in natural bodies for medicinal purposes .. With a copious index to the whole*. Printed for J. Hodges and J. Wood.

Jones, T. W., Lachman, N., & Pawlina, W. (2014). Honoring our donors: A survey of memorial cermonies in United States anatomy programs. *Anatomical Sciences Education, 7*, 219–223.

Journet, C. C. (1974). Transubstantiation. *The Thomist: A Speculative Quarterly Review, 38*(4), 734–746.

Kanter, S. L. (2010). A silent mentor. *Academic Medicine, 85*, 389.

King, P. (2017). *Punishing the criminal corpse, 1700–1840: Aggravated forms of the death penalty in England*. Palgrave Macmillan.

Ladenheim, J. C. (1950). The Doctors' Mob of 1788. *Journal of the History of Medicine and Allied Sciences, 5*(1), 23–43.

Le Breton, D. (2017). *Sensing the world: An anthropology of the senses*. Taylor and Francis LTD.

Mac Gregor, G. (1884). *The history of Burke and Hare and of the resurrectionists Times. A fragment from the criminal Annals of Scotland*. Thomas D. Morison.

Martin, D. L., & Osterholtz, A. J. (2020). The poetics of corpse fragmentation and processing in the ancient southwest. In A. J. Osterholtz (Ed.), *The poetics of processing: Memory formation, cosmology, and the handling of the dead* (pp. 85–102). University Press of Colorado.

Matheson, M. A., Gatti, J. R., Reid, L. D., Gallozzi, S. N., & Cooke, S. B. (2023). Unclaimed bodies in anatomical education: Medical student attitudes at One U.S. Medical Institution. *Teaching and Learning in Medicine, 37*, 1–13. https://doi.org/10.1080/10401334.2023.2277843

McCouat, P. (2019). The life and death of Mummy Brown. *Journal of Art in Society*.

McCue, J. F. (1968). The doctrine of transubstantiation from Berengar through Trent: The point at issue. *Harvard Theological Review, 61*(3), 385–430.

University of Minnesota (2016). *Procuring, using, and disposing of donated human bodies used for research and teaching*. University Policy Library.

Mitford, J. (2000). *The American way of death revisited.* Vintage.

Mufson, B. (2018). This guy served his friends tacos made from his own amputated leg. *Vice*. Retrieved from https://www.vice.com/en/article/legal-ethical-cannibalism-human-meat-tacos-reddit-wtf/

Mütter Matters. (2024). Retrieved from https://muttermuseum.org/mutter-matters

Noble, L. (2011). *Medicinal cannibalism in early modern English literature and culture*. Palgrave Macmillan.

Nystrom, K. C. (2014). The bioarchaeology of structural violence and dissection. *American Anthropologist, 4*, 1–15.

Nystrom, K. C. (2016). *The bioarchaeology of dissection and autopsy in the United States*. Springer.

Osterholtz, A. J. (2010). *Hobbling and torment at sacred ridge: Perimortem damage to the feet as a mechanism of social and physical control.* Paper presented at the Paleopathology Association Meetings, Albuquerque, NM. Poster retrieved from

Osterholtz, A. J. (2012). The social role of hobbling and torture: Violence in the prehistoric southwest. *International Journal of Paleopathology, 2*(2–3), 148–155. https://doi.org/10.1016/j.ijpp.2012.09.011

Osterholtz, A. J. (2013). Hobbling and torture as performative violence: An example from the prehistoric southwest. *Kiva, 78*(2), 123–144.

Osterholtz, A. J. (2014a). Extreme processing at Mancos and Sacred ridge: The value of comparative studies. In A. J. Osterholtz, K. M. Baustian, & D. L. Martin (Eds.), *Commingled and disarticulated human remains: Working toward improved, theory, method, and data* (pp. 105–128). Springer.

Osterholtz, A. J. (2014b). *Patterned processing as performative violence at sacred ridge*. Paper presented at the 79th annual meeting of the Society for American Archaeology, Austin, TX.

Osterholtz, A. J. (2016a). *Each one the same: Performance, demography, and violence at Sacred Ridge*. Paper presented at the 2016 Meeting of the American Association of Physical Anthropologists, Atlanta, GA.

Osterholtz, A. J. (2016b). Patterned processing as performative violence. In A. J. Osterholtz (Ed.), *Theoretical approaches to analysis and interpretation of commingled human remains* (pp. 125–138). Springer.

Osterholtz, A. J. (2018). Interpreting and reinterpreting sacred ridge: Placing extreme processing in a larger context. *Kiva, 84*(4), 461–479.

Osterholtz, A. J. (2020). Pain as power: Torture as a mechanism for social control. In S. G. Sheridan & L. A. Gregoricka (Eds.), *Purposeful pain: The bioarchaeology of intentional suffering* (pp. 215–252). Springer.

Osterholtz, A. J., & Martin, D. L. (2017). The poetics of annihilation: On the presence of women and children at Massacre Sites in the ancient southwest. In C. Tegtmeyer & D. L. Martin (Eds.), *Bioarchaeology of women and children in times of war: Case studies from the Americas*. Springer.

Overholtzer, L., & Argueta, J. R. (2018). Letting skeletons out of the closet: The ethics of displaying ancient Mexican human remains. *International Journal of Heritage Studies, 24*(5), 508–530.

Pan, L. (2022). Cannibalism and the Eucharist: The ethics of eating the human and the divine. *Sophia, 61*(4), 869–885.
Plastination Technique. (2020). Retrieved December 3, 2020, from https://bodyworlds.com/plastination/plastination-technique/
Pawlina, W., Hammer, R. R., Strauss, J. D., Heath, S. G., Zhao, K. D., Sahota, S., Feeley, M. A., et al. (2011). The hand that gives the rose. *Mayo Clinic Proceedings, 86*(2), 139–144. https://doi.org/10.4065/mcp.2010.0625
Peeler, M. (2020). An unexpected education – Unclaimed bodies in the anatomy lab. *New England Journal of Medicine, 383*(21), 2002–2004. https://doi.org/10.1056/NEJMp2015975
Pires, T. ([1516] 1944). *Letter to King Manuel from Chochin. Letter. The Suma Oriental of Tomé Pires* (Ser II, Vol. 90). Hakluyt Society.
Prakash, P., Rai, R., D'Costa, S., Jiji, P. J., & Singh, G. (2007). Cadavers as teachersin medical education: Knowledge is the ultimate gift of body donors. *Singapore Medical Journal, 48*, 186–190.
Ramsey-Stewart, G., Burgess, A. W., & Hill, D. A. (2010). Back to the future: Teaching anatomy by whole-body dissection. *Medical Journal of Australia, 193*(11–12), 668–671.
Riederer, B. M. (2014). Plastination and its importance in teaching anatomy. Critical points for long-term preservation of human tissue. *Journal of Anatomy, 224*, 309–315.
Riederer, B. M., & Bueno-López, J. L. (2014). Anatomy, respect for the body and body donation—A guide for good practice. *European Journal of Anatomy, 18*, 361–368.
Rosner, L. (2009). *The anatomy murders: Being the true and spectacular history of Edinburgh's notorious Burke and Hare, and of the man of science who abetted them in the commission of their most heinous crimes*. University of Pennsylvania Press.
Sakai, T. (2008). Body donation: An act of love supporting anatomy education. *Japanese Medical Association Journal, 51*, 39–45.
Savitt, T. L. (1982). The use of Blacks for medical experimentation and demonstration in the old south. *Journal of Southern History, 48*, 331–348.
Sheppard, K. L. (2012). Between spectacle and science: Margaret Murray and the tomb of the two brothers. *Science in Context, 25*(4), 525–549.
Shupe, E., Karim, S., & Sledge, D. (2023). Unclaimed bodies and medical education in Texas. *JAMA, 330*(12), 1189–1190. https://doi.org/10.1001/jama.2023.15132
Sugg, R. (2016). *Mummies, cannibals, and vampires: The history of corpse medicine from the renaissance to the Victorians*. Routledge.
Tarle, L., Nicholas, G., & Cardoso, H. F. (2020). What next?: Changing ethical protocols for human remains in museums. In C. H. Meloche, L. Spake, & K. L. Nichols (Eds.), *Working with and for ancestors* (pp. 219–232). Routledge.
Tseng, W. T., & Lin, Y. P. (2016). "Detached concern" of medical students in a cadaver dissectio course: A phenomenological study. *Anatomical Sciences Education, 9*, 265–271.
Turner, C. G., II. (1993). Cannibalism in Chaco Canyon: The charnel pit excavated in 1926 at Small House Ruin by Frank H.H. Roberts, Jr. *American Journal of Physical Anthropology, 91*, 421–439.
Turner, C. G., II, & Turner, J. A. (1992). The first claim for Cannibalism in the Southwest: Walter Hough's 1901 discovery at Canyon Butte Ruin 3. *Northeastern Arizona. American Antiquity, 57*(4), 661–682.
Turner, C. G., II, & Turner, J. A. (1999). *Man corn: Cannibalism and violence in the prehistoric American Southwest*. University of Utah Press.
University of Ediburgh. (2024). "Burke and Hare." Retrieved March 5, 2025, Retrieved from https://medicine-vetmedicine.ed.ac.uk/about/history/burke-andhare#:~:text=Hare%20escaped%20by%20turning%20King's,Anatomical%20Museum%20to%20this%20day
Vesalius, A., & van Calcar, J. S. (1555). *De humani corpus fabrica libri septem* (Seven books on the structure of the human body) (LOC Control Number 2021667096 ed.). Joannes Oporinus.

Williams, S. E., & Ross, A. H. (2021). Ethical dilemmas in skeletal collection utilization: Implications of the Black lives matter movement on the anatomical and anthropological sciences. *The Anatomical Record, 305*, 860–868.

Williamson, R. (2007). "Otinontsiskiaj ondaon" ("The house of cut-off heads") The history and archaeology of northern Iroquoian trophy taking. In R. J. Chacon & D. H. Dye (Eds.), *The taking and displaying of human body parts as trophies by Amerindians* (pp. 190–221). Springer.

Ziller Camenietzki, C. (2001). Jesuits and alchemy in the early seventeenth century: Father Johannes Roberti and the weapon-salve controversy. *Ambix, 48*(2), 83–101. https://doi.org/10.1179/amb.2001.48.2.83

Zuckerman, M. K., Austin, R. M., & Hofman, C. A. (2021). Historical anatomical collections of human remains: Exploring their reinterpretation as representations of racial violence. *The Annals of the American Academy of Political and Social Science, 694*(1), 39–47.

Chapter 16
Relics and Reliquaries

Abstract This chapter explores the multifaceted roles of human remains and other physical objects as relics, examining their cultural, religious, and political significance across various contexts. It draws on Gillingham's and Knight's definitions of relics as objects imbued with charismatic power, capable of inspiring veneration, solidifying group identities, and sometimes even performing miraculous acts. The chapter begins by discussing the Christian tradition, where human remains were transformed into sacred relics, often stored in elaborate reliquaries to preserve their spiritual power and foster connections between the living and the dead. It examines the processes of authentication, veneration, and the complex rituals surrounding the use of these relics within Christian worship. Expanding beyond religious contexts, the chapter also investigates secular and nationalistic relics. One key example is the contested remains of Mexico's "niños héroes" (cadet heroes), which were recontextualized in the twentieth century to serve as national symbols of sacrifice and patriotism, despite the uncertainties surrounding their historical authenticity. This case illustrates how material remains can be manipulated to forge and legitimize national identities. Scientific relics and personal relics are also discussed, showing how human remains can serve as poignant links between the living and the deceased. Finally, the concept of war trophies is explored, with a focus on the leg of Mexican General Santa Anna, highlighting how human body parts are transformed into symbols of conquest, memory, and identity. Ultimately, the chapter underscores the enduring power of relics to shape cultural, political, and emotional relationships across time and space.

Gillingham (2010, 203) gives a definition of relics as "artefacts of widely accepted charismatic power, whether they serve as a sign for a famous individual or as the sign of a major, transformative idea." Knight (2010, 230) argues that relics should be defined functionally according to their purpose to inspire veneration, promote solidarity, provide solace, and sometimes work miracles. This functional definition allows for the creation of relics of nationalistic or secular creation, playing into what Gillingham (2010) identified as monarchic, nationalistic and scientific relics. In

A. Osterholtz, *The Bodies Poetic*, Bioarchaeology and Social Theory,
https://doi.org/10.1007/978-3-031-85998-4_16

some places and times, the line between sacred and secular is blurred, such as Latin American veneration of political leaders and actions leading to national identity building. Playing upon the embodiment theory discussed previously in this volume, these relics serve then to embody the cultural and religious intersectional identities and practices important to individuals within these cultures. The importance of a relic is the power and reflected relationships embodied within the material object. Also important is the portability of relics. They can be moved, traded, and exchanged. The presence of a relic can legitimize space as sacred, be used to spread knowledge, and spark interest in the relationships embodied within them—leading to group identity formation.

16.1 Religious Relics

Possibly best known in the western European tradition are the reliquaries associated with the Christian faiths. These should be seen as having a central position within Medieval culture as an embodiment of public and collective understandings of piety and social relationships (Visa & Jussi-Pekka, 2011). Reliquaries are simply the boxes created for the purpose of housing relics and can themselves vary quite a bit. Some are small enough to be worn on the body as jewelry or portable talismans (Lutz, 2015), and become powerful objects in their own right due to their association with the relic within them (Fig. 16.1). Relics and their reliquaries acted as repositories for meaning. According to Lutz (2015, 20), "the center of meaning resided in stuff and substance, in these historical fragments around which elaborate stories were spun." Essentially, these objects are embodied with the acts which are culturally and historically meaningful, drawing the dead and the living together through the manipulation of the physical body of the dead. "The material of the body became a text, and texts became representations of body parts. The miracle can reside in the intimacy of flesh with writing or in the way that inscriptions can make the body a thing not subject to the decay of substances, but rather "eternal"." (Lutz, 2015, 21). Relics were important components of the Catholic church, since every church alter was supposed to contain the remnants of a saint for the church structure to be sanctified (Geary, 1986).

The transformation of the body of the saint into a relic was formulaic and ritualistic within the medieval Catholic church. Geary (1986, 177) suggests that the "career or biography" of objects as part of their transition from "ordinary remains to treasured relics." Essentially, he is arguing that the life history of these objects alters or transforms remains from one social state to another, where agency of the remains develops due to the transformational processes. It was rare for the living or recently dead to be the object of sanctification. More often, the identification of a specific set of remains with a long-dead saint was needed. The location of these remains needed to be identified, typically the location of the body was revealed to those looking via supernatural intervention of the saint him or herself. "Thus the initial impetus for the consideration of a possible relic often came in the form of a vision in which the saint appeared to a holy person and revealed where his remains

Fig. 16.1 Monstrance with a Relic of Saint Sebastian, 1484. Germany, Lower Saxony, Brunswick, Gothic period, fifteenth century. Gilded silver, rock crystal, bone, parchment; overall: 47 cm (18 1/2 in.). The Cleveland Museum of Art, Gift of Julius F. Goldschmidt, Z. M. Hackenbroch, and J. Rosenbaum in memory of the Exhibition of the Guelph Treasure held in the Cleveland Museum of Art from 10 January to 1 February 1931 1931.65

could be found" (Geary, 1986, 177). This could be done through examination of extrinsic or intrinsic criteria. Extrinsic criteria involved examination of the tomb or reliquary, and an examination of documents found either in the tomb or reliquary itself, or in the descriptions of the interment of the saint in hagiographical texts. These examinations were carried out by the local bishop in public sessions attended by both lay and clerical elites.

Once the authenticity of the relic was established, a public ritual known as the *Elevation* was conducted in which the newly authenticated relics were offered to the public for veneration. "Thus corpses passed from the status of mere human remains to that of sacred relics through a public ritual emphasizing both the identity of the remains with those of a saint and the actual miraculous power exercised by that saint through those particular remains" (Geary, 1986, 178). The initial elevation was necessary but not sufficient for the relic to maintain importance through time. Visa and Jussi-Pekka (2011, 163) instead describe a "continuous material process in which relics were framed and made into objects of veneration" in which the objects needed to be handled and venerated continually to maintain their cultural importance. An important part of this was the placing of the relics into reliquaries and their display. The reliquary, then becomes an important house for the relic and its embodied collective acceptance of the miracles produced by the saint. Relics and reliquaries are repositories for the symbolic capital and express agency in their abilities to facilitate miracles for both individuals and communities in need.

The veneration of relics in the medieval church was a principal complaint of Martin Luther in his theses, arguing that relics were merely dead things and shouldn't be worshiped as gods themselves (i.e., they lacked any inherent agency). This argument suggests that the communal understanding of the importance of relics and reliquaries had altered significantly by the sixteenth century; in effect, there was no collective understanding of the embodied meaning, and so *all* meaning for at least some part of the population was lost.

Relics were also used to promote political autonomy and power (Gillingham, 2010). The wrapping of relics in exotic fabrics such as silk procured through trade from China speak to the economic and social power of the Catholic hierarchy as well as the interplay between the monarchical governments and their ecclesiastical counterparts. The skull relic of St. Henry or St. Eric (attribution is disputed) at the sacristy of Turku Cathedral in Southwest Finland is wrapped in Chinese red silk damask and embroidered with a figurative scene depicting a knight wearing chainmail and holding a sword. The knight depicted has beheaded a kneeling man whose hands are raised in prayer; the head lies on the ground between the two men. "Placed in front of the skull's face, the pictorial scene weaves together the narrative of martyrdom with the bones underneath the silk. In effect, the relic and its reliquary form a highly expressive entity" (Visa & Jussi-Pekka, 2011, 164).

Relics are not a phenomenon restricted to western European cultures. Within Buddhist traditions, relics (particularly relics made from human remains) is foundational to belief and ritual. Sharf (1999) argues that the movement of Buddhism throughout Asia should be viewed in terms of the diffusion of sacred objects and the esoteric knowledge that accompanied them. "The relic is a potent vestige of the death of an enlightened being, a memento of his or her abiding liberation" as well as "a physical locus for the saint's enduring charisma, apotropaic power, and grace" (Sharf, 1999, 77–78). Relics have a very special place within Buddhism, they should not be seen as symbolic, but are treated as "presences pure and simple" (Sharf, 1999, 79). This description also highlights the agency of the objects themselves to change how people interact with them, each other, and the spaces in which they reside.

16.2 Monarchic Relics

Monarchic relics contained the bones of royalty. In central America, these proliferated in the mid-seventeenth century. According to Gillingham (2010), monarchic relics draw on the "charismatic sway of kings, expressed in beliefs such as the royal touch..." but also could have material benefits to their possessors in terms of land grants during times of fluctuating property rights. Due to their importance, they were frequently forged. The power was in the object, and even when it wasn't authentic, it still held power. This brings to mind Sahlins' (1983) question for us: how *real* does a process have to be for it to be viewed or interacted with as a *real* object. The agency of monarchic relics is not necessarily based on their firm attribution to specific royal persons but to the concepts, powers, and responsibilities seen to be embodied in those bones.

16.3 Nationalistic Relics

Nationalistic relics may be of particular interest to us in Anthropology. Gillingham (2010, 215) ties the rise of nationalistic relics to the increase in archaeological exploration in Latin America. As he states," archaeologists provided many of the skeletons on which the flesh of a modern nation-state could hang. Ruins and artefacts bore witness to the antiquity and urbane sophistication of Prehispanic culture, which lent Latin American nations legitimate identities in a post-Colonial third space. Bones—if they could be found—would provide secular, nationalist relics to tie contemporary populations to would-be founding fathers. And when neither ruins, artefacts, nor bones were forthcoming they could always be manufactured." Modern nations are fundamentally founded on myths as Vandiver (2000, 4) defines them as "traditional stories a society tells itself that encode or represent the world-view, beliefs, principals, and often fears of that society." Formative stories that may or may not exist in fact are important for social identity.

An excellent example of how this myth can be created and reinforced with material culture is that of Mexico's *niños heroes* (cadet heroes). A group of six cadets were long believed to have been the last defenders of Mexico City against the US Army in 1847, they fought with bayonets after their ammunition was spent and died one after another until the final cadet wrapped a flag around him and jumped to his death from the castle walls (Gillingham, 2010). Moerer (2013, p. 86) documents the rise of myth and public commemoration of this event and their "spiritual victory," beginning in the 1870s. This has become "a central plank in the strongly sacrificial metanarrative of Mexican nationalism" (Gillingham, 2010, p. 216). It wasn't until 1944 that the grave containing the bodies of the *niños heroes* was identified by a Mexican General. Part of a mass grave was excavated to obtain the necessary six skeletons but there was no way to identify the specific individuals involved. Despite the somewhat fuzzy circumstances surrounding the context of the mass grave, the National Institute of Archaeology and History (INAH) produced a report concluding the skeletons were those of young males and *might* be those of the *niños heroes* (Plasencia de la Parra, 1995). The Secretary of Education ultimately removed the 'might' and declared that the remains were those of the *niños heroes*, ever since these remains have been instrumental in commemorative events (Fig. 16.2). The bones recovered from the mass grave are most likely not those of the *niños heroes*, if the *niños heroes* ever existed as real individuals. They have gained a new type of

Fig. 16.2 Obelisk of the niños heroes in Chapultepec Park. ProtoplasmaKid, CC BY 4.0 <https://creativecommons.org/licenses/by/4.0>, via Wikimedia Commons. https://upload.wikimedia.org/wikipedia/commons/5/5c/Obelisco_a_los_Ni%C3%B1os_H%C3%A9roes.jpg

identity, their remains disinterred, manipulated by a nationalistic entity to promote a new concept of identity centered around nation and sacrifice.

In this way, the material culture found within a modern nation-state's boundaries help to deepen the relationship between a (possibly) mythically constructed past to modern leadership, giving legitimacy. This can be thought of as the political power of material remains, particularly within a restricted physical space, harkening back to the post-processual approaches to mortuary treatment. The current political structure takes advantage of material remains to reinterpret historical control over space or resources through a modern interpretation of the past. This philosophical manipulation of the dead and material culture is for the benefit of the living. Verdery (1999, 17) takes a similar view of the preserved bodies of politically important individuals in the former Soviet Union, calling them "quasi-religious relics", with the view of the state taking the place of religion in this case.

16.4 Scientific Relics

Scientific relics are those objects that "radically reshape scientific understandings of the world" (Gillingham, 2010, 217). Important fossil finds such as Lucy (AL 288-1, the *Australopithecus afarensis* skeleton discovered in 1974 by Don Johanson and Tim White) would be an excellent example of a scientific relic. Pyne (2016, n.p.) ascribes the importance of Lucy as a "cultural connection between fossil and audience… when a fossil is no longer an 'it'—an object that is merely a museum catalogue number, a thing on display. When a fossil has become a 'her'—an anthropomorphized ancestor complete with a paleo-persona—the fossil can connect with audiences on any number of levels."

One such example, and the discussion around its use to promote scientific ideals, are the South African fossils that were launched into space in 2023 (effectively giving them a space burial). Lee Berger, the discoverer of the fossil and a Research Professor at the University of Witwatersrand, advocated for the sending of the fossils into space. Lee Berger's goals were two-fold, first as a celebration of African ancestors and second as a way to bring increased awareness of South Africa's role in early human evolution. In his letter of request to the South African Heritage Resources Agency (SAHRA), Berger (2023, n.p.) writes that carrying two fossils into space (one *Australopithecus sediba* and one *Homo naledi*) would be the first instance of "extinct humans" in space (the place Of *H. naledi* within the human tree of these two species is still a matter of debate, but there is a general acceptance that they were important species whether they were on the direct line to anatomically modern humans or on a side branch). Berger follows that this would be "a sign of respect by humanity to our African ancestors who gave us the technology, skills and mind that allows this perhaps greatest expression of human technological achievement—the exploration of space."

Berger recognized that this would be controversial, and that this controversy in itself would bring attention to South African paleoanthropology. Berger's request was granted by SAHRA, and those remains were launched in the fall of 2023. The fossils entered low-earth orbit and then returned to the ground where they were collected thanks to the GPS tracking of the capsule. Callaway (2023) documented the reaction of fellow scholars, noting that some, such as Robyn Pickering, a geologist at the University of Cape Town called this action "unethical". Yonatan Sahle, an archaeologist at the University of Cape Town saw this as a perpetuation of colonial acts, "a perpetuation of past, very ugly aspects of paleoanthropological research" (scholars quoted in Callaway, 2023). Central to the concern was the safety record of the vehicles used to transport the fossils. A previous version of the vehicle had resulted in a catastrophic failure (Malik, 2015), leading to concerns about the potential loss of the fossils. To mitigate this concern, Zipfel, the curator of collections at the University of Witwatersrand, noted that the remains chosen for inclusion in the capsule had been thoroughly documented in 3D scans, casts, and photographs (Callaway, 2023).

The American Association of Biological Anthropologists (AABA) published a statement concerning the launch that highlighted the authority of governmental agencies to permit items of cultural heritage to be used for the promotion of science (Hlusko & Leigh, 2023). In recognizing the controversy, the AABA statement concludes "we hope that this incident will focus the attention of the scientific community on the responsibility that we, as scientists, have to a variety of entities (current and future), when it comes to the study and preservation of objects of great scientific and cultural significance." Would such a controversy have developed if authorities from South Africa had requested permission to send stone tools of similar age into space? The importance of these remains was tied, at least in part, to the fact that these were once the bones of potential human ancestors. The transformation via the fossilization process has turned them into something else, but they still retain the properties of being bone, and therefore can be seen as a tie to an ancient past the focused on the importance of Africa (and South Africa in particular) in the human story. Given the history of racism within paleoanthropology (hinted at by Sahle above), physical remains (i.e., bones) that show the centrality of Africa within the human narrative and are concrete objects that can be held in the hand, filled with agency that centers around nationalistic, scientific, and cultural identity.

16.5 Cultural and Personal Relics

Visitors to the Keats-Shelley Memorial House in Rome see, in addition to other things, parts of Shelley's jaw on display. Pieces of his bones were also collected after his seaside cremation and are also on display with his hair (Lutz, 2015). Relics need not be religious, as the example of Shelley's jaw and hair attest. Lutz (2011, 128) argues that initially relics were restricted to the physical remains of saints, but eventually transition into a secular space for the Victorians. These new secular relics "still held some of the sublime, fetishistic magic of those outmoded holy relics of bygone days." The Victorian era can be seen as a "relic culture" in which "every substantial library or archive has a body part, a book bound in human skin, or fragments of hair stored away" (Lutz, 2015, 15) (e.g., Fig. 16.3). But the relic culture of

Fig. 16.3 Untitled (Case with portraits of a man and woman and hair ornaments), c. 1870s. nineteenth century. Photographs: photographs on enamel, framed in copper alloy metal frames; hair lockets: hair with pen and black and ink on cream-colored backing; case: leather with metal push-clasp and velvet-lined interior; image: 3.5 × 3 cm (1 3/8 × 1 3/16 in.); case: 9 × 5 × 1 cm (3 9/16 × 1 15/16 × 3/8 in.). The Cleveland Museum of Art, John Cook Memorial Fund 2017.9

the Victorians was not the first time that human remains were used to convey social meaning. They, in fact, built upon a long tradition of this.

In his patent application "Deciduous dentition jewelry and method of Manufacture", Fischer (2012) writes that the loss of deciduous teeth are milestones in childhood, and that these milestones are often lost in the course of every-day life. He notes that storage solutions for deciduous teeth are "biologically contaminated" (Fischer, 2012, 4 of 8) may become lost or combined with the teeth of other children within the same household. The solution to this problem, according to the patent, is the creation of jewelry made from the teeth after they have been cleaned, crushed and set in epoxy. Wilbert, the creators of the modern Vault, also provide a variety of services, including the production of memorial jewelry, some of which is designed to hold cremains as "a truly meaningful way of holding memories of a loved one close" (Wilbert, 2020). These jewelry pieces are designed to hold a lock of hair or a small amount of cremated remains.

The creation of jewelry made from human remains (e.g., teeth, bones, hair, and even glass eyes) has a long history, this is discussed at greater detail in the chapters of this volume centered on cremations and the dispersal of cremains (Chaps. 10 and 12). One form of personal relic that has gained notice in the past few years is the preservation of tattooed skin. Kyle and Michael Sherwood, funeral directors in the US set up a system called "Save my Ink Forever" to preserve tattooed skin after the wearer has died. Kyle Sherwood sat for an interview to discuss the process and the reasoning behind it, stating "With this, you have a piece of that person, something that was truly important to them. There's also the fact that some tattooists are modern-day Michelangelos… Up until now, their legacy was through photographs, tattoos usually die with the person, and you wouldn't burn or bury a Picasso. So these tattoos can be physically seen for generations to come" (Snape, 2022). For their service, families don't have to arrange the service ahead of time, but some do. Interested family members need to get in touch with the Sherwoods within 48–72 h after death to start the paperwork involved in collecting the tattoos (this can be done before or after embalming). The Sherwoods will stabilize the preserve the tattooed skin as well as frame the piece (complete with UV protective glass).

When asked what people do with the preserved tattoos, Kyle Sherwood stated "people will sometimes make memorials within their home. Some people just hang them up with other photos and you don't even know the difference." The preserved tattoos are missing the elements that make them connected to the body and therefore appear as static art pieces. The preserved tattoos resemble parchment. These can be important touchstones for those who choose this service. As Kyle said "One of the tattoos that I liked was a man who passed away and he had the handprint of his child tattooed on him. The child was still young, so his wife thought that having the handprint tattoo preserved would be something that his child would like to have. They didn't get to spend much time together, so it's a testament to that dad's love for their child." When clients request this service from their funeral director, the Sherwoods send a package of materials with instructional videos to the funeral home. The excised skin is then shipped to Ohio for the Sherwoods to preserve and frame. Families choose the frame and matting, the preserved tattoo is sewn to canvas and the frame is pumped with nitrogen to help with preservation behind museum-grade UV blocking glass. The Sherwoods charge between around $1700 for a 5 × 5 inch tattoo to more than $120,000 for a full body suit (Oxenden, 2022).

With tattoos becoming more commonplace (Ipsos, 2019), this type of memorial and personal relic is likely to become more common. The process is legal in all states except for Washington. A New York Times article from 2022 notes that Walker Posey, the spokesperson for the National Funeral Directors Association, said that more than half of his clientele ask about the service (Oxenden, 2022). As Kyle Sherwood said, "The gratification people have and that connection I'm able to make, you can't explain it… It's very humbling and powering to have that impact on someone" (quoted in Oxenden, 2022). The preservation of the skin links the living to the dead by creating a stable relic of the dead, one that would normally disappear when its wearer died. It is an interesting thought experiment to think what will ultimately happen to these personal relics when the personal connection between the living and the dead is lessened by generations. The majority of people interviewed for their perspective on preserving tattooed skin spoke of connection to the dead and how they knew the tattoos were important to the deceased; this was the reason for preserving this particular relic. In general, this connection should become lessened in great grandchildren or those generations more removed. Will we see preserved tattoos in antique stores in 50 years, just like we see mortuary art made from the hair of the deceased as was common in the nineteenth century?

16.6 War Trophies

The taking of war trophies is not uncommon. Case studies within Chacon and Dye's (2007b) edited volume *The Taking and Displaying of Human Body Parts as Trophies by Amerindians* provide a wealth of data and interpretation about the power that these particular objects can hold. In their Introduction, Chacon and Dye (2007a, 5) note that "it would be naïve in this context to propose a single or even primary cause or set of variables that underlie human trophy taking" and acknowledge that "human motivations for taking trophies in any one context may vary dramatically and or change accordingly through time." In the context of the current volume, war trophies should be viewed as a kind of relic, an embodiment of the cultural performance of violence that created the trophy. This is indicated by the differential life history of the trophy as an artifact (along the lines of Walker, 1996). In fact, it is always a difficult task during fieldwork to determine how to catalog and analyze these remains. Should they be given a burial number? Or a special finds number? Or both?

An excellent example of this would be the many legs of José Antonio López de Santa Anna. Santa Anna was born in 1794 and served as the president of Mexico 11 times between 1833 and 1855. He first came to prominence as an officer in the Mexican War of Independence (1810–1821). Initially a royalist, he became a fighter for Independence in 1821. In the fluid and experimental political system of a newly independent Mexico, Santa Anna was a key figure. In 1829, Santa Anna became a national hero and was promoted to General after helping to turn away a Spanish invasion force at Tampico. After his victory at the Alamo in 1835, he was defeated at San Jacinto, a low point in his career. This defeat stripped him of the accolades he'd won at Tampico. The 1830s were a time of flux politically, complete with foreign intervention and the seizing of foreign property and interests. One victim of this was a French pastry chef whose bakery was destroyed by Mexican soldiers. The French played up this claim and ultimately blockaded the Gulf Coast and bombarded Veracruz. During this Pastry War, Santa Anna took command of Mexican troops and attempted to counter-attack the disembarking French. During the attack he was struck by grapeshot (ammunition made from packing many small iron or lead balls between 1 and 1.5 inches in diameter into a bag, shot from a canon, that acts similarly to a shotgun blast with the shot spreading out during impact), killing his horse underneath him and severely injuring his left leg. When gangrene set in, they amputated the leg. Unfortunately, the amputation was not surgically well done, and he suffered pain and periodic opening of the wound with exposure of the bone. The leg was initially buried at his Veracruz home. When he ascended to the presidency again in 1842, Santa Anna exhumed his leg, paraded it to Mexico City in a coach and buried it beneath a cemetery monument in an elaborate state funeral. In 1844, when public opinion turned against Santa Anna, rioters tore down the monument, tied the desiccated leg to a rope and drug it through the streets. Langford (1945, p. 73) writes that it was presented to Santa Anna in 1876, perfectly preserved "even to the toenails!". The ultimate disposition of the amputated leg is unknown.

Numerous protheses were produced for the again-lauded general, including wooden pegs and several that approximated the shape of legs. One set of "mock legs" was purchased from a specialist company in New York. After recuperation at his home, Santa Anna was brought to Mexico City where he was named as provisional president. Though Santa Anna was not a particular astute politician for day-to-day governing ("he liked gaming and cockfights better than wrestling with the nation's intractable finances" (Knight, 2010, 235)), his popularity increased with "sporadic deeds of derring-do, which reinforced his role as a national hero in moments of crisis" (Knight, 2010, 235)). He did do some governing such as setting up an administration, introduced currency reform and planning the re-conquest of Texas. "Santanista ritual now became elaborate, systematic and personalist; indeed, as the government's fortunes declined, and as a combative liberal Congress confronted the executive, ritual became an antidote to mounting unpopularity" (Knight, 2010, 235). Retelling of past glories, military parades, and the raising of statues became common elements. Ultimately, though, this cult of personality waned due to failure of the currency reforms (and subsequent tax increases and confiscations of property) and a lack of progress in retaking Texas. In response to protests to his rule, he became more despotic, took frequent absences from the capital and forcibly closed Congress. Revolt against his rule was inevitable. In the time since Santa Anna's escapades in the independent country of Texas, Texas had joined the United States in 1836 and the US was eyeing land to the south as a way to expand.

The Mexican government in some disarray called on Santa Anna to repulse the Americans in 1846, when he again assumed the presidency and took command of the military. He fought two major battles, at Buena Vista and Cerro Gordo, Veracruz in 1847. Buena Vista was a draw, while Cerro Gordo was a major loss. Santa Anna escaped to his hacienda with Americans in pursuit. On the night of the battle, the fourth Illinois Infantry nearly captured the defeated general who was escaping on a mule, appropriated everything in his mule train (including his dinner, a large amount of money, and his artificial left leg, Fig. 16.4). The Illinois irregulars claimed the leg as booty of war. The leg was held as a trophy, and was used as a bat in what has been described as the first time baseball in any form was played in a foreign country (Angus MacFarland to Illinois State Mulitary Museum (19 July 2004), cited in Knight, 2010). Santa Anna resigned later that year, but this was not the end of his political career. He had numerous other military campaigns, he rose and fell within the political hierarchy, and was finally exiled in 1867. He was allowed to return 7 years later under amnesty. He returned a different figure, but still a figure that a portion of the population idolized. Upon his return to Mexico City after his exile, he was gifted with a box containing the remnants of his amputated leg, which the gifter claimed he had rescued from a mob in 1844. He is said to have told the former

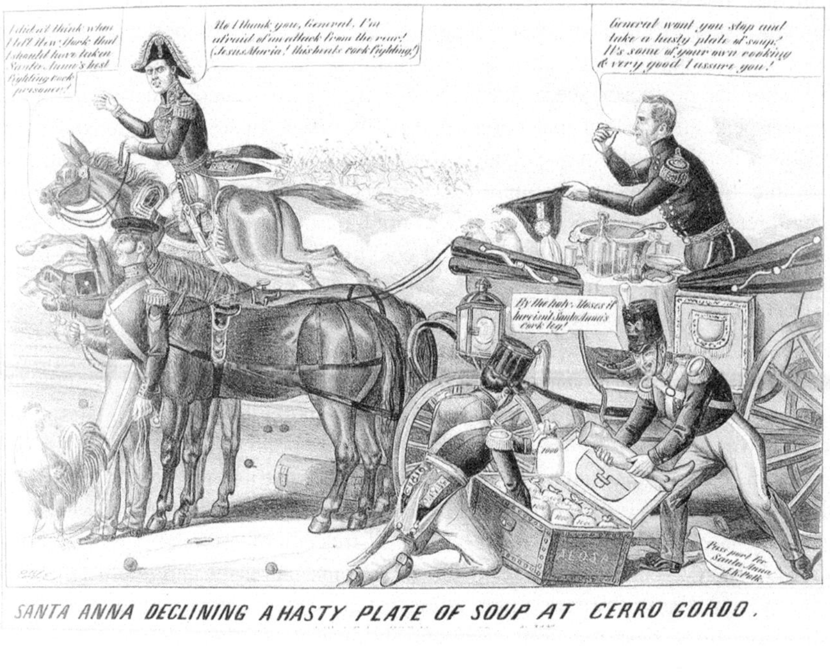

Fig. 16.4 Print by Edward Williams Clay (1799–1847) depicting the retreat of Santa Anna while soldiers discover his cork leg in a box. Edward Williams Clay (1799–1857), Public domain, via Wikimedia Commons. https://upload.wikimedia.org/wikipedia/commons/6/67/Santa_Anna_declining_a_hasty_plate_of_soup_at_Cerro_Gordo.jpg

general that he had his wife look after the leg as a sacred trust. He died in June 1876, no mention is made of whether his severed leg was buried with him (Knight, 2010).

The prosthetic leg appropriated in 1847, however, continues to have cultural resonance. This leg has been on display in the Illinois State Military Museum since 2006. The soldiers who procured the leg signed it. It was part of private collections until 1882, but was never displayed until 1938 at the Illinois State Fair. During World War II, three Illinois State Representatives proposed returning the leg to Mexico as a sign of hemispheric solidarity, but it was never returned. The leg was used as a military recruiting tool during the Vietnam conflict, when it travelled all over the state of Illinois with the National Guard. How to interpret the use of the leg? Knight (2010, 244–245) notes that the leg "had several labels slapped on it: for some it was, indeed, a 'relic'; for others, a 'souvenir'; but most commonly, it was called a 'trophy', specifically a 'war trophy'."

War trophies are embodiments of the actions that occur during war. Bender (2009) views war trophies as transgressive objects. "Trophies transgress boundaries between different bodies, different owners taking possession of them, and different cultures trying to frame their agency and make sense of them" (Bender, 2009, p. 503). Essentially, in being transgressive, they blur boundaries and by blurring those boundaries, embody the actions of war. They are taken, not given. They are an embodiment of the actions of the victor and because of this exhibit significant post mortem agency.

16.7 Summary

Bodies or parts of bodies can come to embody or represent large social movements, including revolutions, but they can also come to embody scientific revolutions and personal attachments. The phenomenological elements of the body allow for its use. When preserved, human remains (whether soft tissue or bone) or the representation of those limbs (in the case of Santa Anna's leg) allow for these relics to be passed down and passed around within a community. The passing of these remains helps to force social ties. Some relics have tremendous staying power, maintaining their power for hundreds of years (such as religious relics), while others may lose their importance after several generations as the individual attachment to the donor of the materials fades in local memory (such as the many personal relics that wind up in antique stores). The scale of connections represented by the relic may be the key factor in how long these relics hold power: if the connection is personal, it may fade after memory of the individual donor passes out of living memory. For religious relics, they take on agency of their own, and the attribution to specific individuals may become somewhat fuzzy. In visiting reliquaries throughout Europe, it is not uncommon to find "relics" that are not human bone, but obviously faunal elements within the containers. This doesn't lessen their cultural or religious significance, because even though they aren't *real human bones*, they have the cultural association with being *real*, in line with Sahlin's (1983) observation that something doesn't have to be *real* to have real influence.

References

Bender, C. (2009). "'Transgressive Objects' in America: Mimesis and Violence in the Collection of Trophies during the Nineteenth Century Indian Wars." Civil Wars 111(4): 502-513

Berger, L. R. (2023). Letter to South African Heritage Resources Agency. In terms of Section 32(19) of the National Heritage Resources Act (Act 25 of 1999). Retrieved from https://sahris.sahra.org.za/sites/default/files/casedecisions/21821%20letter_export_hominin_space.pdf

Callaway, E. (2023). Ancient-human fossils sent to space: Scientists slam 'publicity stunt'. *Nature* Retrieved July 16, 2024, from https://www.nature.com/articles/d41586-023-02882-1#:~:text=Precious%20bones,%2Dyear%2Dold%20Homo%20naledi

Chacon, R. J., & Dye, D. H. (2007a). Introduction to human trophy taking: An ancient and widespread practice. In R. J. Chacon & D. H. Dye (Eds.), *The taking and displaying of human body parts as trophies by Amerindians* (pp. 5–31). Springer Science and Business Media.

Chacon, R. J., & Dye, D. H. (Eds.). (2007b). *The taking and displaying of human body parts as trophies by Amerindians, interdisciplinary contributions to archaeology*. Springer.

Fischer, J. G. (2012). Deciduous dentition jewelry and method of manufacture. *United States of America, 8226877*, B2.

Geary, P. (1986). Sacred commodities: The circulation of medieal relics. In A. Appadurai (Ed.), *The social life of things: Commodities in cultural perspective* (pp. 169–194). Cambridge University Press.

Gillingham, P. (2010). The strange business of memory: Relic forgery in Latin America. *Past & Present, 206*(suppl_5), 199–226. https://doi.org/10.1093/pastj/gtq018

Hlusko, L. J., & Leigh, S. R. (2023) AABA statement concerning the hominin fossils carried on virgin galactic spaceflight. *American Association of Biological Anthropologists*. Retrieved July 17, 2024. https://bioanth.org/about/aaba-statement-concerning-the-hominin-fossils-carried-on-virgin-galactic-spaceflight/

Ipsos. (2019). *More Americans have Tattoos Today than Seven Years Ago*. https://www.ipsos.com/en-us/news-polls/more-americans-have-tattoos-today

Knight, A. (2010). The several legs of Santa Anna: A Saga of secular relics. *Past & Present, 206*(suppl_5), 227–255. https://doi.org/10.1093/pastj/gtq019

Langford, W. M. (1945). "Santa Anna and the Pastry War." Southwest Review 31(1): 70-73.

Lutz, D. (2011). The dead still among us: Victorian secular relics, hair jewelry, and death culture. *Victorian Literature and Culture, 39*(1), 127–142. https://doi.org/10.1017/S1060150310000306. https://www.cambridge.org/core/article/dead-still-among-us-victorian-secular-relics-hair-jewelry-and-death-culture/FDF70B02AB55F656CF073BEE383DE0A4

Lutz, D. (2015). *Relics of death in Victorian literature and culture*. Cambridge University Press.

Malik, T. (2015). Deadly SpaceShipTwo Crash Caused by Co-Pilot Error: NTSB. Space.com. Retrieved from https://www.space.com/30073-virgin-galactic-spaceshiptwo-crash-pilot-error.html

Moerer, A. K. (2013). *Changing Chapultepec: Construction, consumption, and cultural politics in a Mexico City Forest, 1934–1944*. PhD, University of Minnesota.

Oxenden, M. (2022). Their loved ones died. Preserved tattoos offer a way to keep them close. *New York Times*, October 8, 2022. https://www.nytimes.com/2022/10/08/arts/save-my-ink-forever-tattoo-preservation.html

Plasencia de la Parra, E. (1995). Conmemoracio'n de la hazan˜a e'pica de los nin˜os he'roes: su origen, desarrollo y simbolismos'. *Historia Mexicana, 45*(2), 241–279.

Pyne, L. (2016). *Why we still love 'Lucy': A compelling mix of science, history and culture continues to draw the public to teh world's most famous hominin*. History Today.

Sahlins, M. (1983). Other times, other customs: The anthropology of history. *American Anthropologist (New Series), 85*(3), 517–544.

Sharf, R. H. (1999). On the allure of Buddhist relics. *Representations, 66*, 75–99.

Snape, A. (2022). Ever thought about having your tattoos preserved when you die? *Journal* (blog), Stories&Ink. November 7, 2022. https://www.storiesandink.com/en-us/blogs/journal/tattoo-preservation-after-death

Vandiver, E. (2000). *Classical mythology part I*. The Teaching Company.

Verdery, K. (1999). *The political lives of dead bodies: Reburial and Postsocialist change*. Columbia University Press.

Visa, I., & Jussi-Pekka, T.. (2011). Finger of a saint, thumb of a priest: Medieval relics in the diocese of Turku, and the archaeology of lived bodies. *Scripta Instituti Donneriani Aboensis, 23*(0). https://doi.org/10.30674/scripta.67401. https://journal.fi/scripta/article/view/67401

Walker, W. H. (1996). Ritual deposits: Another perspective. In E. C. Adams (Ed.), *River of change: Prehistory of the middle Little Colorado River Valley, Arizona* (pp. 75–91). Arizona State Museum, University of Arizona.

Wilbert. (2020). *Memorial jewelry*. Wilbert Funeral Services Inc. Accessed December 2, 2020. Retrieved from https://www.wilbert.com/store/cremation/memorial-jewelry/

Chapter 17
Wrapping Up: The Poetics of Mortuary Performance and Us

Abstract This chapter summarizes the arguments made throughout the chapters of this book, focusing on how mortuary practices reflect and shape cultural identities. By expanding Whitehead's Poetics model to mortuary processes, the diverse ways societies memorialize the dead, both historically and in modern contexts, are approached. I argue that such practices serve as forms of social communication between the individual, the Social Body, and the Body Politic. The chapter highlights the cultural and historical contingency of death rituals, showing how the meaning of "respect" for the dead has evolved over time and is shaped by broader societal changes. Through case studies ranging from Roman burial practices to contemporary acts like turning ashes into jewelry, the chapter briefly recaps how death rituals can express both collective national identities (e.g., U.S. National Cemeteries) and individual identities (e.g., simple cremations). By analyzing these diverse mortuary performances, the chapter underscores the importance of the dead in defining social bonds, collective memory, and cultural values. Ultimately, it argues that the ways we interact with the dead are instrumental in constructing identity and navigating the complexities of social and cultural change.

I began this volume by discussing the boxes of certificates that sit in my spare room. Those boxes are still there, periodically shifted from one side of the room to the other. They were given to me by my mother, likely as a way of honoring the part of my father that was a doctor—his dedication to patient care and his desire to help others. It was also a quiet acknowledgment of the importance of education in the environment where I was raised. For now, those certificates will remain framed, waiting. I'm in no rush.

This book focuses on the relationship between the living and the dead, and how these interactions shape our understanding of group identity across different times and places. Biocultural bioarchaeology, as a discipline, uses multiple lines of evidence to reconstruct past lives and lived experiences. It helps us write the stories of people who may not have left written records or were never deemed important enough to be remembered—such as Roman women and children, whose lives rarely appear in the works written by Roman men. My goal here has been to expand

A. Osterholtz, *The Bodies Poetic*, Bioarchaeology and Social Theory,
https://doi.org/10.1007/978-3-031-85998-4_17

Whitehead's Poetics model to mortuary processes, applying it to both archaeological case studies and modern mortuary practices. The case studies I've selected are those I find most intriguing and instructive for this purpose.

One thing I've discovered is that the sheer variety of ways modern societies memorialize the dead cannot possibly be captured in a single volume. Historical and contemporary accounts of mortuary practices reveal that social messages about the living's responsibility to the dead are central to how we understand ourselves. Honoring the dead has always been important, but what 'honoring' means has changed over time. In the nineteenth century, sending ashes into space would have been seen as profoundly disrespectful. Yet today, such an act may be understood as a tribute to the deceased's social identity, reflecting respect. There is no single "right" way to honor the dead in modern society—this reflects the growing importance of individuality. Whether it's a grand, formal ceremony or a simple, unattended cremation with ashes scattered at sea, what matters is that the memorialization aligns with the deceased's identity. For example, when Jessica Mitford died in 1996, her friends and family honored her with a simple cremation, costing only $475 (Mallon, 2006), which was entirely consistent with her persona as a writer and critic of the funeral industry. Her death was memorialized in a way that was deeply appropriate to who she was.

Putnam's (2020) book *Bowling Alone: The Collapse and Revival of American Community* explores the role of civic engagement and identity. In this book, he examines how the different forms of social connection (bridging and bonding social capital) reflect different understandings of the individual to other individuals and their larger communities. In examining these relationships, he gets at some key elements of identity formation but utilizes a different approach (sociology) as well as different data sets. One of the central arguments of the book is that civic engagement has declined in the second half of the twentieth century, something tied to an increase in bonding social capital and a decrease in bridging social capital. An overemphasis on reinforcing social ties amongst those who already think like you may make an individual less likely to form bonds between those who don't share your own niche interests. The emphasis on bonding social capital is reflected in the increasing focus on individualized mortuary performance and is a reflection of the particular historical and cultural context of the late twentieth and twenty-first century.

Just as Whitehead teaches us about the Poetics of violence, I've come to see the Poetics of Mortuary Performance as equally culturally and historically contingent. Throughout this volume, I've threaded themes of post-mortem agency and embodiment, which allow us to use the same theoretical frameworks to understand the relationship between culture change and cremation in the Roman world, and the rise of the virtual cemetery in the late twentieth century. I've also explored the material culture that emerges from these practices, demonstrating how these objects take on meaning of their own. The changing ways we interact with the body—whether in art, medicine, or memorialization—reveal shifting cultural understandings of who has responsibility to the dead, and how structural violence plays a role in those decisions. Bodies can serve as powerful symbols, both in how we see ourselves as cultural groups and in their geopolitical significance.

Consider the example of National Cemeteries in the United States. These sites not only function as places of individual remembrance but also as tools of nation-building. They represent one expression of collective identity that can unite diverse groups—whether along lines of gender, race, ethnicity, or religion—into a shared identity: the soldier. The Tomb of the Unknown Soldier at Arlington National Cemetery, with its iconic changing of the guard ceremony, exemplifies this. The performance, conducted every hour of every day, regardless of the weather, symbolizes a steadfast promise to bring every soldier home and care for them after death. These spaces, though contested, become expressions of national unity and identity. Through burial, display, and repatriation, bodies help to create and solidify national identity and sovereignty.

Mortuary performance also provides a platform for the expression of individual identity. This can be seen in everything from fandom (e.g., honoring scientific achievement) to religious or ethical expressions of remembrance. How a person is memorialized—whether through an elaborate funeral or a modest cremation—reveals a preference for certain aspects of their identity. This personalization in death reflects broader trends in modern society toward individuality, where the process of dying and remembering is increasingly seen as a deeply personal, rather than communal, experience.

Regardless of the scale of the ceremony, mortuary performance serves as a form of social communication between the Individual Body, the Social Body, and the Body Politic. It is the use of the body to convey meaning—whether through traditional rituals or modern memorial innovations. Human cultures have always used the body as a material for communication, shaping it into symbolic forms of memory. We see this in the evolution of coffins, urns, and other memorial objects, which change not only as expressions of personal identity but also in response to larger social shifts. These objects, like the performative rituals they accompany, can be both destructive (acknowledging loss and the reality of death) and constructive (creating new social bonds and forms of collective identity). Whether it's the ancient practice of mummification in the Outer Hebrides or the modern act of turning cremated remains into diamonds, we can interpret these diverse practices through the same lens. The way the living interact with the dead always carries meaning, not only for the deceased but also for those who remain.

References

Mallon, T. (2006). Red Sheep. *The New Yorker*. Accessed October 3, 2024.

Putnam, R. D. (2020). *Bowling alone: The collapse and revival of American community (Revised and Updated)*. Simon & Schuster Paperbacks.

Index

A. Osterholtz, *The Bodies Poetic*, Bioarchaeology and Social Theory,
https://doi.org/10.1007/978-3-031-85998-4

The manufacturer's authorised representative in the EU is Springer Nature Customer Service Centre GmbH, Europaplatz 3, 69115 Heidelberg, Germany. If you have any concerns regarding our products, please contact ProductSafety@springernature.com

Printed and bound by CPI Group (UK) Ltd, Croydon, CR0 4YY
01/10/2026
02225909-0012